Orbital mechanics and formation flying

Related titles:

Aerodynamic Measurements: From physical principles to turnkey instrumentation
(ISBN 978-1-84569-992-5)
Aerodynamic Measurements will present to its reader a comprehensive treatment of the theoretical bases which experimental techniques used in aerodynamics are based on. Limitations of each method in terms of accuracy, response time and complexity will also be addressed. This book will serve as a guide in choosing the most pertinent technique for each type of flow field including: 1D, 2D, 3D, steady or unsteady, subsonic, supersonic or hypersonic. This is currently no book that presents as many techniques as those presented in this volume, with such a critical review of the various methods offered. Aerodynamic Measurements will help guide the reader to choose the most appropriate in each case and for each technique the evolution from old-fashioned mechanical process to the modern computerized version is reported in order to allow students and practitioners to understand what is behind the results shown on the coloured screen of their computers.

Leveraging Information Technology for Optimal Aircraft Maintenance, Repair and Overhaul
(ISBN 978-1-84569-982-6)
Aimed at professionals in the aviation industry and students of Aircraft Maintenance, this book covers the use and management of technology in the Aviation Maintenance, Repair, and Overhaul (MRO) context. This book will provide a backdrop of current trends in the industry where Airlines are tending to retain their aircraft longer on the one hand, and rapidly introducing a new genre of aircraft such as the A380 and B787 into service. This book aims to provide industry professionals and students of Aviation MRO (at all levels) with the necessary principles, approaches and tools to respond effectively and efficiently to the constant development of new technologies - both in general and within the Aviation MRO profession. It highlights the unique information requirements for Aviation MRO and delves into the detailed aspects of information needs from within the industry such as Engineering and Maintenance Planning, Aircraft/Engine/Component Configurations, Materials Management, Logistics, and Technical Documentation. It will highlight the difference between on-wing and off-wing maintenance with their unique information processing needs. This book is designed as a primer on IT services for Aircraft Engineering professionals and a handbook for IT professionals servicing this niche industry.

Modelling and Simulation of Integrated Systems in Engineering: Issues of methodology, quality, testing and applications
(ISBN 978-1-85709-078-2)
This book places particular emphasis on issues of model quality and ideas of model testing and validation. Mathematical and computer-based models provide a foundation for explaining complex behaviour, for decision-making, for engineering design and for real-time simulators for research and training. Many engineering design techniques depend on suitable models. Assessment of the adequacy of a given model for an intended application is therefore critically important. Generic model structures and dependable libraries of sub-models that can be applied repeatedly are increasingly important. Applications are drawn from the fields of mechanical, aeronautical and control engineering, and mainly involve non-linear lumped-parameter models described by ordinary differential equations. Topics considered repeatedly in the context of these applications include: system identification and parameter estimation; parameter sensitivity analysis; model optimisation; and inverse simulation.

Details of these and other Woodhead Publishing books can be obtained by:

- visiting our web site at www.woodheadpublishing.com
- contacting Customer Services (e-mail: sales@woodheadpublishing.com; fax: +44 (0) 1223 832819; tel.: +44 (0) 1223 499140; address: Woodhead Publishing Limited, 80 High Street, Sawston, Cambridge CB22 3HJ, UK)

If you would like to receive information on forthcoming titles, please send your address details to Customer Services, at the address above. Please confirm which subject areas you are interested in.

Orbital mechanics and formation flying

A digital control approach

PEDRO A. CAPÓ-LUGO

AND

PETER M. BAINUM

WOODHEAD PUBLISHING

Oxford Cambridge Philadelphia New Delhi

Published by Woodhead Publishing Limited, 80 High Street, Sawston, Cambridge CB22 3HJ, UK
www.woodheadpublishing.com

Woodhead Publishing, 1518 Walnut Street, Suite 1100, Philadelphia, PA 19102-3406, USA

Woodhead Publishing India Private Limited, G-2, Vardaan House, 7/28 Ansari Road, Daryaganj, New Delhi – 110002, India
www.woodheadpublishingindia.com

First published 2011, Woodhead Publishing Limited

British Library Cataloguing in Publication Data
A catalogue record for this book is available from the British Library.

Library of Congress Control Number: 2011919191

Woodhead Publishing ISBN: 978-0-85709-054-6 (print)
ISBN: 978-0-85709-387-5 (online)

Typeset by RefineCatch Ltd, Bungay, Suffolk

To my wife, my newborn son, and my family, thanks for their patience and support.

Pedro A. Capó-Lugo

To my wife, my son, and his wife for their support and understanding.

Peter M. Bainum

Contents

List of figures

List of tables

List of symbols

$\mathbf{A}$	State matrix
$\hat{\mathbf{A}}$	Discrete state matrix
$\mathbf{A}_{CL}$	Closed loop matrix
A	Area of the ellipse
A_F	Axial force
A_{VO}	Area of the valve opening
ARE	Adjoint Riccati equation
$\vec{a}$	Acceleration of the body
$\vec{a}_{0/1}$	Acceleration of the body seen from the reference point
$\vec{a}_{2/0}$	Acceleration of the body seen from the observer 2 near point O
a	Semimajor axis
a_t	Semimajor axis for the transfer orbit
a_N	Coefficient of the polynomial
a	Semi-conjugate axis for the hyperbola
$\bar{\alpha}$	Angle of attack
α	Angle of the incident light
$\boldsymbol{B}$	Control matrix
$\hat{\boldsymbol{B}}$	Discrete control matrix
B_2	Constant describing the Earth oblateness
B_C	Ballistic coefficient
$\bar{B}$	Unit vector applied along the tangential direction
B_0	Earth's magnetic field intensity
$\vec{B}_m$	Earth's magnetic field vector
$\vec{B}_h$	Horizontal component of the Earth's magnetic field vector
b	Semiminor axis
$\hat{b}_i$	Body coordinates
β	Gimbal angle of the motors about the Y axis
β_{max}	Maximum gimbal angle

C	Controllability matrix
$\hat{C}$	Discrete output matrix
CG	Center of gravity
CM	Center of mass
CP	Center of pressure
c	Number of constraints
$c(t)$	Surface constraint
C_D	Drag coefficient
$\bar{c}$	Velocity of light in vacuum
DOF	Degrees of freedom
D	Magnetic declination
D_E	Earth's magnetic dipole moment
$\vec{D}$	Magnetic dipole moment
D_{max}	Maximum magnetic dipole moment
d	Distance from the center of the Earth to the directrix line
d_{RW}	Damping of the reaction wheel
$\vec{d}$	distance between parallel coordinate systems
$d(\ )$	Change or variation of a variable
δ	Gimbal angle for the motors about the Z axis
δf	Difference in the true anomaly angle
Δt	Change in time or sampling time
Δf	Sampling time in the true anomaly angle
$\Delta V_{P,SB}$	Change in velocity at the perigee point for SB
$\Delta V_{P,SA}$	Change in velocity at the perigee point for SA
$\Delta V_{P,SC}$	Change in velocity at the perigee point for SC
$\Delta V_{P,SH}$	Change in velocity at the perigee point for SH
ΔV_a	Change in velocity at the apogee point
∇f	Gradient of a function
$E(z)$	Auxiliary variable in the discrete domain
E	Eccentric anomaly
ECI	Earth centered inertial frame
E_0	Light energy arriving at the infinitesimal surface
E_b	Voltage across the rotor
$E_a(t)$	Applied voltage
E_{max}	Maximum applied voltage
E_r^2	Mean square error
e	Eccentricity of the orbit
ϵ_a	Absorptivity coefficient

ϵ_d	Diffuse reflectance coefficient
ε	Reflectivity coefficient
ξ, ζ, η	Coordinate system of the center of mass
e_x	Actual state error
$e_{\dot{x}}$	Time derivative of the actual state error
F	Final state weighting matrix
FS	First stage of Ares V
FL	Fuzzy logic
F_A	Aerodynamic force
$\vec{F}$	Forces applied to the body
$\bar{F}$	Solar radiation force
$\vec{F}_i^b$	Body forces
$\vec{F}_i^S$	Surface forces
$\vec{F}^+$	Solar radiation force for absorbing surface
$\vec{F}^-$	Solar radiation force for reflective surface
$\vec{f}_{J2,M}$	J2 perturbation force for the maneuvering spacecraft
$\vec{f}_{J2,R}$	J2 perturbation force for the reference spacecraft
$\vec{f}_{P,M}$	Perturbation force for the maneuvering spacecraft
$\vec{f}_{P,R}$	Perturbation force for the reference spacecraft
f	True anomaly angle
f_L	Final true anomaly angle
f_0	Initial true anomaly angle
f_M	Frequency of the magnetic natural response function
f_v	Damper constant
$G_P(S)$	Transfer function for the process or plant
$G_A(S)$	Transfer function for the actuator
$G_C(S)$	Transfer function for the controller
$G_S(S)$	Transfer function for the sensors
$G_{OL}(S)$	Open loop transfer function
$G_{CL}(S)$	Closed loop transfer function
G	Universal gravitational constant
GA	Genetic algorithms
GP	Gimbal point
$\vec{G}$	Adjoint Riccati equation
$\vec{G}_\infty$	Steady-state adjoint Riccati equation
$\vec{\hat{G}}_\infty$	Discrete steady-state adjoint Riccati equation
g	Gravitational acceleration for the Earth
γ	Flight path angle
γ_L	Learning parameter

$\vec{\gamma}$	Error vector
$\bar{\vec{\gamma}}_\infty$	Steady-state error vector
H	Density scale height
H_m	Flux density in the rod
$\mathcal{H}$	Hamiltonian equation
$\vec{H}$	Angular momentum of the body
$\vec{H}_T$	Total angular momentum
$\vec{H}_{RW}$	Reaction wheel angular momentum
h	Angular momentum per unit mass
$\bar{h}$	Power of electromagnetic wave per unit surface
$\bar{h}_0$	Solar radiation constant
$\mathbf{I}$	Identity matrix
$\hat{\mathbf{I}}$	Inertia matrix
I	Magnetic inclination angle
I_{max}	Maximum applied current
I_{ij}	Moments of inertia, i=1,2,3 and j=1,2,3
I_a	Armature current
I_{SP}	Specific impulse for any motor
I_S	Current source
i	Inclination angle
i_S	Inclination angle of the Sun with respect to the Earth
$\mathbf{J}$	Principal moments of inertia
J_{RW}	Reaction wheel moment of inertia
$\mathcal{J}$	Cost function or system performance
$\mathbf{K}_\infty$	Steady-state control gain matrix
$\mathbf{K}$	Control gain matrix
K	Momentum of light
K_b	Back electromotive force constant
K_t	Motor torque constant
K_P	Proportional control gain
K_I	Integral control gain
K_D	Derivative control gain
$\tilde{K}_P$	Proportional adaptive control gain
$\tilde{K}_d$	Derivative adaptive control gain
K	Spring constant
$\vec{K}^{P}_\infty$	Steady-state perturbation gain vector
k	Sample taken in the discrete control process

L	Lagrange equation or Lagrange function
$\mathcal{L}\{\ \}$	Laplace transform
L	Number of runs in the hierarchical control scheme
L_a	Armature inductance
L_{CP}	Distance from the CG to the CP
L_{GP}	Distance from the CG to the GP
L_{SRB}	Distance from the CG to the center of the nozzle of the SRB
LQR	Linear quadratic regulator
LHS	Left hand side
l	Angular momentum
l_C	Length of the cylinder
$\vec{\lambda}$	Co-state variable
λ	Longitude
$\lambda_{ecliptic}$	Mean longitude of the ecliptic plane
λ_{sun}	Mean longitude of the Sun
$\mathcal{M}$	End conditions in the cost function
M	Mean anomaly angle
$\bar{M}$	Total system mass
M_b	Mass of the body
M_E	Mass of the Earth
M_{sun}	Mean anomaly of the Sun
m	Mass of the body
m_0	Initial mass
m_i	Mass of a particle
$\dot{m}$	Mass ratio at the exit nozzle of any motor
$\dot{m}_{max}$	Maximum mass flow rate
μ	Earth gravitational constant which is equal to GM_E
N	Number of particles
N_F	Normal force
N_f	Last sample obtained in the control solution
N_{STALL}	Stall torque
$\bar{N}$	Unit vector applied along the normal to the orbital plane
$\vec{N}$	Torques
$\vec{\mathcal{N}}$	End constraints
$\vec{N}_A$	Torque due to the atmospheric pressure
$\vec{N}_{AR}$	Torque due to the satellite angular momentum
$\vec{N}_E$	Gravity-gradient torques
$\vec{N}_m$	Magnetic torques

$\vec{N}_P$	Torque due to perturbing forces
$\vec{N}_{RW}$	Reaction wheel torques
$\vec{N}_{SP}$	Torque due to solar pressure
$\vec{N}_T$	Torque due to the thrusters
N_{max}	Maximum torque due to the reaction wheels
n	Mean motion
NN	Neural network
n_T	Number of thrusters for the RoCS
n_t	Number of turns of the coil
$\hat{n}$	Normal vector to the surface
υ	Angular velocity of the reaction wheels
$\upsilon\_$	No load angular velocity
Ω	Anti-symmetric or skew-symmetric matrix
Ω_E	Earth angular velocity about the polar axis
Ω	Right ascension of the ascending node
ω	Argument of perigee
ω_n	Natural frequency
$\vec{\varpi}$	Angular velocity of the satellite
$\vec{\varpi}_R$	Reference angular velocity of the satellite
$\vec{\varpi}_n$	Mean motion of the angular velocity of the satellite
P	P_∞ Steady-state Riccati matrix
$\hat{P}_\infty$	Steady-state discrete Riccati matrix
$\%OS$	Percent of overshoot
p	Semi-latus rectum
ϕ	Latitude
φ	Roll angle
φ_m	Damping function for the magnetic dipole system
φ_n	Natural response function for the magnetic dipole system
ξ	Damping coefficient
ψ	Yaw angle
$\vec{\psi}$	Perturbation column vector
$\vec{\psi}_A$	Perturbation column vector for atmospheric density force
$\vec{\psi}_{SP}$	Perturbation column vector for solar pressure force
$\vec{\psi}_{Unknown}$	Perturbation column vector for unknown forces
$\mathbf{Q}$	State weighting matrix
$\tilde{\mathbf{Q}}$	State weighting matrix in the true anomaly angle
$\mathcal{Q}$	Dynamic pressure
$\mathcal{Q}_j$	Generalized forces

q	Quaternion, and term in the Legendre polynomial expansion of the Earth's potential function
q_i	Generalized coordinates
q_e	Error quaternion
q_c	Commanding quaternion
$\mathbf{R}$	Rotational matrix
$\boldsymbol{R}$	Control weighting matrix
$\tilde{\boldsymbol{R}}$	Control weighting matrix in the true anomaly angle
$\boldsymbol{R}_J$	Direction cosines matrix for the principal moments of inertia
$\bar{R}$	Unit vector applied along the radial direction
R_A	Armature resistance
R_E	Radius of the Earth
R_C	Radius of the cylinder
$RoCS$	Roll control system
RHS	Right hand side
RE	Riccati equation
R or $\vec{R}$	Radius that defines the distance from the center of the Earth to the satellite
$\hat{\vec{R}}$	Unit vector associated with $\vec{R}$
$\vec{R}_m$	Distance from the center of mass of the satellite to a point in the body of the satellite
$\vec{R}_{CM}$	Distance from the center of the Earth to the center of mass of the satellite
r_a	Radius of apogee
r_0, R_{SUN}	Mean distance of the Earth to the Sun
r_p	Radius of perigee
$\vec{r}$	Position of the maneuvering satellite
$\vec{r}_i$	Position vector of the body
$\vec{r}_2$	Distance from point O to the particle
$\vec{\rho}(t)$	Separation distance between a pair of satellites
$\vec{\rho}_{Traj}(t)$	Separation distance between a pair of satellites from the trajectory system
$\vec{\rho}_{NAV}(t)$	Separation distance between a pair of satellites from the navigation system
ρ_{Prop}	Density of the propellant
S	Satellite surface area
$\bar{s}$	Altitude of the orbit
SRB	Solid rocket booster motors

$sgn(\ )$	Signum function
$\hat{\sigma}$	Direction of the light flux
ς	Function in terms of the true anomaly angle
T	Period of the orbit
T_t	Period of the transfer orbit
T_{JC}	Time in Julian centuries
T_{JD}	Time in Julian dates
$T_{J2\text{-}X}$	Thrust force for the J2-X engines
T_{RoCS}	Thrust force for the RoCS
$T_{RS\text{-}68b}$	Thrust force for the RS-68b motors
T_{SRB}	Thrust force for the SRB motors
T_{max} or T_m	Maximum thrust force
T_K	Kinetic energy
T_P	Time of perigee passage
T_r	Rise time
T_s	Settling time
T_p	Peak time
TPBVP	Two point boundary value problem
tol	Tolerance value
$\vec{\Gamma}(k)$	Solar pressure force disturbance vector
θ	Pitch angle
θ_A	Momentum parameter
ϑ	End conditions co-state variables
U	Potential energy
US	Upper stage flight for Ares V
U_L	Magnetic potential function
$U_e(k)$	Unit value for the error containing the orbital elements
$U_X(k)$	Unit value for the state vector containing the orbital elements
$u^*(t)$	Particular control function
$u_A(t)$	Actuator control function
$u_m(t)$	Maximum input acceleration
$\vec{u}(t)$	Control vector function
$\vec{u}_A(t)$	Atmospheric force per unit mass
$\vec{u}_C(t)$	Control vector function containing adaptive and baseline controller
$\vec{u}_{SP}(t)$	Solar pressure force per unit mass
$\tilde{\vec{u}}(t)$	Adaptive control function
υ_f	Relative angular velocity of the satellite

V	Energy equation
V_L	Lyapunov homogeneous function
$\mathbb{V}$	Volume
V_{Prop}	Velocity of the propellant
$V_{P,SB}$	Velocity at the perigee point for SB
$V_{P,SA}$	Velocity at the perigee point for SA
$V_{P,SC}$	Velocity at the perigee point for SC
$V_{P,SH}$	Velocity at the perigee point for SH
V_A	Velocity at the apogee point
$\vec{V}$	Velocity of center of mass or velocity vector
$\hat{V}$	Unit vector of the velocity vector
v	Velocity equation
v_{ESC}	Escape velocity
v_p	Velocity at the perigee point
v_i	Velocity of a particle
W	Work done on a system
W_V	Weight of the vehicle
w	Sum of the true anomaly angle and the argument of perigee
$\bar{w}$	Light energy per volume
$\vec{w}$	Angular velocity
X_B, Y_B, Z_B	Body coordinate frame
X_R, Y_R, Z_R	Reference coordinate frame
X_N, Y_N, Z_N	Nominal separation distance for the satellites in the ECI frame
X_S, Y_S, Z_S	Initial separation distance for the satellites in the ECI frame
$\vec{X}$	State vector for the orbital elements
$\vec{X}_D$	State vector for the desired orbital elements
x_{EM}, y_{EM}, z_{EM}	Magnetic field lines
$x_C(s)$	Transfer function for the command function
$x_A(s)$	Transfer function for the actuator function
$\vec{x}$ or x_i	State vector ($i = 1,2,3, \ldots$)
$\vec{x}_0$	Initial condition
$\vec{x}_1$	Final condition
$\vec{x}_D$	Desired state vector
$\bar{y}$	Linear transformation in the true anomaly angle of the separation distance

$Z\{\}$	Z-transform
$\vec{z}$	State variables with integral terms
$\int_{\mathbb{V}} ( )\, d\mathbb{V}$	Volumetric integral

Acknowledgements

I want to thank my colleagues from EV-41, ED-04, and ET-40 for their help and guidance when I was working in the Constellation program. I want to thank Mr. Mark West, EV-41 Branch Chief, and Mr. Mark E Jackson, EV-41 Deputy Branch Chief that led me in the work for the Constellation program, the Fast Affordable Science and Technology Satellite, and the heavy lift vehicle. Also, I want to thank Mr. Reginald Alexander, ED-04 Branch Chief, for letting me work in the guidance, navigation, and control for the heavy lifter vehicle, Ares V. I want to thank Mr. Timothy E. Driskill, ET-40 Branch Chief, for allowing me to work in the environmental testing laboratory to gain knowledge about the systems hardware. Specially, I want to thank Dr. Tannen Van Zwietten, from EV-42 Branch, for her input and help in the adaptive control scheme explained and demonstrated in Chapters 7 and 11.

Pedro A. Capó-Lugo

First I would like to thank all of my Master's and Ph.D. students over more than a 40 year time span. I have learned more from all of you than you ever realized.

In addition, much appreciation goes to my Visiting Scholars and Visiting Professors who have inspired my graduate students, other faculty, and myself.

Thirdly, I would like to thank to my colleagues in the Department of Mechanical Engineering at Howard University and other Departments such as Physics, Electrical Engineering, Computer Sciences, and Mathematics. The financial and other support provided by the Mechanical Engineering Department Chairs, Dr. M. Lucius Walker, Jr., Dr. Charles B. Watkins, Dr. Robert Reiss, Dr. Lewis Thigpen, and Dr. H. A. Whitworth have been invaluable. Special thanks to Dean James Johnson, Dean Orlando Taylor, Dean Edward Hawthorne, and Dean William Sadler.

And to other supervisors who directed my early assignments at the Martin Company, IBM Federal Systems Division, The Johns Hopkins University – Applied Physics Laboratory, and NASA Goddard Spaceflight

Center. This experience provided a firm basis for much of my practical real world experience that is reflected throughout the book and in many of the suggested homework problems that appear at the end of each chapter.

Lastly, my thanks go to the many organizations that have invited me to present short courses and lectures which include: AIAA National Capital Section; AIAA Hampton Roads Section; European Space Agency; INPE (Brazil); ISAS/JAXA – Japan; LAPAN – Indonesia; the Indian Space Research Organization; Beijing Institute of Control Engineering; Shanghai Satellite and Launch Vehicle Center; The Martin Company Evening Program; The Applied Physics Laboratory Evening Program. In addition numerous universities in many parts of the world have provided many ideas.

And I should include thanks to many funding agencies that have provided support for students and Visiting Scholars and thesis topics including: NASA Goddard; NASA Headquarters; NASA Langley; NASA Marshall Space Flight Center; NASA Lewis (Glenn); AFOSR; AFOSR – Cooperation with NRL; DARPA; USAF Wright Laboratory; and Nippon Telephone and Telegraph (NTT) – Japan.

Peter M. Bainum

Preface

This preface attempts to summarize the objective of this text. The authors have more than 54 years of practical real-world experience and attempt to encompass this experience in the chapters and suggested problems. Many of these problems have been adapted from these actual space projects and cannot be found in many contemporary text books.

This book will be used by practicing engineers and other texts may provide a deeper theoretical background. The reader can refer to more than 170 historical references, beginning with Arthur C. Clarke's 1945 reference describing his ideas for formation flying of three communication stations in a geosynchronous orbit in order to provide, through relay technology, instantaneous communication to/from any location on the Earth's surface, and also to/from various orbiting spacecraft.

Although the text begins with a review of orbital mechanics, the emphasis here is placed on digital control techniques and specific application for formation flying, deployment, station keeping, and reconfiguration. Part of the formation flying is based on research contracts with the Applied Physics Laboratory, and DARPA, focusing on on-track (string of pearls) and 3-D tetrahedron configurations in highly elliptical orbits. The tetrahedron was suggested as part of a design challenged by NASA Goddard.

Advanced modern control techniques, including intelligent control applications such as fuzzy logic, hierarchical control, and adaptive control, are proposed for the first time for formation flying applications. Some of these methods have been previously implemented for industrial applications, such as in the use of stepper motors.

Where possible, problems with analytical or closed-form solutions are suggested; in some cases a background in MATLAB and ability to solve nonlinear differential equations and in discrete form is assumed, as well as the ability to produce results in graphical format. In some cases answers and/or hints are provided. The philosophy here is that most of us learn by practicing, which means by working problems of various degrees of difficulty. Suggested problems attempt to follow the material covered in the preceding chapter.

Some of the suggested exercises came from the author's own classes taken at MIT, the Catholic University of America, Howard University, various short courses at UCLA, and evening classes taken at the Applied Physics Laboratory and elsewhere. And many others were adapted from the authors' practical experience.

Peter M. Bainum

About the authors

Dr. Pedro A. Capó-Lugo has a B.S. in Electrical Engineering from the University of Puerto Rico at Mayaguez, Puerto Rico. He concentrated in analog and digital electronics and is certified in the analysis of satellite photos using remote sensing and pattern recognition. After completing his B.S., he joined Dr. Peter M. Bainum as his graduate student at Howard University in Washington, DC. At Howard University, he completed an M.E. in Aerospace Engineering and a Ph.D. in Mechanical Engineering. In his graduate studies, he mainly researched space flight dynamics and control, and formation flying.

In 2008, Dr. Capó-Lugo was employed by NASA George C. Marshall Space Flight Center in Huntsville, Alabama to work as an aerospace engineer. Currently, he is working for the Flight Mechanics and Analysis Division (EV-40) in the Control Systems Design and Analysis Branch (EV-41). In this division, he works with launch vehicles and satellites. In launch vehicle systems, he works with the Ares I and V of the Constellation program. In the Constellation program, he develops the simulation for the manual steering function for the Ares I rocket such that the astronauts can control the vehicle in the upper stage flight of the rocket; for the Ares V, he analyzes the stability and control of the vehicle during the first and upper stage flight of the vehicle. In more recent work, he is analyzing the stability and control of other versions of the heavy lift vehicle. In satellite systems, he works in the concept designs of satellites for lunar, asteroid, and Earth scientific missions. In addition, he works in the control design of the Fast Affordable Science and Technology Satellite (FASTSAT). FASTSAT was launched on November 20, 2010 for a mission period of approximately one year.

As part of NASA's educational outreach, Dr. Capó-Lugo is an adjunct faculty in the Calhoun Community College – Huntsville Campus. In this college he teaches pre-engineering courses with their respective laboratories. Also, he and his colleague, Dr. Hien Vo, are supervising a group of undergraduate and graduate students from the Department of Electrical Engineering, Mechanical Engineering, Computer Sciences in the InterAmerican University – Bayamón Campus, the InterAmerican University – Ponce Campus, University of Puerto Rico – Mayaguez Campus, and the Polytechnic University in Puerto Rico. The objective of this group is to design, fabricate, and launch the first cube satellite by the Hispanic institutions. The cube satellite is called the Space Weather using Ion spectrometer and Magnetometers (SWIM) project. With this same group of students and his colleague, he is supervising the design of an attitude determination and control system board that will be implemented in the High Altitude Student Platform (HASP) program. This project uses the basis of the attitude determination of a satellite to control the orientation of the payload during the ascent flight of a balloon.

Dr. Capó-Lugo has published approximately 20 articles in international and national journals and in conference proceedings.

Dr. Peter M. Bainum is Distinguished Professor of Aerospace Engineering, Emeritus, in the Department of Mechanical Engineering, Howard University, Washington, DC, USA. His education includes a B.S. in Aeronautical Engineering from Texas A&M University; an S.M. in Aeronautics and Astronautics from MIT, and a Ph.D. from the Catholic University of America. His appointment at Howard extends from 1967 to the present.

He has been responsible for directing 20 Masters students, 14 Ph.D. students, and 10 visiting scholars and visiting professors from China, Korea, Brazil, Indonesia, India, Mali, and the USA. Grant/contract work has included the initial Tethered Shuttle Subsatellite Project (NASA/Italian Space Agency), dynamics analysis of the Small Astronomy Project for the San Marco Program, and development of control strategies for the Large Space Structures Program (NASA Langley), and autopilot control laws for expendable launch vehicles (NASA -Glenn). Current research focuses on orbital formation flying.

Dr. Bainum's research has concentrated on the dynamics and control of large structural systems for aerospace applications. He has been the Principal Investigator/Co-Investigator of 38 research grants/contracts sponsored by NASA, US AFOSR, US Air Force Wright Laboratory, INTELSAT, Ball Brothers Research Corp., Martin Marietta, The Johns Hopkins University/Applied Physics Laboratory, and Nippon Telegraph and Telephone Corp. (Japan). Before joining academia he had 10 years industrial experience with Martin-Orlando, IBM, and The Johns Hopkins University/Applied Physics Laboratory. Projects included the Gemini Project and various communication and experimental gravity-gradient satellites.

Dr. Bainum is the author/co-author of more than 200 journal articles, conference proceedings, and technical reports, and he is the editor or co-editor of 21 books. He has accepted more than 220 invitations to serve as lecturer, panelist, or session chairman/organizer worldwide.

He is a licensed Professional Engineer in the District of Columbia.

Service to Professional organizations and honors include:

AIAA- Fellow; former members of Astrodynamics TC; Space Transportation TC; International Activities Committee; recipient International Cooperation Award, 2008.

American Astronautical Society (AAS)-Fellow; officer and member board of directors since 1976; member, International Programs Committee; recipient Dirk Brouwer Award, 1989.

International Astronautical Federation (IAF)-Former Chair, Astrodynamics TC; current member, Materials & Structures TC; AAS Chief/Alternate Delegate to the IAF General Assembly since 1980; recipient, IAF Frank Malina Education Award, 2007.

Fellow AAAS; Fellow British Interplanetary Society; Honorary Member, Japanese Rocket Society; Sen. Spark Matsunaga Memorial Award for International Cooperation in Space, 2001 Member, International Academy of Astronautics.

The authors may be contacted at:

Dr. Pedro A. Capó-Lugo
NASA George C. Marshall Space Flight Center
Mail Stop: EV-41
Huntsville, AL 35812, USA
E-mail: pedro.a.capo-lugo@nasa.gov

Dr. Peter M. Bainum
Howard University
Department of Mechanical Engineering
2300 Sixth Street, NW
Washington, DC 20059, USA
E-mail: pbainum@howard.edu

1

Introduction

1.1 Introduction to the book

For the past 50 years, the orbital motion of a satellite has been studied and proved by the launch of different satellites into Space such as the Sputnik, the Gemini mission, and the Apollo mission. In addition, the spacecraft mechanics and control have been developed and used in classical, modern, intelligent, and adaptive control schemes. These advances in the areas of orbital mechanics and control allow the engineer to perform different maneuvers with a single satellite.

Using this knowledge, more complex and larger satellites are developed to satisfy certain mission requirements; also, these satellites require a rocket large enough that can carry them into Space. On the other hand, the invention of smaller satellites, weighing less than a 100 kg, can be used to perform the same mission goals as the larger satellites. These satellites carry a single instrument and are placed in the same orbit with a required separation distance. This group of satellites is known as a constellation or formation flying. A constellation can have any type of geometrical configuration. The advantage of a constellation is that the satellites can take different measurements at various points in the same orbit at the same time. This advantage allows the scientist to do more experiments at a single point; in addition, there is an opportunity to fit more than one satellite in a single rocket. Although, the sensors proposed for use in smaller satellites are not flight tested yet; it is possible to use them to navigate a constellation into the required formation.

The objective of this book is to take the reader into the development of the orbital and attitude motion of a satellite; in this way, the background information can be used to develop the equations of motion for a constellation. This book uses a practical approach to explain the orbital mechanics and control equations; in addition, all the control techniques applied to formation flying are specified in the discrete domain. Using the digital formulation, the code can be easily implemented in the computer onboard the satellite. There are books [1] [2], journal articles [3] [4] [5], and conference articles [6] that provide a more complete mathematical background for solving the equations and controlling a single satellite and a group of satellites in formation flying. The intention of this book is not to add more mathematical depth into the control system of a spacecraft; on the other hand, it shows the practical aspects of controlling a single satellite and formation flying.

1.2 Book division

Figure 1.1 shows the conceptual map explaining the book division. Before developing any type of formulations, Chapter 2 provides an explanation of orbital mechanics. In this chapter, the motion of a single spacecraft in the orbit is explained. In Chapter 3, the orbital perturbations that affect the motion of the satellite in orbit are explained. The main perturbations that are defined are the perturbations due to the Earth and Sun. Chapters 2 and 3 explain the translational motion of the vehicle without and with disturbances.

To begin explaining the attitude motion of the vehicle, the frame transformations are explained in Chapter 4. These frame transformations are important to explain the rotation from a reference to a body frame in the satellite. In Chapter 5, the attitude dynamics are explained based on the formulation of the angular momentum equations, Lagrange equations, quaternions, and the quasi-coordinates. All these equations explain the orientation of the spacecraft about the center of mass of the satellite. After describing its motion, there are two types of applied torques acting on the satellite; and in Chapter 6, the environmental and actuator torques are explained. The environmental torques are the outside forces that cause disturbances to the orientation of the satellite. The actuator torques define the forces applied along the body of the satellite to cause a rotation to correct the orientation of the vehicle.

These environmental and actuator torques provide a formal introduction to control systems. Control systems are used to correct the translational and attitude motion of a vehicle. These control systems are divided into two big sections which are the classical and modern control systems. The classical control systems are described by known techniques that are continuous in the time domain. Once the control system is expressed in the computer, the control system is defined in the discrete

Figure 1.1 Conceptual map of the book

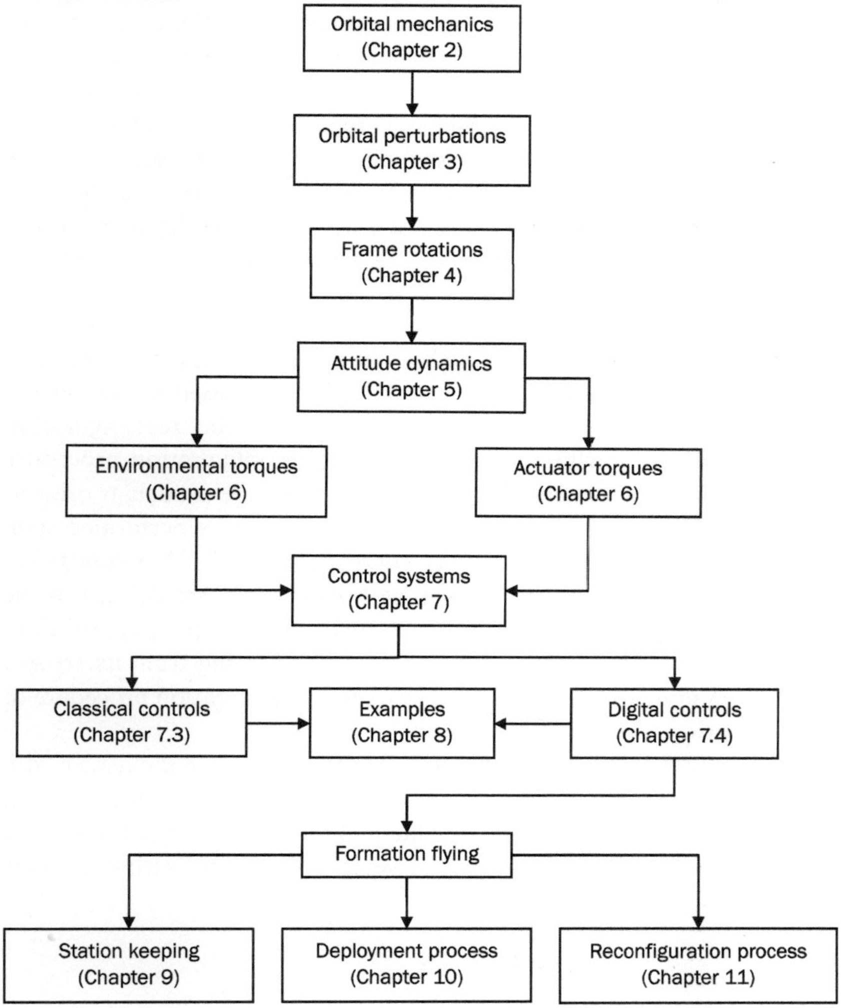

domain which is known as the modern controls. In classical and modern controls, the techniques are similar; but the main difference is a sampling time used to describe how long the system holds to collect one sample of data. In this chapter, the difference in continuous and discrete control is expressed in terms of this sampling time. Before wrapping up this chapter, Chapter 8 takes on a single example based on a small satellite to explain many of the techniques in Chapter 7.

Chapters 1 through 6 explain the motion of a single satellite in orbit; but Chapters 9 through 11 explain formation flying. Chapter 9 explains the translational dynamics of a pair of satellites in an elliptical orbit. This problem can be solved for a circular orbit, but the circular is one particular solution of the elliptical orbits. In this chapter, a control system is presented to take care of nonlinearities in the equations of motion defined by the disturbances due to the Earth oblateness and the solar pressure effects. Chapter 9 shows the station-keeping process of a satellite. To show the complexity of launching a constellation, the deployment and reconfiguration process is explained. The deployment procedure is shown in Chapter 10. This chapter uses very simple techniques to cause the satellite to depart from a near-Earth orbit to a highly elliptical orbit. Once the satellite reaches the apogee point of the transfer orbit, the station-keeping techniques shown in Chapter 9 are used to correct the orbital motion of the satellite. In Chapter 11, the reconfiguration procedure of the satellite is performed. The reconfiguration procedure changes the orbital dimensions of a satellite from one particular orbit to another. In this case, the reconfiguration procedure is performed with intelligent controllers and adaptive control schemes. These controllers can be easily implemented in real time because they are defined in the discrete domain; but these controllers provide an interesting approach to control the satellite by 'taking decisions' and 'learning from its errors'. These techniques can be used to achieve the reconfiguration procedure of a constellation.

As shown in Figure 1.1, the book is divided to provide the reader with clear information about the application of orbital mechanics, actuators, and control systems to a single satellite and formation flying. Hence, this information allows the reader to develop and control either a single satellite or a constellation.

2

Two body orbital motion

Abstract: Orbital mechanics is the study of the motion of planets and satellites in Space. In order to understand this motion, the Lagrange's equations are explained to determine the equations of motion for a body. In addition, the motions of a single and a system of particles are defined such that they can be used for the development of the motion of a satellite about a planet. From these equations of motion, the orbital elements, the angles, and other parameters that describe an orbit are defined and explained. In summary, the main purpose of this chapter is to explain the orbital motion of a satellite about a planet which is the Earth.

Key words: two body orbital motion, degrees of freedom, Lagrange's equation, orbital elements, center of gravity, system of particles, inertial coordinate system.

2.1 Introduction to orbital motion

The objective of this chapter is to introduce the equations of motion for a satellite in any orbit about the Earth. In many references [7] [8] [9], this study of the celestial bodies (or planets) is known as astrodynamics or orbital mechanics. The intention of this chapter is not to explain in depth the motion of the planets and satellites, but is to provide the general concepts for the motion of a satellite or planet; in this way, the reader can understand a single body motion before developing more complex definitions about the motion of multiple satellites. Also, this book treats different control problems that use this chapter as a basis for the development of the problem.

The chapter is decomposed into different topics to provide the basic knowledge of dynamics. Dynamics refers to the motion of a body in free space. In topics 2.2 through 2.4, the Lagrange equations of motion are explained for single and multiple mass systems. In addition, the definition of constraints and generalized coordinates are explained because the equations of motion can be written for specific independent variables. The topics mentioned do not show examples to solve different dynamical problems; but in topics 2.5 through 2.8, the equations of motion for two body central force motion are developed and explained.

It is important to determine the type of orbit in which the satellite is orbiting. In topic 2.9, the Earth Coordinate Inertial (ECI) frame is explained. This frame is commonly used to explain the orbital elements associated with the motion of a satellite. In Chapter 4, this coordinate system can be used to explain the frame rotations. As part of the equations of motion, Kepler's equations are explained in topics 2.10 through 2.11. These equations are based on the period and mean motion of the satellite about the Earth. These topics are important because they relate the time with respect to the location of the satellite. When topics 2.10 through 2.11 are compared with topics 2.5 through 2.8, the main difference is how the location of the satellite is determined.

In summary, this chapter explains the orbital motion of the satellite based on the solution of Lagrange's equations. As explained before, this is an important chapter before the motion of multiple satellites is explained.

2.2 Constraints and generalized coordinates

Before developing the equations of motion of a vehicle, it is necessary to understand the significance of degrees of freedom and how the degrees of freedom are related to the constraints of a body. Imagine a particle moving in free space; also, locate a Cartesian coordinate system in the center of gravity (CG) of the particle. The particle can translate along any of the three axes of the Cartesian coordinate system. In addition, the particle can rotate about any of the three Cartesian axes. The degrees of freedom (DOF) explain the minimum number of independent coordinates specifying the motion of a body. A particle moving in free space has six DOF which contain three DOF for translation and three DOF for rotation. The motion of a body is not only constrained to move in a Cartesian coordinate system but can be explained in cylindrical and spherical coordinate systems to simplify the equations of motion.

On the other hand, the motion of a particle can be constrained to follow a path or rotate about a single point. If the motion of the particle is constrained, the numbers of DOF are reduced. Assuming that the particle is no longer moving in free space, the particle is constrained to move in a plane and cannot rotate about any of the three axes; then, the number of DOF is two because the particle can only move in the directions of two of the Cartesian axis. If the same particle is constrained to move along a straight line, the number of DOF is one because the particle can travel along one (Cartesian) axis. The motion of a satellite, launch vehicle, or car, among others, can be constrained to reduce the number of DOF. In order to understand the definition of DOF, it is necessary to study small dynamical problems before developing complex problems containing more than 2 DOF.

As a general case, the constraints can be classified as follows,

Constraint to a surface

$$F(x, y, z) = 0$$

Constraint to a surface and time

$$C(x, y, z, t) = 0$$

When a particle is constrained to move over the surface of a sphere, the constraint can be written as $F(x, y, z) = x^2 + y^2 + z^2 - r^2 = 0$ where r is the radius of the sphere. A constraint due to the surface and time can happen with a satellite when the thrusters are fired at certain periods of time in certain locations about the Earth. This constraint is shown in the reconfiguration procedure for a constellation in Chapter 11. A simple equation is used to determine the number of DOF associated to the particle in motion and is written as follows,

$$DOF = 3N - c \tag{2.1}$$

where N is the number of particles, and c is the number of constraints.

The variables that are independent and free of any constraint equation are called generalized coordinates. The number of generalized coordinates corresponds to the number of DOF associated with the motion of a body. The advantage of knowing the generalized coordinates for a system is that the number of independent variables can be reduced. In general, the generalized coordinates can be expressed as,

$$f_i = f_i\,(q_1, q_2, q_3, \ldots, q_i)$$

where q_i are the generalized coordinates. These generalized coordinates are used to determine the number of equations of motion for a vehicle.

2.3 Lagrange's equation

Newton's laws are commonly used to explain the motion of a body [10]. The simplified second law says that the sum of the forces equals the (constant) mass of the body times its acceleration and is written as follows,

$$\Sigma \vec{F} = m\vec{a} \tag{2.2}$$

where $\vec{F}$ are the forces acting on the body, $\vec{a}$ are the accelerations of the body, and m is the mass of the body. Equation (2.2) explains the translational motion of a body in three-dimensional space. These equations can be expressed for any number of independent variables which are reduced by the number of constraint equations. It is necessary to redefine equation (2.2) in a more general form such that the generalized coordinates are used to obtain the necessary equations of motion.

As shown in Goldstein [11], D'Alembert's principle can be used to obtain the impulsive form of Lagrange's equations. This form of Lagrange's equation can be written as,

$$\frac{d}{dt}\left(\frac{\partial T_K}{\partial \dot{q}_j}\right) - \frac{\partial T_K}{\partial q_j} = Q_j \tag{2.3}$$

where,

$$Q_j = \Sigma_i \vec{F}_i \frac{\partial \vec{r}_i}{\partial q_j} \tag{2.4}$$

T_K is the kinetic energy associated with the motion of the particle, and Q_j is the generalized force. The generalized forces are expressed in terms of the applied forces on the body ($\vec{F}_i$), and the location in which the forces are applied ($\vec{r}_i$). In equation (2.3), the kinetic energy can be expressed in terms of the position and velocity of the body. There are two types of forces that can be applied to the body: 1) body forces ($\vec{F}_i^b$), and 2) surface forces ($\vec{F}_i^S$). The body forces can be derived from a potential function as follows,

$$\vec{F}_i^b = \nabla U = \frac{\partial U}{\partial \vec{r}_i} \tag{2.5}$$

The commonly used body force is the gravity, because the gravity force is derivable from a potential function. Also, electromagnetic forces are another form of body forces that are derivable from a potential function. The surface forces, $\vec{F}_i^S$, also known as contact forces, are applied over an infinitesimal area on the body. Examples of surface forces are the thrusts provided by the vehicle, tension in a cable, and friction forces. The body forces are only functions of the position of the body while the surface forces can be functions of the position and velocity of the body. Also, the surface forces are functions of time depending on the problem constraints. Substituting the definitions of body and surface forces into equation (2.3), the equations of motion are represented as,

$$\frac{d}{dt}\left(\frac{\partial L}{\partial \dot{q}_j}\right) - \frac{\partial L}{\partial q_j} = Q_j^S \tag{2.6a}$$

where,

$$Q_j^S = \Sigma_i \vec{F}_i^S \frac{\partial \vec{r}_i}{\partial q_j} \tag{2.6b}$$

$$L = T_K - U \tag{2.6c}$$

In the practical sense, equation (2.3) is commonly used to determine the applied forces over a body. Lagrange's equations shown in equation (2.6) are also known as the energy method for the development of equations of motion. Equation (2.6) is a special form of Lagrange's equation [**11**], but $\vec{F}_i^S$ is normally zero. For some of the problems solved here, the surface forces are non-zero.

2.4 System of particles

The study of the orbital motion between a planet and satellite is concerned with the motion of a system of particles. In a system of particles, there is more than one mass traveling as a whole body. The applied forces associated with a group of particles are defined in terms of the weak and strong forces. The weak forces are the forces exerted on two particles in which the forces are equal and opposite. The weak forces refer to Newton's third law of motion. The strong forces are the forces exerted on two particles in which the forces are equal and opposite; but the forces applied to the two particles lie on the same line joining the particles [**11**].

Figure 2.1 provides a more comprehensive visualization and understanding of weak and strong forces. The black particles have a mass and are moving with some velocity. The gray point is the center of mass of the system of particles. In Position 1, particles 1, 2, and 3 have strong forces because the applied forces of the particles lie on the line connecting

Figure 2.1 System of particles

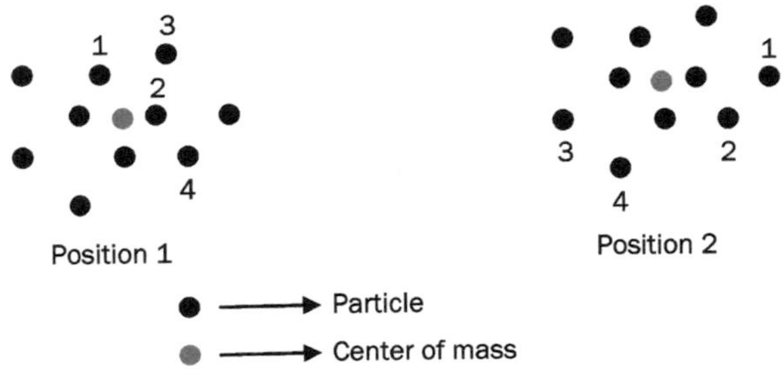

each other. Instead, particles 1 and 4 can be connected by different paths, and the forces applied to each other are weak. The main question is the following: what happens when the particles travel from Position 1 to Position 2? When the particles travel from one position to another, the weak and strong forces are interchanged as shown in the Position 2 of Figure 2.1, but the traveling of the particles from one point to another is always about the center of mass of the system of particles.

Because of the motion of a system of particles in Figure 2.1, the motion (kinetic energy) of a system of particles can be represented mathematically [11] as follows:

$$T_K = \bar{M}\ |\vec{V}|^2 + \Sigma_i m_i\ |\vec{v}_i|^2 \tag{2.7a}$$

where $\bar{M}$ is the total system mass, $\vec{V}$ is the velocity of the center of mass, m_i is the mass of a single particle, and $\vec{v}_i$ is the velocity of each particle. The position of the center of mass can be represented as follows,

$$\vec{R} = \frac{\Sigma_i m_i \vec{r}_i}{\Sigma_i m_i} \tag{2.7b}$$

In Equation (2.7a), the motion of a system of particles is described by the motion of the center of mass (as if all the mass were concentrated there) and the motion of the particles about the center of mass as shown in Figure 2.1. Goldstein [11] provides a more detailed development of equation (2.7a). The potential function is described for the system in terms of the potential for each particle and is defined as follows,

$$U = \Sigma_i U_i \tag{2.7c}$$

The definitions in equations (2.7) can be substituted into equations (2.6) to obtain the equations of motion for a system of particles. This formulation is mostly used for the solution of the two body orbital problem.

2.5 Two body orbital motion problem

Up to this point, an example of Lagrange's equation has not been used to demonstrate how the equations of motion can be obtained. In this section, the explanation of the two body problem for the orbital motion of a body is performed. Lagrange's equation is used to determine the equations of motion. As shown in Figure 2.2, assume two point masses, m_1 and m_2, moving about some central point. The only applied forces affecting the motion of these two masses are due to the interaction with a potential function, U. Also, assume that the system is conservative which means that the work done by the applied force is equal to zero. Using equations (2.6c) and (2.7), the Lagrange equations for the motion of the particles in Figure 2.2 can be described as,

$$L = T_K(\dot{\vec{R}},\dot{\vec{r}}) - U(\vec{r},\dot{\vec{r}}) \tag{2.8}$$

The kinetic energy associated to the motion of a system of particles in Figure 2.2 is equal to,

$$T_K(\dot{\vec{R}},\dot{\vec{r}})= \frac{1}{2}(m_1+m_2)\left|\dot{\vec{R}}\right|^2 + \frac{1}{2}m_1\left|\dot{\vec{r}}_1\right|^2 + \frac{1}{2}m_2\left|\dot{\vec{r}}_2\right|^2 \tag{2.9}$$

where $\vec{r}_1$ and $\vec{r}_2$ are the radii vectors of two particles measured from the center of the coordinate system. $\vec{R}$ is the distance of the center of mass relative to the coordinate system. The center of mass equation can be written as,

$$\vec{R} = \frac{m_1\vec{r}_1 + m_2\vec{r}_2}{m_1 + m_2} \tag{2.10a}$$

Figure 2.2 **Two body motion**

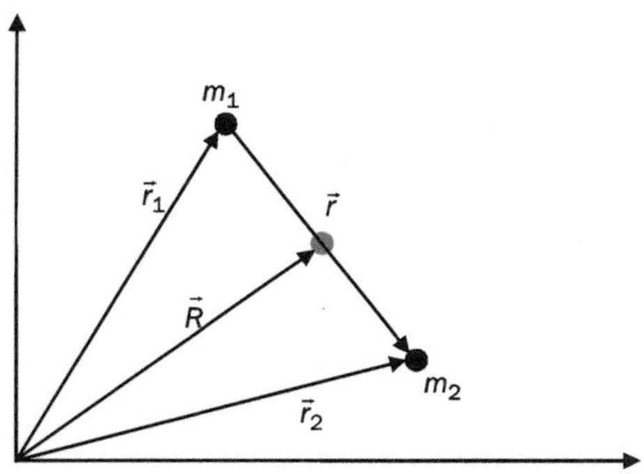

The distance between the two masses can be described as follows,

$$\vec{r} = \vec{r}_2 - \vec{r}_1 \tag{2.10b}$$

Using equations (2.10), the radii vectors of the two particles can be written as follows,

$$\vec{r}_1 = \vec{R} - \frac{m_2}{m_1 + m_2}\vec{r} \qquad \vec{r}_2 = \vec{R} + \frac{m_1}{m_1 + m_2}\vec{r} \tag{2.11}$$

If it is assumed that $\vec{R} = 0$, the center of mass is either at rest or moving uniformly; then, equation (2.11) becomes,

$$\vec{r}_1 = -\frac{m_2}{m_1 + m_2}\vec{r} \qquad \vec{r}_2 = \frac{m_1}{m_1 + m_2}\vec{r} \tag{2.12}$$

And, the kinetic energy equation in equation (2.9) becomes as,

$$T(\dot{\vec{R}},\dot{\vec{r}}) = \frac{1}{2}\frac{m_1 m_2}{m_1 + m_2}\left|\dot{\vec{r}}\right|^2 \tag{2.13}$$

Thus,

$$L = \frac{1}{2}\frac{m_1 m_2}{m_1 + m_2}\left|\dot{\vec{r}}\right|^2 - U(\vec{r},\dot{\vec{r}}) \tag{2.14}$$

The Lagrange equation does not contain terms involving the motion of the center of mass (M) relative to the location of $\vec{R}$. From the Lagrangian formulation and Figure 2.2, it can be observed that a fixed central force at particle 1 will act on particle 2 located at distance $\vec{r}$ with a reduced mass equal to,

$$\mu = \frac{m_1 m_2}{m_1 + m_2} \tag{2.15}$$

For the Earth-satellite problem, $m_1 = M_E \gg m_2$ where M_E is the mass of the Earth, and m_2 is the mass of the satellite. The reduced mass (μ) can be approximated to the mass of the satellite.

The solutions shown in equation (2.14) and (2.15) refer to the central force motion between a planet and a satellite. The central force motion of two bodies about their center of mass can always be reduced to an equivalent one body problem. For the Earth-satellite motion, the central force is exerted on the satellite by the Earth. As shown in equation (2.11) and in Figure 2.2, the center of mass for the system of particles is not necessarily at the center of the coordinate system.

2.6 Orbital equations of motion

From the previous section, it is shown that the two body problem can be expressed as a reduced single body problem as shown in Figure 2.3 for the Earth-satellite system. The potential function for the inverse square force field law defined by Newton can be written as [10],

$$U(R) = -\frac{GM_E m}{R} \tag{2.16a}$$

where G is the universal gravitational constant, and R is the distance from the center of the Earth to the satellite. The kinetic energy of the relative motion of the satellite about the Earth can be defined as,

$$T\tilde{K} = \frac{1}{2}m(\dot{R}^2 + R^2\dot{f}^2) \tag{2.16b}$$

where $\dot{R}^2 = \dot{\vec{R}} \cdot \dot{\vec{R}} = |\dot{\vec{R}}|^2 \quad R^2 = \vec{R} \cdot \vec{R} = |\vec{R}|^2$. The Lagrangian equation becomes as,

$$L = \frac{1}{2}m(\dot{R}^2 + R^2\dot{f}^2) + \frac{GM_E m}{R} \tag{2.17}$$

For this formulation, there are no applied surface forces on the satellite. This problem has two DOF because the satellite is constrained to move about the Earth in a plane. The two generalized coordinates for this system are $q_1 = R$ and $q_2 = f$. The two equations of motion for the satellite can be obtained from the solution of Lagrange's equation and are written as,

Figure 2.3 Earth-satellite system

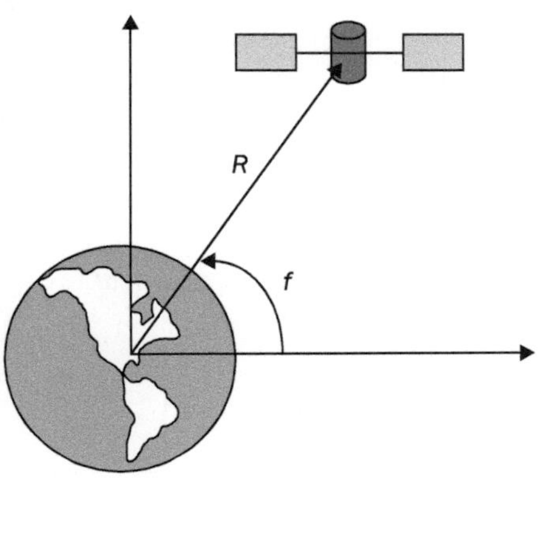

$$\ddot{R} - R\dot{f}^2 = -\frac{\mu}{R} \tag{2.18a}$$

$$mR\ddot{f} + 2m\dot{R}\dot{f} = 0 \tag{2.18b}$$

where $\mu = GM_E = 398600.4418$ (km^3/sec^2) [10]. From equation (2.18b), the following equation can be written associated with the momentum of the vehicle,

$$mR\ddot{f} + 2m\dot{R}\dot{f} = \frac{d}{dt}(mR^2\dot{f}) = 0$$

Integrating the previous equation,

$$l = mR^2\dot{f} \Rightarrow h = \frac{l}{m} = R^2\dot{f} \tag{2.19}$$

l is the magnitude of the angular momentum associated with the orbital motion of the vehicle, and h is the angular momentum magnitude per unit mass. If $h = 0$, the particle is moving in a straight line because the angular momentum vector is perpendicular to the radius of the orbit motion. If $h \neq 0$, the particle is subjected to move along an orbit path [11]. Figure 2.4 shows the orbit path of the satellite about the Earth. In a period of time (dt), the satellite sweeps through an angle df. As shown from Figure 2.4, the differential area associated with the motion of the satellite from one point to another can be written as,

$$dA = \frac{1}{2}R(R\,df) = \frac{1}{2}R^2\frac{df}{dt}dt$$

$$\frac{dA}{dt} = \frac{1}{2}R^2\frac{df}{dt} \Rightarrow 2\frac{dA}{dt} = h \tag{2.20}$$

Equation (2.20) demonstrates the conservation of angular momentum which is equivalent to saying that the areal velocity of the satellite is

Figure 2.4 **Area covered by the satellite orbit path**

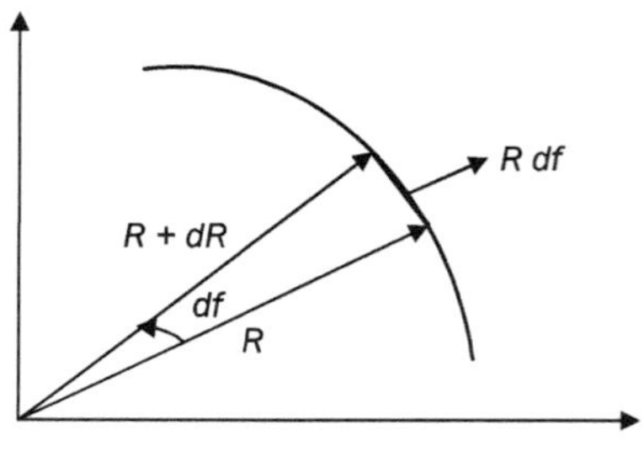

constant. In the early 1600s, Kepler developed three laws of planetary motion [**10**]. These three laws were developed by observing the motion of the planets. Kepler's second law states that under the action of a gravitational field, the motion of the radius vector to the orbiting body sweeps out equal areas in equal times. Equation (2.20) demonstrates mathematically Kepler's second law of planetary motion. Kepler's second law was developed in 1609.

In 1609, Kepler developed his first law that explains the orbit path of the planets. Kepler's first and second laws of planetary motion were developed at the same time. Kepler's first law says that all planets describe elliptical orbits which have the Sun at one focus. He discovered the elliptical orbits by observing Mars and by selecting the correct geometrical figure that could fit the planet path. Kepler's first law is demonstrated by solving equation (2.18a). Substituting equation (2.19) into equation (2.18a), the radial equation can be written as follows,

$$\ddot{R} = \frac{h^2}{R^3} - \frac{\mu}{R^2} \tag{2.21}$$

Let $u = 1/R$ and $\dot{f} = hu^2$; then, the following transformations can be performed,

$$\frac{dR}{dt} = \frac{d}{dt}(u^{-1}) = -\frac{1}{u^2}\dot{f}\frac{du}{df} = -h\frac{du}{df} \tag{2.22a}$$

$$\frac{d}{dt}\frac{dR}{dt} = -h^2u^2\frac{d^2u}{d^2f} \tag{2.22b}$$

Using equations (2.22), equation (2.21) is transformed as,

$$\frac{d^2u}{d^2f} + u = \frac{\mu}{h^2} \tag{2.23}$$

The solution of equation (2.23) with the variation of parameters [**12**] is,

$$u = B\cos f_0 \cos f + B\sin f_0 \sin f + \frac{\mu}{h^2} \tag{2.24}$$

Using the following trigonometric identity [**13**],

$$\sin x \sin y = \cos(x - y) - \cos x \cos y$$

Equation (2.24) is reduced to,

$$u = B\cos(f_0 - f) + \frac{\mu}{h^2} \tag{2.25}$$

Without loss of generality, let $f_0 = 0$ and substitute the definition for μ; then, the radial equation can be written as,

$$R = \frac{h^2/\mu}{1+\epsilon\cos f} \tag{2.26}$$

where,

$$\epsilon = \frac{Bh^2}{\mu}$$

Equation (2.26) is the equation of a conic section in polar coordinates. Before explaining and defining the unknown variables in equation (2.26), the significance of B should be determined. A set of curves called conic sections are defined by the situation shown in Figure 2.5. A conic section is defined as the locus of a point whose undirected distance from a fixed point (focus) divided by its undirected distance from a fixed line (called directrix) is a constant. In Figure 2.5, D is the directrix line shown by the dashed line, d is the undirected distance from the focus point to the directrix line, and SD is the undirected distance from the satellite to the directrix line. Using the definition of a conic section, the following equation is obtained,

$$\epsilon = \frac{R}{SD} = \frac{R}{d - R\cos f} \Rightarrow \frac{1}{R} = \frac{1}{\epsilon d} + \frac{1}{d}\cos f \tag{2.27}$$

Comparing equation (2.27) to equation (2.25), $B = 1/d$. B is used to determine the location of the directrix line with respect to the focal point. In equation (2.26) and (2.27), ϵ is known as the eccentricity and determines the nature (shape) of the curve. Table 2.1 shows the different orbits that can be produced with different eccentricity values. Figure 2.6

Figure 2.5 Definition of a conic section

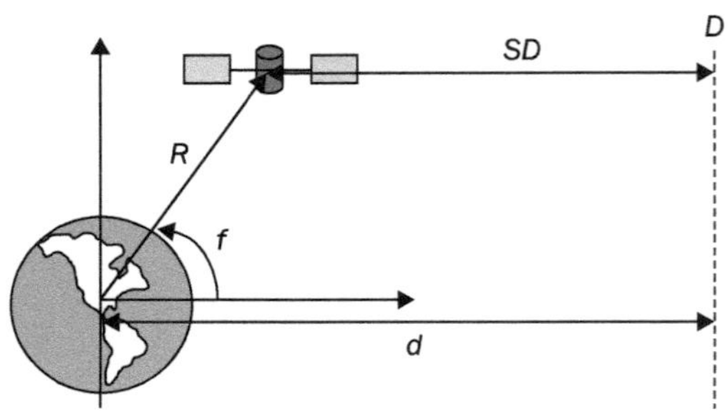

Table 2.1 **Nature of the curves**

Nature of the curve	Eccentricity value
Circular orbit	0
Elliptical orbit	$0 < \epsilon < 1$
Parabolic orbit	1
Hyperbolic orbit	$\epsilon > 1$

shows a graphical visualization of the different curves for the angular momentum per unit mass (h) equal to 73,000 km/sec^2.

There are other properties that can be determined from the elliptical orbit. If $f = 0°$ *or* $180°$, there are two radii that describe the size of the orbit. If $f = 0°$, the satellite is located at the perigee point of the ellipse which is the point closest to the Earth. On the other hand, the point farthest from the Earth is known as the apogee point, and the radius is determined when $f = 180°$. Evaluating these angles in equation (2.26), the radius of perigee (r_p) and radius of apogee (r_a) are equal to,

Figure 2.6 **Satellite orbit paths**

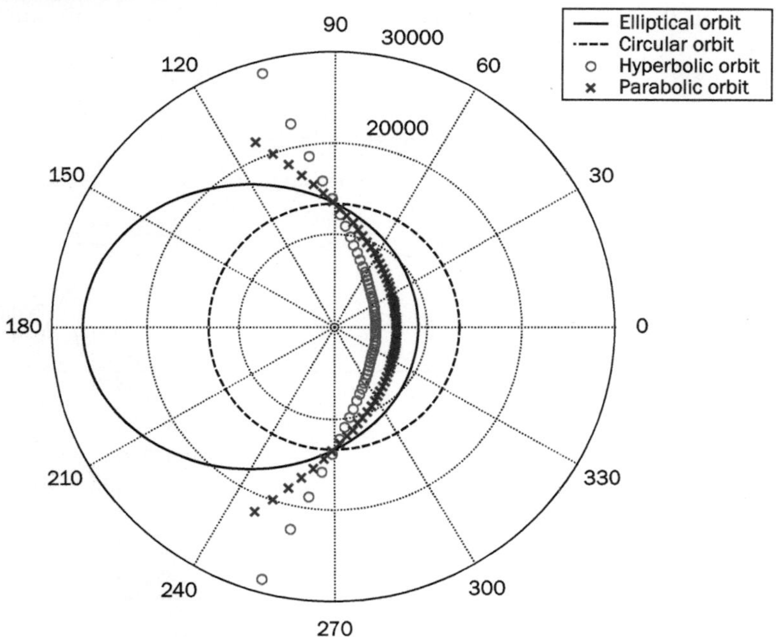

$$r_p = R(0) = \frac{h^2}{\mu}\frac{1}{1+\epsilon} \qquad r_a = R(180) = \frac{h^2}{\mu}\frac{1}{1-\epsilon} \tag{2.28}$$

In the ellipse shown in Figure 2.7, the longest distance from perigee to the apogee point passing through the focus point is known as the major axis. The focus point is located at the center of the Earth. The major axis is calculated as twice the semimajor axis (a) and is described as,

$$2a = r_p + r_a \Rightarrow a = \frac{r_p + r_a}{2} \tag{2.29}$$

Using equation (2.28) and (2.29), the numerator in equation (2.26) can be calculated as,

$$\frac{h^2}{\mu} = a(1-\epsilon^2) \tag{2.30}$$

And, the equation of a conic section can be rewritten as,

$$R = \frac{a(1-\epsilon^2)}{1+\epsilon\cos f} \tag{2.31}$$

In equation (2.31), R defines the distance from the focus point to the position of the satellite in the orbit. f determines the position of the satellite in the orbit; this variable is known as the true anomaly angle.

Figure 2.7 **Elliptical orbit size specifications**

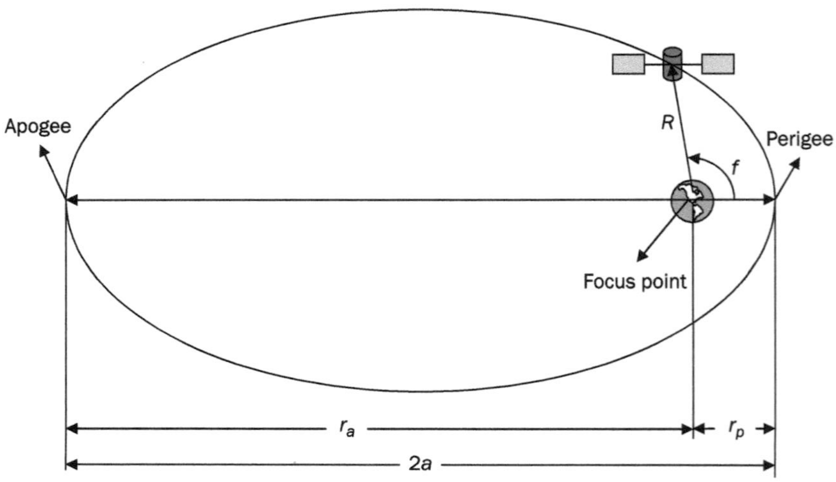

The equation of a conic section is a function of the true anomaly angle or the location of the satellite in the orbit path.

There is one more distance associated with the 90 degree angle. When $f = 90°$, equation (2.31) equals,

$$R\left(\frac{\pi}{2}\right) = \frac{a(1-e^2)}{1} = p \tag{2.32}$$

p is known as the semi-latus rectum and is also used to describe the size of the orbit, in addition to the semimajor axis.

2.7 Energy and velocity of orbiting bodies

The different orbits of the satellite are determined in terms of the size and nature of the curve. In addition to size and type of the orbit, the energy and velocity of the satellite can be used to describe the orbit path. Multiplying equation (2.18a) by $\dot{R}$ and equation (2.18b) by $R\dot{f}$, the addition of both equations results in,

$$m(\dot{R}\ddot{R} + R\dot{R}\dot{f}^2 + R^2\dot{f}\ddot{f}) + \frac{\mu m}{R^2}\dot{R} = 0 \tag{2.33}$$

In the left hand side (LHS) of equation (2.33), the first three terms can be written as,

$$\dot{R}\ddot{R} = \frac{1}{2}\frac{d}{dt}(\dot{R}^2) \qquad R\dot{R}\dot{f}^2 + R^2\dot{f}\ddot{f} = \frac{1}{2}\frac{d}{dt}(R^2\dot{f}^2) \tag{2.34}$$

Using equation (2.33) and multiplying the final equation by dt, equation (2.33) is equal to,

$$\frac{m}{2}\left(\frac{d}{dt}(\dot{R}^2)dt + \frac{d}{dt}(R^2\dot{f}^2)dt\right) + \frac{\mu m}{R^2}dR = 0 \tag{2.35}$$

Integrating equation (2.35), the energy equation for the satellite motion is,

$$\frac{m}{2}(\dot{R}^2 + R^2\dot{f}^2) - \frac{\mu m}{R} = V \tag{2.36}$$

The first two terms on the LHS of equation (2.36) resemble the kinetic energy of the vehicle along the orbit path, and the third term of equation (2.36) is the potential energy described by the inverse square law. V is the energy associated with the motion of the satellite in the orbit and is a constant. Equation (2.36) can be rewritten as follows,

$$\frac{m}{2}v^2 - \frac{\mu m}{R} = V \tag{2.37}$$

where $v^2 = \dot{R}^2 + R^2\dot{f}^2$. In equation (2.37), the square of the velocity varies inversely with the distance from the center of the Earth and inversely proportional to the energy. Also, the sum of the kinetic energy and the potential energy is equal to the total energy of the orbit which is equal to a constant. Equation (2.37) may also be written as,

$$v^2 = -2\int_\infty^R \frac{\mu}{r^2}dr + \frac{V}{m} \tag{2.38}$$

Equation (2.38) is referred to as the vis-viva ('life energy') integral.

At the perigee and apogee points, the radial velocity is equal to zero; the only non-zero component is the tangential velocity. At the perigee point, the velocity component can be written as,

$$v_p = r_p\dot{f}_p = \frac{h}{r_p} \tag{2.39}$$

Remember that the angular momentum is a constant as shown in equation (2.19). The radius of perigee can be written as,

$$r_p = \frac{h^2}{\mu}\frac{1}{1+\dfrac{Bh^2}{\mu}} \Rightarrow \frac{1}{r_p} = \frac{\mu}{h^2} + B \tag{2.40}$$

Then, equation (2.39) is equal to,

$$v_p = \frac{\mu}{h} + Bh \tag{2.41}$$

Substituting equation (2.40) and (2.41) into equation (2.37), the energy equation is defined as,

$$V = -\frac{m}{2}\left(\frac{\mu}{h}\right)^2\left(1-\left(\frac{Bh^2}{\mu}\right)\right)$$

From equation (2.30) and the definition of eccentricity in equation (2.26), the energy equation is reduced to,

$$V = -\frac{m\mu}{2a} \tag{2.42}$$

The total energy in an elliptical orbit is inversely proportional to the semimajor axis. It is possible to follow the same calculations to determine the energy associated with other types of orbits.

For the parabola, the semimajor axis tends to infinity because the vacant focus in the ellipse tends to $-\infty$. Thus, the energy associated with the parabola is equal to zero. If the kinetic energy is larger than the potential energy of the satellite, the energy equation (2.37) is larger than zero which implies that the eccentricity of the orbit is larger than one. This eccentricity refers to the hyperbolic orbit in which the total energy is equal to,

$$V = \frac{m\mu}{2a} \tag{2.43}$$

where a is the semi-transverse axis for the hyperbola.

Substituting equation (2.42) into equation (2.37), the velocity at any point in the elliptical orbit is equal to,

$$v = \sqrt{\mu\left(\frac{2}{R} - \frac{1}{a}\right)} \tag{2.44}$$

A similar analysis can be performed to determine the velocity at any point for any orbit. Table 2.2 shows the final results between the energy and velocity of the orbit. In Table 2.2, the energy associated with the circular orbit is lower than the energy associated with the hyperbolic orbit. This shows that an increment in energy is required to transfer from a circular orbit to any other orbit. The velocity does not show the same variation because it is not a constant in the elliptical, hyperbolic, and parabolic orbits. The velocity is only constant for the circular orbit. In practice, the circular orbit is used as a transitory orbit to transfer to other orbits.

Table 2.2 **Energy and velocity relations for the orbits**

Type of orbit	Energy	Velocity
Circular orbit	$-\frac{m\mu}{2R}$	$\sqrt{\frac{\mu}{R}}$
Elliptical orbit	$-\frac{m\mu}{2a}$	$\sqrt{\mu\left(\frac{2}{R} - \frac{1}{a}\right)}$
Parabolic orbit	0	$\sqrt{\frac{2\mu}{R}}$
Hyperbolic orbit	$\frac{m\mu}{2\mathrm{a}}$	$\sqrt{\mu\left(\frac{2}{R} + \frac{1}{\mathrm{a}}\right)}$

2.8 Escape velocity

The escape velocity, v_{esc}, is the necessary velocity to escape the Earth's gravitational attraction. In theory, this means that the initial kinetic energy plus the gravitational potential is equal to zero when the distance approaches infinity. In practice, the escape velocity is associated with the necessary velocity to send a rocket into Space.

Using Figure 2.8, the work done against the gravitational force for a moving body over a distance dr can be written as,

$$W_{12} = -\mu m \int_{R_e}^{\infty} \frac{dr}{r^2} = -\frac{\mu m}{R_E} \tag{2.45}$$

where R_E is the radius of the Earth which is equal to 6,378,136.49 (m) [10]. For a conservative system, the energy equation can be written as,

$$W_{12} = T_2 - T_1 = -\frac{1}{2} m v_{esc}^2 \tag{2.46}$$

then,

$$v_{esc} = \sqrt{\frac{2\mu}{R_E}} \tag{2.47}$$

Equation (2.47) is the necessary velocity to escape from the Earth's surface. It is possible to generalize equation (2.47) to escape from any altitude as,

$$v_{esc} = \sqrt{2} v_c(R) \tag{2.48}$$

where,

$$R = R_E + \overline{s} \qquad v_c(R) = \sqrt{\frac{\mu}{R}} \tag{2.49}$$

Figure 2.8 Diagram for the escape velocity calculation

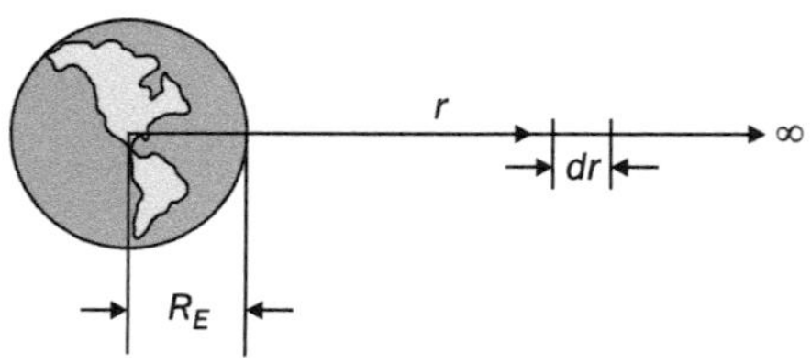

Figure 2.9 **Escape velocity and circular velocity vs. altitude**

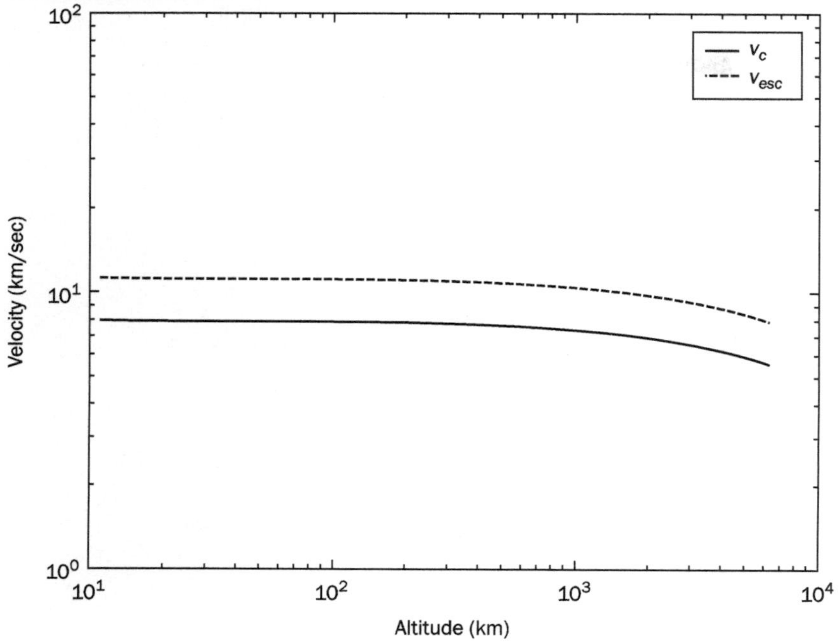

$\bar{s}$ is known as the altitude of the orbit and is measured from the Earth's atmosphere to the satellite orbit. Figure 2.9 shows the graph of the escape velocity and circular velocity versus altitude. The difference between the two graphs provides the necessary change in velocity to cause the body to be injected from an initial circular orbit into an escape trajectory.

2.9 Earth Coordinate Inertial (ECI) system

The Earth Coordinate Inertial system (ECI) is an imaginary Cartesian coordinate system placed at the center of the Earth. The ECI can be used to describe the orbit path of a satellite. To describe the orbit path, six orbital elements are used in which four elements explain the planar motion of the satellite and two elements specify the orientation of the plane. Figure 2.10 is used to describe the six orbital elements, but there are other definitions of different planes and lines needed for the explanation of the orbital elements.

Figure 2.10 **Earth Centered Inertial (ECI) frame**

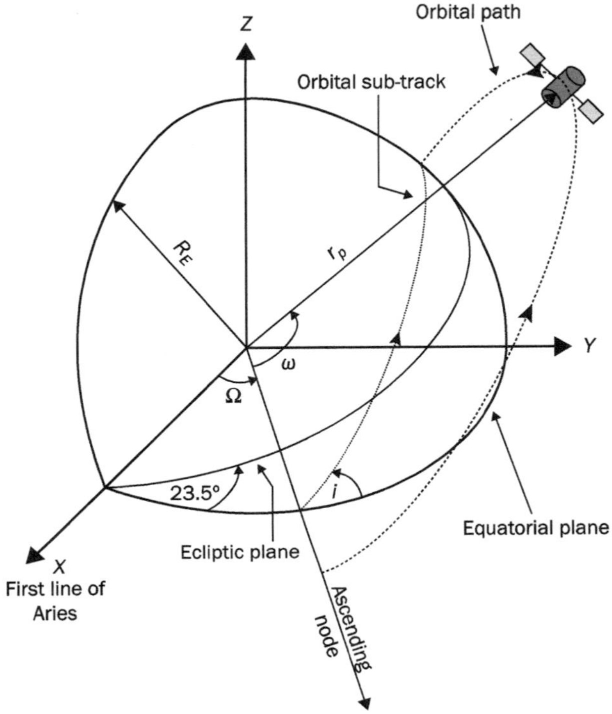

In Figure 2.10, the X direction of the Cartesian frame is known as the First Line of Aries. The First Line of Aries is the line from the Earth's center to the direction of the Sun at the time when the Sun crosses the Celestial Equator moving North at the vernal (Spring) equinox. The Celestial Sphere is an imaginary sphere of infinite radius upon which the celestial bodies are assumed to be fixed. This is the analogy with planetariums where the Earth's center is at the center of the celestial bodies that are projected upon the inner walls of the planetarium. The Earth's radius is not significant in comparison, and the calculations are based upon the assumption that the terrestrial observations are made from the center of the Earth rather than at the surface. The Celestial Equator is the great circle of the Celestial Sphere formed by extending the plane of the Earth's equator until it cuts the celestial sphere (equinoctial equator). The Ecliptic plane is the plane which describes the relative motion of the Sun about the Earth (or vice-versa). The Ecliptic plane is inclined 23.5 degrees with respect to the Celestial Equator. The orbit plane is the extended plane of the orbit path of the satellite.

As explained in the previous sections, the semimajor axis (a) and the eccentricity (e) explain the size and shape of the orbit, respectively. The intersection between the orbital plane and the Celestial Equator is known as the Line of Nodes. The angle measured from the First Line of Aries to the Line of Nodes is known as the Right Ascension of the Ascending Node (RAAN or Ω). This angle explains the location of the line of nodes when the satellite is moving from South to North. There is a second angle 180 degrees apart from the RAAN known as the Right Descension of the Descending Node. The argument of perigee (ω) is the angle measured from the Line of Nodes to the perigee point. The inclination angle (i) is the measured angle from the line perpendicular to the Celestial Equator to the line perpendicular to the orbital plane. The inclination angle also describes the coverage pattern of the satellite about the Celestial Sphere. The Time of Perigee (T_p) passage is the time that the satellite takes to move from the Line of Nodes to the perigee point.

In summary, the four orbital elements describing the planar motion are: the semimajor axis, eccentricity, argument of perigee, and time of perigee passage. The inclination angle and the RAAN are used to describe the orientation of the orbital plane about the ECI frame. These parameters are important for the explanation of frame rotations in Chapter 4.

2.10 Period of an orbit

Another parameter that can be used to describe different orbits is the period of an orbit, T. The orbital period describes the amount of time that the satellite takes to complete one full orbit. From the conservation of angular momentum, the areal velocity (equation (2.20)) is constant.

The area of the orbit (A) can be found by integrating equation (2.20) over one complete period as,

$$\int_0^T \frac{dA}{dt} dt = A = \frac{h}{2} T \tag{2.50}$$

The area of an ellipse is equal to [13],

$$A = \pi ab \tag{2.51}$$

where a is the semimajor axis, and b is the semiminor axis of the ellipse. The semiminor axis of an ellipse is described as [13],

$$b^2 = a^2(1 - \epsilon^2) = ap = a\frac{h^2}{\mu} \tag{2.52}$$

Equation (2.52) is obtained from the definition of the semi-latus rectum and equation (2.30). Substituting equations (2.50) and (2.52) into equation (2.51), the orbit period can be written as,

$$T = 2\pi\sqrt{\frac{a^3}{\mu}} \tag{2.53}$$

Kepler developed his third law of planetary motion [10] in 1618. The third law states that the square of the period of revolution of a planet is proportional to the cube of its semimajor axis. The period of an orbit can be used to classify some orbits. One example is the geosynchronous orbit which has the same period as the Earth's rotation about its polar axis. When a geosynchronous satellite is observed from the Earth, the satellite appears to be stationary but is moving at the same rate as the Earth. Substituting $T = 24$ hrs, the semimajor axis equals to 2,241.10 km. The satellites in geosynchronous orbits are placed in circular orbits. Figure 2.11 shows the relation between the semimajor axis and the orbit period.

Figure 2.11 **Orbital period**

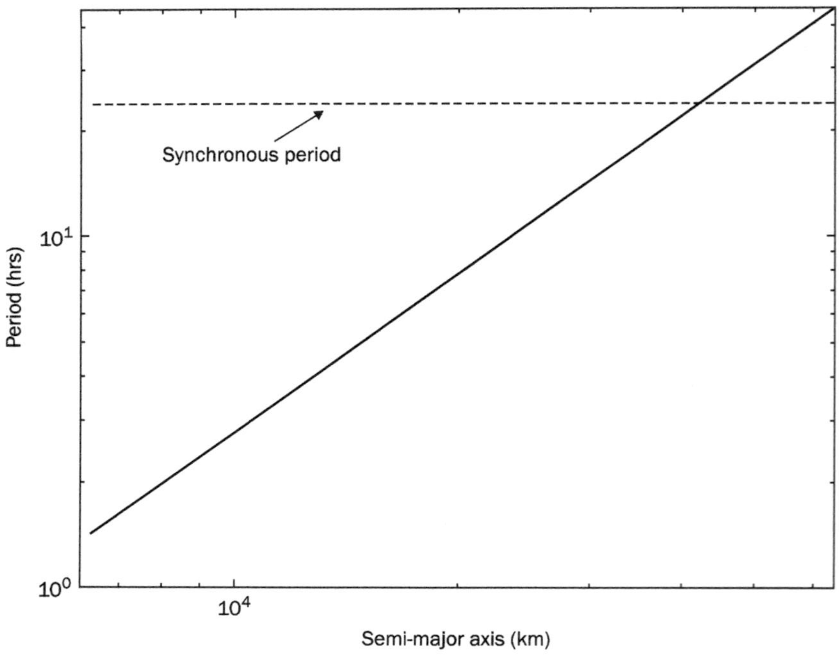

2.11 Development of Kepler's equation

In Section 2.6, the equation of a conic section is defined as a function of the true anomaly angle, R(*f*). This section develops the equation of a conic section and the true anomaly as a function of time. In reality, a relation between the time and the true anomaly angle for eccentric orbits is required. The true anomaly angle has different time relationships for circular and elliptical orbits. Figure 2.12 shows the elliptical orbit of a satellite. In Figure 2.12, there is an outer circle of radius a centered at the geometrical center of the ellipse. The angle called E is known as the eccentric anomaly and is used to describe the projection of the satellite position on the outer circle. The distance from the focus to the center of the ellipse is described by c which equals to ϵa. From the triangle ABD, the following relationship can be written,

$$R \cos f = a \cos E - \epsilon a \tag{2.54}$$

Using equation (2.32) and the previous equation, the equation of a conic section (equation (2.31)) can be defined as a function of the eccentric anomaly as,

$$R = p - \epsilon R \cos f = a(1 - \epsilon \cos E) \tag{2.55}$$

Figure 2.12 Geometry of the ellipse about the Earth

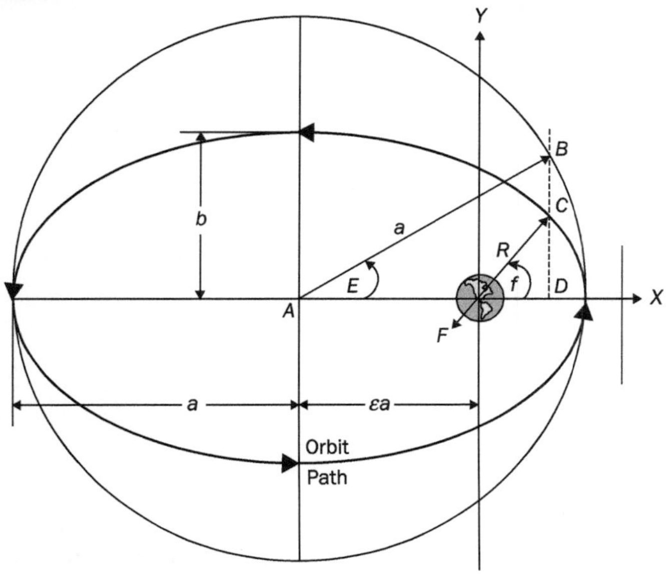

The eccentric anomaly is also a function of time. The time derivative of equation (2.55) determines the radial velocity equation as,

$$\dot{R} = a\epsilon\dot{E}\sin E \tag{2.56}$$

Using the triangle FCD and equations (2.54) and (2.55), a relationship for the sin E can be obtained as follows,

$$R^2 \sin^2 f = R^2 - R^2 \cos^2 f$$

$$R \sin f = a\sqrt{1-\epsilon^2}\sin E \tag{2.57}$$

Substituting equation (2.57) into equation (2.56), the radial velocity can be related to the time derivative of the eccentric anomaly as,

$$\dot{E} = \frac{\dot{R}}{R}\frac{\sqrt{1-\epsilon^2}}{\epsilon \sin f} \tag{2.58}$$

The time derivative of the inverse of the equation of the conic section equals to,

$$\frac{\dot{R}}{R} = \frac{\epsilon \sin f}{a(1-\epsilon^2)}\frac{R^2\dot{f}}{R}$$

Using the definition of the angular momentum and equation (2.55), the previous equation is defined as follows,

$$\frac{\dot{R}}{R} = \frac{\epsilon h \sin f}{a^2(1-\epsilon^2)(1-\epsilon\cos E)}$$

Then, equation (2.58) equals to,

$$\dot{E} = \frac{h}{a^2(1-\epsilon^2)(1-\epsilon\cos E)} \tag{2.59}$$

Separating the variables, equation (2.59) can be integrated as,

$$\int_0^E (1-\epsilon\cos E)\,dE = \int_{T_p}^t \frac{h}{a^2(1-\epsilon^2)}dt$$

$$E - \epsilon\sin E = \frac{h}{a^2(1-\epsilon^2)}(t-T_p) \tag{2.60}$$

The RHS of equation (2.60) is known as the mean anomaly angle (M) and is a function of time. The mean anomaly angle is written as,

$$M(t) = n(t - T_P) \tag{2.61}$$

where n is the mean motion or average angular velocity of the satellite. The mean motion can be written as,

$$n = \frac{h}{a^2(1-\epsilon^2)} = \frac{2\pi}{T}$$

Then,

$$M(t) = n(t - T_P) = E - \epsilon \sin E \tag{2.62}$$

Equation (2.62) is Kepler's equation which is transcendental in E. There are several methods to solve equation (2.62), but Newton's method [**14**] can be easily applied to obtain the solution for equation (2.62). Kepler's equation can be solved as follows:

a. Find $M(t)$ given $t - T_P$ and the orbital elements.

b. Solve Kepler's equation for E with Newton's method.

c. Solve for R in equation (2.55).

d. Determine the true anomaly angle with the equation of a conic section.

There are other relations that can be used to obtain more equations describing relations between the eccentric anomaly and the true anomaly angle. With equations (2.62) and (2.55), the location of the satellite in an elliptical orbit can be obtained. For circular orbits, the use of these equations is meaningless. Instead, the true anomaly angle in a circular orbit is described as $f = nt$. This shows that the eccentric anomaly and the mean anomaly are the same.

2.12 Suggested problems

Problem 2.1. Show that the general Lagrange's equation (equation (2.3)) can be also written as,

$$\frac{\partial \dot{T}_K}{\partial \dot{q}_j} - 2\frac{\partial T_K}{\partial q_j} = Q_j$$

Problem 2.2. A particle of mass m hangs by an extensionless string of length l from a point P on the circumference of a rigid disk of mass $2m$ and radius a. The disk can rotate about the fixed axis passing through its

Figure 2.13 Problem 2.2

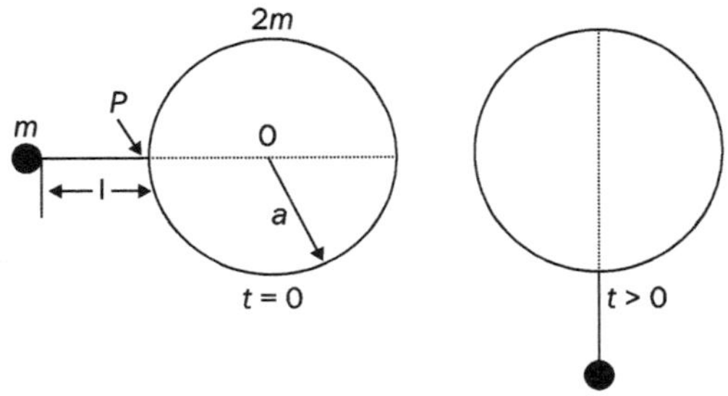

center O. At $t = 0$, the system is released from rest when the string is horizontal with the diameter of the circular; in other words, the mass is released when the string is parallel with the dotted line. Derive the equation of motion using Lagrangian formulation and find the path of the particle shortly after the system is released in a linear system.

Problem 2.3. Suppose that a particle of mass m is in the motion describing the circle r and height z in a conservative force field in which the potential energy is $V(r, z)$, where $r^2 = x^2 + y^2$.

a. Find the equations of motion.
b. Consider the steady motion of mass m in which $\dot{\theta}$ is constant. Find the condition of radial stability of the motion. [Hint: Study the perturbation of the radius of the circle with a perturbation term, ϵ_p.)

Problem 2.4. Formulate the equations of motion for a mass which is held by a force of the form

$$F = -\frac{a}{r^2} + \frac{b}{r^3}$$

where $\vec{F} = -\nabla V_r$, that is directed only along the radial (refer to Figure 2.3). Assume that there are no other external forces or torques acting on the mass.

a. Find the potential function from $\vec{F} = -\nabla V_r$.
b. Derive the Lagrange's equation of motion in the generalized coordinates (r, f).
c. Integrate the equations of motion analytically.
d. Discuss the physical significance of your result.

Problem 2.5. In the Gemini Mission, the Gemini vehicle was to rendezvous with the target Agena vehicle which was initially in a higher (altitude) orbit.

Figure 2.14 Problem 2.3

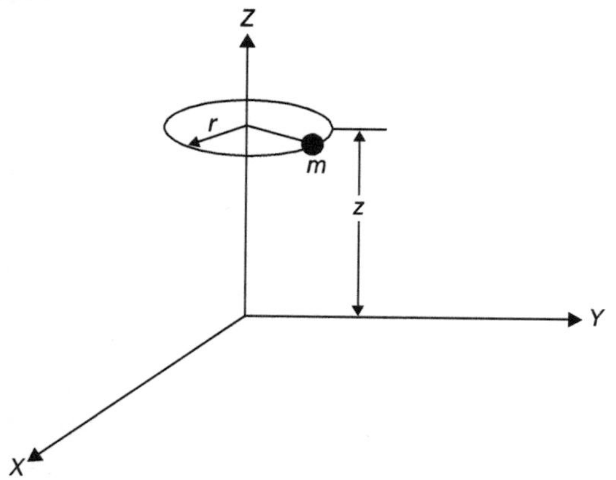

The intermediate portion of this mission was accomplished by maneuvering the Gemini vehicle into a transfer orbit between its initial 87 nautical mile altitude circular orbit and the 161 nautical mile altitude circular orbit of the target. Consider the transfer orbit with the following characteristics:

Perigee altitude = 87 nautical miles
Apogee altitude = 161 nautical miles

a. Calculate the semimajor axis in feet, the eccentricity, and the orbital period in seconds for the transfer orbit.
b. Calculate the orbital period (sec) for the target vehicle in the 161 nautical mile altitude circular orbit. What is the mean value of the Agena's orbital angular velocity?
c. At the initial time, $t = 0$, consider the Gemini to be at the perigee position in the transfer orbit. It is desired to complete rendezvous at the first available time after two complete revolutions in the transfer orbit. What is the total time required?
d. Assuming an impulsive velocity correction capability, compute the velocity increment and direction in order to affect a soft rendezvous at the time indicated in (c).

Constants for this problem,

$$\mu = GM_E = 1.407654 \times 10^{16} \text{ ft}^3/\text{sec}^2$$

$$R_E = 20925874 \text{ ft}$$

1 nautical mile = 6076.1033 ft

Problem 2.5. A Venus probe departs from Earth on June 21 following a transfer orbit of the following characteristics,

$a = 384 \times 10^9$ ft

$e = 0.277$

$\Omega = 90°$ East of Aries

Assume that the orbits of Venus and the Earth about the Sun are coplanar and circular.

a. Sketch the transfer orbit, roughly showing its relationship to the orbits of the Earth and Venus. Label the aphelion and perihelion points in this orbit. Indicate the Earth's position in its orbit on March 21 and June 21.

b. What velocity increment is required to leave the Earth's orbit on June 21 and establish the transfer orbit?

c. Calculate the time required in the transfer orbit from June 21 to the first contact with the orbit of Venus. Indicate this point on your sketch.

d. Without a numerical calculation, explain the impulsive maneuver required to enter the orbit of Venus at this point. (i.e. will this impulse involve a change in velocity magnitude, direction or both?)

Constants for the use in this problem,

Venus orbital radius 355×10^9 ft; velocity 114,800 ft/sec

Earth orbital radius 491×10^9 ft; velocity 97,600 ft/sec

With respect to the heliocentric (sun-centered coordinate system)

$GM_{SUN} = 4.679 \times 10^{21}$ ft^3/sec^2

3

Orbital perturbations in the two body motion

Abstract: The two body orbital motion is considered as an undisturbed movement of a body about a planet. In Space, there is an unexpected environment that can cause disturbances to the motion of a satellite. The objective of this chapter is to present the perturbations due to the Earth and Sun that highly affect the orbital motion. These effects can cause variations in the orbital elements that are explained through Lagrange's planetary equations. Throughout this chapter, the disturbances by the Earth and the Sun are explained in detail such that there is an understanding of how the satellite motion can be affected.

Key words: Earth and Sun disturbances, Lagrange's planetary equations for potential function and impulsive forces, oblateness of the Earth (J2 perturbation), atmospheric force, solar pressure force.

3.1 Introduction to disturbance effects

Kepler's equations are used as the base formulation for the description of the satellite motion without any disturbances. In reality, this orbital motion of the satellite does not maintain the desired orbit for a long period of time as shown in Chapter 2. The orbit begins to deform after some period of time. The effects that cause these variations to the orbital motion of a satellite are known as perturbations. The principal perturbations affecting a satellite are due to the Earth and the Sun.

The objective of this chapter is to present the formulation for the perturbations due to the Earth and the Sun. Before the perturbations are formulated, the Lagrange planetary equations are defined [**15**]. These equations describe the variations of the orbital elements in terms of a potential function or applied surface forces along the orbit. Using these equations, the perturbations are analyzed to determine how the satellite's orbit will behave. The disturbances can be classified as secular and short and long period variations. A secular perturbation is a linear variation in the orbital element that keeps increasing its effects through time [10]. Short period variations are perturbations that show its effects in less than one orbital period; otherwise, the perturbations that happen after one period are known as long period variations. Many of the perturbations due to the Earth and Sun cause secular effects to the orbital elements. Throughout this chapter, some of the perturbations are explained in terms of the mean of a function. The mean of a function is obtained to determine the effects of the perturbations on the orbital elements in one orbital period.

The Earth has disturbance effects due to the oblateness of the Earth and the atmosphere. The Earth is not a spherical body but has a bulge about the Equator. The gravitational potential function of the Earth is expanded in terms of Legendre polynomials to explain different shapes of the Earth. On the other hand, the Earth's atmosphere also causes a disturbance on the satellite. Because of the interaction with the Earth's atmosphere, the satellite orbit keeps decreasing its energy in every orbit and moves toward the center of the Earth. The Sun also causes effects on the satellite motion due to the interaction with the surface of the satellite. The reflective effects of the material and the area of contact of the satellite with the Sun create additional forces that can affect the satellite motion. The combination of these effects can cause the satellite to move away from the desired orbit.

In summary, the following topics explain the principal disturbance effects due to the Earth and Sun. Also, the Lagrange planetary equations are used to explain the effect of the perturbations on the orbital elements; in this way, the reader can have an idea of what is happening to the satellite orbit.

3.2 Lagrange planetary equations

The orbital elements can be perturbed by single impulses applied to the satellite or a potential function. Also, the orbital elements can be propagated in time to determine the effects of the perturbing forces on the satellite. Battin [16] presented a set of equations that are perturbed by a conservative field. These equations are developed with the Lagrange brackets [11] for which the name of the equations are derived. The Lagrange planetary equations can be written as follows,

$$\frac{da}{dt} = \frac{2}{na}\frac{\partial U}{\partial M} \tag{3.1a}$$

$$\frac{d\epsilon}{dt} = \frac{1-\epsilon^2}{na^2\epsilon^2}\frac{\partial U}{\partial M} - \frac{\sqrt{1-\epsilon^2}}{na^2\epsilon}\frac{\partial U}{\partial \omega} \tag{3.1b}$$

$$\frac{di}{dt} = \frac{-1}{na^2\sqrt{1-\epsilon^2}\,\sin(i)}\left[\frac{\partial U}{\partial \Omega} + \cos(i)\frac{\partial U}{\partial \omega}\right] \tag{3.1c}$$

$$\frac{d\Omega}{dt} = \frac{1}{na^2\sqrt{1-\epsilon^2}\,\sin(i)}\frac{\partial U}{\partial i} \tag{3.1d}$$

$$\frac{d\omega}{dt} = \frac{\sqrt{1-\epsilon^2}}{na^2\epsilon}\frac{\partial U}{\partial \epsilon} - \frac{\cos(i)}{na^2\sqrt{1-\epsilon^2}\,\sin(i)}\frac{\partial U}{\partial i} \tag{3.1e}$$

$$\frac{dM}{dt} = n - \frac{2}{na}\frac{\partial U}{\partial a} - \frac{1-\epsilon^2}{na^2\epsilon}\frac{\partial U}{\partial \epsilon} \tag{3.1f}$$

where U is a potential function.

There is a special form of the Lagrange planetary equation in which the effects of small impulses can cause variations to the orbital elements. This form of the planetary equations is shown by Danby [15]. In Figure 3.1, the effects of the impulses on the orbital elements are shown,

$$\cos^2\lambda = \frac{1+\cos 2\lambda}{2} \qquad \sin^2\lambda = \frac{1-\cos 2\lambda}{2} \tag{3.2a}$$

$$\frac{d\epsilon}{dt} = \frac{na^2\sqrt{1-\epsilon^2}}{\mu}\sin(f)\bar{R} + \frac{na^2\sqrt{1-\epsilon^2}}{\mu}\big(\cos(f)+\cos(E)\big)\bar{B} \tag{3.2b}$$

$$\frac{di}{dt} = \frac{naR}{\mu\sqrt{1-\epsilon^2}}\cos(w)\bar{N} \tag{3.2c}$$

Figure 3.1 Lagrange planetary equations diagram

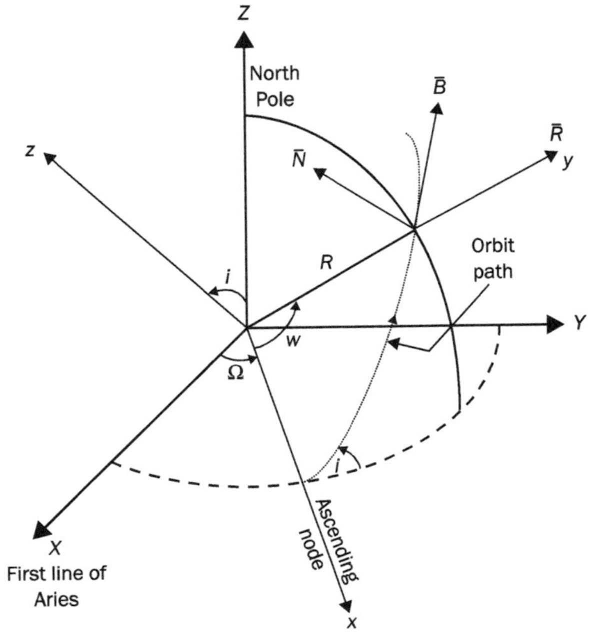

$$\frac{d\Omega}{dt}=\frac{naR}{\mu\sqrt{1-\epsilon^2}}\sin(w)\,csc(i)\bar{N} \tag{3.2d}$$

$$\begin{aligned}\frac{d\omega}{dt}=&-\frac{na^2\sqrt{1-\epsilon^2}}{\mu\epsilon}\cos(f)\bar{R}+\frac{na^2\sqrt{1-\epsilon^2}}{\mu\epsilon}\left(1+\frac{R}{p}\right)\sin(f)\bar{B}\\&-\frac{naR}{\mu\sqrt{1-\epsilon^2}}\sin(w)\csc(i)\cos(i)\bar{N}\end{aligned} \tag{3.2e}$$

As shown in Figure 3.1, the directions of the force impulse per unit mass are explained as: $\bar{R}$ represents the radial direction component, $\bar{B}$ represents the tangential component, and $\bar{N}$ represents the component normal to the orbit plane. w is the sum of the true anomaly angle and the argument of perigee. In equations (3.1) and (3.2), the Lagrange planetary equations have singularity problems. These equations are singular when the eccentricity is equal to zero or one and the inclination angle is equal to 0 and 180 degrees. This equation cannot be applied to circular orbits in order to transfer from one orbit to another; the equations can be only applied to elliptical orbits. For eccentricities higher and/or equal than 1, the solution is also imaginary and/or singular. There is another special form of equations that takes out the effects of the nonlinearities of the Lagrange planetary equations. This set of equations is called the equinoctial equations and are explained in Battin [16].

3.3 Perturbation due to the earth oblateness

There are other perturbations that are not associated with the effects of forces applied to the satellite. The perturbing forces can happen due to the interaction of the satellite with the Earth and the Sun. One of the first perturbations that is analyzed is the perturbation due to the oblateness of the Earth. It is known that the Earth is not a spherical body because it has a bulge around the equator with some symmetry near the poles about the polar axis [11]. The Earth is described as an oblate spheroid due to these characteristics. There are external forces associated with this bulge that can create changes in the orbital motion of the satellite. It is also known that the Earth is a top whose figure axis is precessing about the normal to the ecliptic plane which is a motion known as the precession of equinoxes. To calculate this precession, the potential energy [11] is expanded to find the gravitational potential of a point mass and a non-spherical distribution of matter. This derivation of the gravitational potential is explained in terms of the inertia tensor that leads to the known constant J2, describing the Earth as an oblate spheroid.

Brouwer and Clemence [17] developed the potential function in terms of the J2 coefficient, and, Goldstein [11] shows a similar development which is defined in terms of inertia tensors. The present development follows Brouwer and Clemence [17] because the gravitational potential function is explained in terms of the J2 coefficient. Figure 3.2 shows a schematic in which the Earth is assumed as an arbitrary body. In Figure 3.2, there are two different coordinate systems. The ξ, η, and ζ coordinate system is centered at the center of mass (point O) of the Earth, and the x, y, z coordinate system is parallel to it showing the position of a point mass, P, moving about the Earth. An elementary mass (dm) located at any point in the Earth is denoted by the point D, and the angle S is shown between the points POD. The point P is located in the parallel coordinate system x, y, z.

Figure 3.2 **Schematic for the calculation of the potential function**

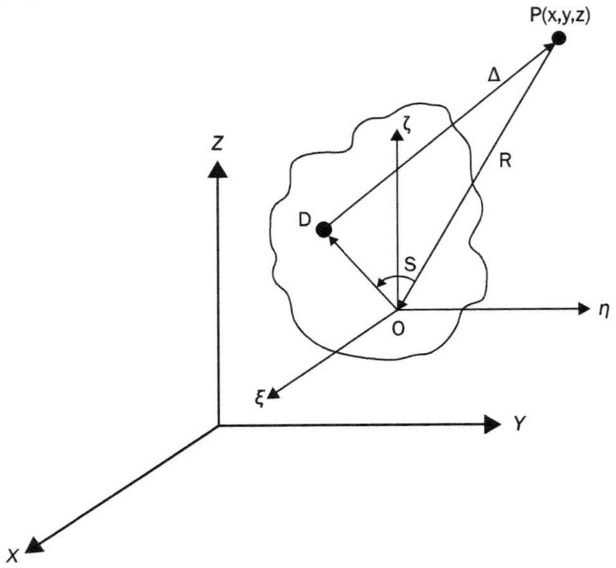

From Figure 3.2, the square of the distance from the center of mass of the Earth to the point P can be described as,

$$R^2 = x^2 + y^2 + z^2 \tag{3.3}$$

and the square of the distance from O to the point D is,

$$\rho^2 = \xi^2 + \eta^2 + \zeta^2 \tag{3.4}$$

The square of the distance from D to P is therefore,

$$\Delta^2 = (x - \xi)^2 + (y - \eta)^2 + (z - \zeta)^2 \tag{3.5}$$

Equation (3.5) can be expanded and reduced to the following form:

$$\Delta^2 = R^2\left[1 + \left(\frac{\rho}{R}\right)^2 - \frac{2(x\xi + y\eta + z\zeta)}{R\rho}\left(\frac{\rho}{R}\right)\right] \tag{3.6}$$

It is assumed that ($\rho/R << 1$), but this ratio is defined as,

$$\frac{\rho}{R} = \vartheta \tag{3.7}$$

Also, the third term on the right hand side of equation (3.6) can be written as,

$$q = \cos S = \frac{\vec{R}\cdot\vec{\rho}}{R\rho} = \frac{x\xi + y\eta + z\zeta}{R\rho} \tag{3.8}$$

where $|q| \leq 1$. Equation (3.5) can be reduced with equations (3.7) and (3.8) as,

$$\Delta^2 = R^2[1 + \vartheta^2 - 2\vartheta q] \tag{3.9}$$

The potential energy expressing the attraction of the mass (M), which is the Earth, and the point mass (m), located at point P, is denoted as,

$$U = G\int\frac{dM}{\Delta} = G\int\frac{dM}{R\sqrt{[1+\vartheta^2 - 2\vartheta q]}} \tag{3.10}$$

where G is the universal gravitational constant. The integrand of equation (3.10) can be expressed in terms of the Legendre polynomials and the binomial expansion. The binomial expansion can be expressed in the following form,

$$(1 + \vartheta^2 - 2\vartheta q)^{-1/2} = P_0(q) + P_1(q)\vartheta + P_2(q)\vartheta^2 + \ldots + P_n(q)\vartheta^n + \ldots \tag{3.11}$$

where P_n are the Legendre polynomials in terms of q of degree n. The left hand side of equation (3.11) can also be expanded using the binomial theorem as follows,

$$(1-(2\vartheta q - \vartheta^2))^l = \Sigma_{k=0}^{\infty}(-1)^k\frac{l(l-1)\ldots(l-k+1)}{k!}(2\vartheta q - \vartheta^2)^k \tag{3.12}$$

where $l = -1/2$. By equating equation (3.11) and equation (3.12), the collection of the coefficients of ϑ^n gives the following general expression,

$$P_n(q) = \frac{1\cdot3\cdot5\ldots(2n-1)}{1\cdot2\cdot3\ldots n}\left[q^n - \frac{n}{2}\left(\frac{n-1}{2n-1}\right)q^{n-2} + \frac{n(n-1)(n-2)(n-3)}{2\cdot4\cdot(2n-1)(2n-3)}q^{n-4} + \ldots\right] \tag{3.13}$$

The first three Legendre polynomials can be written using equation (3.13) as,

$$P_0(q) = 1 \qquad P_1(q) = q \qquad P_2(q) = \frac{3q^2-1}{2} \tag{3.14}$$

With equations (3.11) and (3.14), equation (3.10) is defined as follows,

$$U = U_0 + U_1 + U_2 + \ldots + U_n + \ldots \tag{3.15}$$

where,

$$U_0 = \frac{G}{R}\int P_0 \; dM \tag{3.16a}$$

$$U_1 = \frac{G}{R}\int P_1(q)\left(\frac{\rho}{R}\right) dM \tag{3.16b}$$

$$U_2 = \frac{G}{R}\int P_2(q)\left(\frac{\rho}{R}\right)^2 dM \tag{3.16c}$$

...

$$U_n = \frac{G}{R}\int P_n(q)\left(\frac{\rho}{R}\right)^n dM \tag{3.16d}$$

Throughout the following development, only the first three terms of the Legendre polynomial expansion are considered. With these three terms, the Earth's oblateness is approximated. Higher order terms can be used to define a highly deformed Earth, but they are not considered here. Expanding equation (3.16a) and using P_0 in equation (3.15), the first potential term (U_0) can be defined as follows,

$$U_0 = \frac{G}{R}\int P_0 \, dM = \frac{\mu}{R} \tag{3.17}$$

where $\mu = GM$. Equation (3.17) describes a case where all the mass is concentrated at the center of mass of the Earth. This is the definition used for the central force body motion in which the Earth is assumed as a homogeneous spherical body. Expanding equation (3.16b) and using P_1 in equation (3.15), U_1 is defined as,

$$\begin{aligned} U_1 &= \frac{G}{R}\int q\left(\frac{\rho}{R}\right) dM = \frac{G}{R}\int \frac{x\xi + y\eta + z\zeta}{R^2} \, dM \\ &= \frac{G}{R^2}\left[\frac{x}{R}\int \xi \; dM + \frac{y}{R}\int \eta \; dM + \frac{z}{R}\int \zeta \; dM\right] \end{aligned} \tag{3.18}$$

By the definition of the center of mass [11], the integrals in equation (3.18) are equal to zero so that $U_1 = 0$.

For $n = 2$, equation (3.16c) can be written as,

$$U_2 = \frac{G}{R}\int \frac{3q^2-1}{2}\left(\frac{\rho}{R}\right)^2 dM = \frac{G}{R}\int\left[-\frac{\rho^2}{2R^2}+\frac{3}{2}\left(\frac{x\xi+y\eta+z\zeta}{R^2}\right)^2\right]dM \quad (3.19)$$

Expanding equation (3.19),

$$U_2 = -\frac{G}{R^3}\left\{\left(\frac{1}{2}-\frac{3x^2}{2R^2}\right)\int \xi^2\, dM + \left(\frac{1}{2}-\frac{3y^2}{2R^2}\right)\int \eta^2 dM + \left(\frac{1}{2}-\frac{3z^2}{2R^2}\right)\int \zeta^2 dM\right\}$$
$$-\frac{G}{R^3}\left\{-\frac{3xy}{R^2}\int \xi\eta\, dM - \frac{3xz}{R^2}\int \xi\zeta\, dM - \frac{3yz}{R^2}\int \eta\zeta\, dM\right\} \quad (3.20)$$

Equation (3.20) can be directly related to the moments and products of inertia due to the mass (M), or Earth. Let α, β, and γ be the direction cosines of an axis (n) thru the center of mass as shown in Figure 3.3. From Figure 3.3, it is known that,

$$\rho \cos\theta = \vec{\rho}\cdot\vec{n} = \xi\alpha + \eta\beta + \zeta\gamma \quad (3.21)$$

The moment of inertia about the assumed axis in Figure 3.3 can be written as,

$$I_n = \int \rho^2 \sin^2\theta\, dM = \int \rho^2(1-\cos^2\theta)\, dM \quad (3.22)$$

Substituting equation (3.4) and (3.21) into equation (3.22), the moment of inertia about the assumed axis is defined as,

$$I_n = \int [(\xi^2+\eta^2+\zeta^2) - (\xi\alpha+\eta\beta+\zeta\gamma)^2]\, dM \quad (3.23)$$

Figure 3.3 **Axis used to calculate the moments of inertia related to U_2**

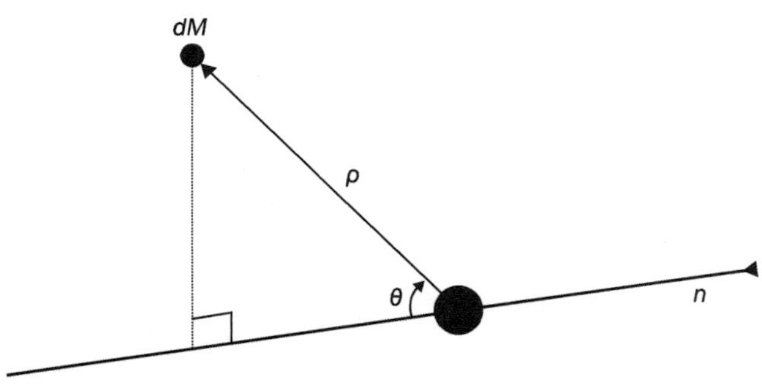

It is known that $\alpha^2 + \beta^2 + \gamma^2 = 1$. Multiplying $\alpha^2 + \beta^2 + \gamma^2$ by the first term on the right hand side of equation (3.23), equation (3.24) can be rewritten as follows,

$$I_n = \int [(\eta^2 + \zeta^2)\,\alpha^2 + (\xi^2 + \zeta^2)\,\beta^2 + (\xi^2 + \eta^2)\,\gamma^2]\,dM$$
$$-2\int (\xi\eta\alpha\beta + \xi\alpha\zeta\gamma + \eta\beta\zeta\gamma)\,dM \quad (3.24)$$

But,

$$I_x = \int (\eta^2 + \zeta^2)\,dM \quad I_{xy} = \int \zeta\eta\,dM$$
$$I_y = \int (\xi^2 + \zeta^2)dM \quad I_{xz} = \int \xi\zeta\,dM$$
$$I_y = \int (\xi^2 + \eta^2)\,dM \quad I_{xz} = \int \eta\xi\,dM$$

Equation (3.24) therefore reduces to,

$$I_n = I_x\alpha^2 + I_y\beta^2 + I_z\gamma^2 - 2I_{xy}\alpha\beta - 2I_{xz}\alpha\gamma - 2I_{yz}\beta\gamma \quad (3.25)$$

Dividing by I_n, the equation for the ellipsoid of inertia for the body (M) can be represented as,

$$1 = I_x x^2 + I_y y^2 + I_z z^2 - 2I_{xy}xy - 2I_{xz}xz - 2I_{yz}yz \quad (3.26)$$

where,

$$x = \frac{\alpha}{\sqrt{I_n}} \quad y = \frac{\beta}{\sqrt{I_n}} \quad z = \frac{\gamma}{\sqrt{I_n}}$$

Selecting x, y, and z to be the principal axis, the cross product terms are equal to zero, and the equation of the ellipsoid of inertia becomes,

$$1 = I_x x^2 + I_y y^2 + I_z z^2 \quad (3.27)$$

where x, y, and z are the coordinates of any point over the surface of the ellipsoid α, β, and $\gamma$are the direction cosines of the line joining the points to the center of the ellipsoid. Also,

$$I_{xy} = I_{yz} = I_{xz} = 0$$

And,

$$\int (\xi^2 + \eta^2 + \zeta^2)\,dM = \frac{I_x + I_y + I_z}{2}$$

The integrals in equation (3.24) are equal to,

$$\int \xi^2\,dM = \frac{-I_x + I_y + I_z}{2} \quad (3.28a)$$

$$\int \eta^2 \, dM = \frac{I_x - I_y + I_z}{2} \tag{3.28b}$$

$$\int \zeta^2 \, dM = \frac{I_x + I_y - I_z}{2} \tag{3.28c}$$

$$\int \xi\eta \, dM = \int \xi\zeta \, dM = \int \eta\zeta \, dM = 0 \tag{3.28d}$$

Substituting equations (3.28) into equation (3.20), U_2 can be written as,

$$U_2 = \frac{G}{R^3}\left\{\left(\frac{3x^2}{2R^2} - \frac{1}{2}\right)\frac{-I_x + I_y + I_z}{2} + \left(\frac{3y^2}{2R^2} - \frac{1}{2}\right)\frac{I_x - I_y + I_z}{2} + \left(\frac{3z^2}{2R^2} - \frac{1}{2}\right)\frac{I_x + I_y - I_z}{2}\right\} \tag{3.29}$$

Equation (3.29) can be expanded and rewritten as,

$$U_2 = \frac{G}{R^3}\left\{\left(\frac{3x^2}{2R^2} - \frac{1}{2}\right)\frac{-2I_x + I_y + I_z + I_x}{2} + \left(\frac{3y^2}{2R^2} - \frac{1}{2}\right)\frac{I_x - 2I_y + I_z + I_y}{2} + \left(\frac{3z^2}{2R^2} - \frac{1}{2}\right)\frac{I_z + I_x + I_y - 2I_z}{2}\right\}$$

$$U_2 = \frac{G}{R^3}\left\{\frac{1}{2}\left(I_x + I_y + I_z\right) - \frac{3}{2}\left(\frac{x^2 I_x + y^2 I_y + z^2 I_z}{R^2}\right)\right\} \tag{3.30}$$

where x, y, and z are the coordinates of a point mass, P, and I_x, I_y, and I_z are the principal moments of inertia of the mass M (defining the mass of the Earth). Introducing spherical coordinates as shown in Figure 3.4, the coordinates of the point mass can be written as,

$$\begin{aligned} x &= R \cos\lambda \cos\phi \\ y &= R \sin\lambda \cos\phi \\ z &= R \sin\phi \end{aligned} \tag{3.31}$$

where λ and ϕ are the longitude and the latitude, respectively. Substituting equations (3.31) into equation (3.30), U_2 is defined as,

$$U_2 = \frac{G}{R^3}\left\{\frac{1}{2}(I_x + I_y + I_z) - \frac{3}{2}(I_x \cos^2\lambda + I_y \sin^2\lambda)\cos^2\phi - \frac{3}{2} I_z \sin^2\phi\right\} \tag{3.32}$$

Using half angle formulas,

$$\cos^2\lambda = \frac{1 + \cos 2\lambda}{2} \qquad \sin^2\lambda = \frac{1 - \cos 2\lambda}{2}$$

Figure 3.4 **Cylindrical coordinate system for the point mass**

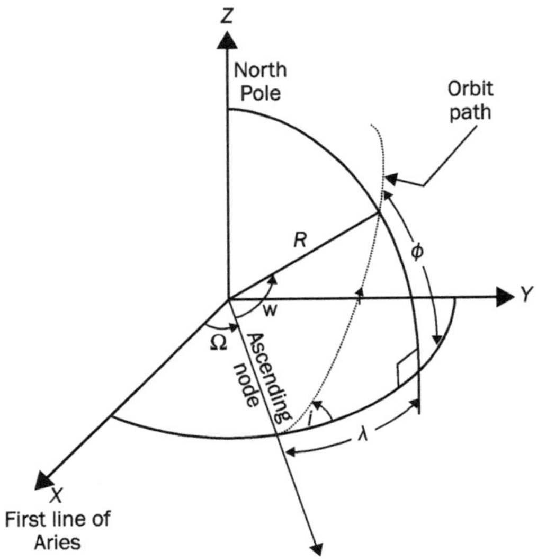

equation (3.32) is reduced to,

$$U_2 = \frac{G}{R^3}\left\{\left(\frac{1}{2}-\frac{3}{2}\ \sin^2\phi\right)\left(I_z - \frac{I_x + I_y}{2}\right) - \frac{3}{4}(I_x - I_y)\cos 2\lambda\ \cos^2\phi\right\} \quad (3.33)$$

If $I_x = I_y$, the Earth is rotationally symmetric; then,

$$U_2 = \frac{G}{R^3}\left(\frac{1}{2}-\frac{3}{2}\ \sin^2\phi\right)(I_z - I_x) \quad (3.34)$$

The potential function defined in equation (3.15) is equal to,

$$U = U_0 + U_1 + U_2$$

$$U_2 = \frac{\mu}{R^3}\left(1 + \frac{B_2}{2R^2}(1 + 3\ \sin^2\phi)\right) \quad (3.35)$$

where,

$$\mu = GM_E \qquad B_2 = -\frac{I_z - I_x}{M_E}$$

The B_2 term has dimensions of length to the square but can be rewritten as,

$$B_2 = -\frac{2}{3} J_2 R_E^2 \tag{3.36}$$

where R_E is equal to the radius of the Earth at the Equator, and,

$$J_2 = \frac{3}{2}\left(\frac{I_z - I_x}{M_E R_E^2}\right) = \frac{3}{2}\left(\frac{k_z^2 - k_x^2}{R_E^2}\right) \approx 1.082x10^{-3} \tag{3.37}$$

where k_z and k_x are the radius of gyration about the z and x axis, respectively. R_E is the radius about the Equator of the Earth which is approximated to 6,378.13649 (km). The J_2 parameter is dimensionless and represents an Earth with symmetry about the polar axis.

From equation (3.35), the Earth's potential function which includes the perturbation due to the oblateness of the Earth can be defined as,

$$U = \frac{\mu}{R} - B_2 \frac{\mu}{2R^3}(1 - 3\sin^2\phi) \tag{3.38}$$

As mentioned earlier, the perturbation due to the oblateness of the Earth causes a precession of the line of nodes about the normal to the ecliptic plane. The second term on the right hand side (RHS) of equation (3.38) can be used to understand the precession of the line of nodes. In Figure 3.4, the sine of the latitude can be rewritten from the right spherical triangle as,

$$\sin\phi = \sin(\omega + f)\sin i$$

Then, equation (3.34) can be rewritten as,

$$U_2 = \frac{1}{3} J_2 \left(\frac{R_E}{R}\right)^2 \frac{\mu}{R}(1 - 3\sin^2(\omega + f)\sin^2 i) \tag{3.39}$$

Equation (3.39) can be used to determine the variation of the orbital elements by propagating the orbit. The orbit propagation with equation (3.39) are known as the osculating orbital elements. If the orbital perturbations are analyzed for one orbit, the average of a function over one period can be used. Rewriting equation (3.39) with different trigonometric identities and the equation of a conic section **[18]** and taking the average over one period **[16]**, the average over one period of equation (3.39) is equal to,

$$\bar{U}_2 = \frac{n^2 J_2 R_E^2}{4(1-\epsilon^2)^{3/2}}(2 - 3\sin^2 i) \tag{3.40}$$

Substituting equation (3.40) into equation (3.1), the mean orbital elements affected by the J2 perturbation are equal to,

$$\frac{d\bar{a}}{dt} = \frac{d\bar{\epsilon}}{dt} = \frac{d\bar{\imath}}{dt} = 0 \tag{3.40a}$$

$$\frac{d\bar{\Omega}}{dt} = -\frac{3}{2}nJ_2\left(\frac{R_E}{\bar{p}}\right)^2 \cos\bar{\imath} \tag{3.40b}$$

$$\frac{d\bar{\omega}}{dt} = -\frac{3}{4}nJ_2\left(\frac{R_E}{\bar{p}}\right)^2 (4-5\sin^2\bar{\imath}) \tag{3.40c}$$

$$\frac{d\bar{M}}{dt} = n + \frac{3}{4}\frac{nJ_2}{(1-\bar{\epsilon}^2)^{3/2}}\left(\frac{R_E}{\bar{a}}\right)^2 (2-3\sin^2\bar{\imath}) \tag{3.40d}$$

$\bar{a}$, $\bar{\epsilon}$, $\bar{\imath}$, $\bar{\Omega}$, $\bar{\omega}$ and $\bar{M}$ are the mean orbital elements for the semimajor axis, eccentricity, inclination angle, RAAN, argument of perigee, and mean anomaly. The semimajor axis, the eccentricity, and the inclination angle are not affected by the J2 perturbation. The only orbital elements affected are the RAAN, the argument of perigee, and the mean anomaly which cause a rotation and precession of the orbit. The rotation and precession can be compared to the spinning of a symmetrical top with an inclined rim. If the rim is nudged in such a way that it can create a torque, the rim also precesses (or regresses) about the polar axis of the top; in other words, the top will show a 'wobble' about the polar axis [**11**].

This precession of the orbit is observed in the rate of change of the RAAN. If the inclination angle is equal to 90 degrees, there is no precession or regression of the orbit. For $0 < i < 90°$, the orbit precesses west and is called a prograde orbit. If $90° < i < 180°$, the orbit precesses east and is called a retrograde orbit. The argument of perigee has a rate of precession about the polar axis which causes a motion of the major axis of the ellipse for $i = 90°$. There are two inclination angles for which there is no precession (or regression) of the argument of perigee, 63.4 degrees and 116.6 degrees. If $i < 63.4°$, the line of apsides precesses in the direction of the orbit motion. For $63.4° < i < 116.6°$, the line of apsides precesses in opposite sense to the orbital motion.

Assume a circular orbit and define

$$L = \frac{R_E + \bar{s}}{R_E}$$

where $\bar{s}$ is the mean altitude of the orbit. Figures 3.5 and 3.6 show the results explained in the previous paragraph. As the mean altitude keeps

Figure 3.5 Nodal precession rate for a circular orbit

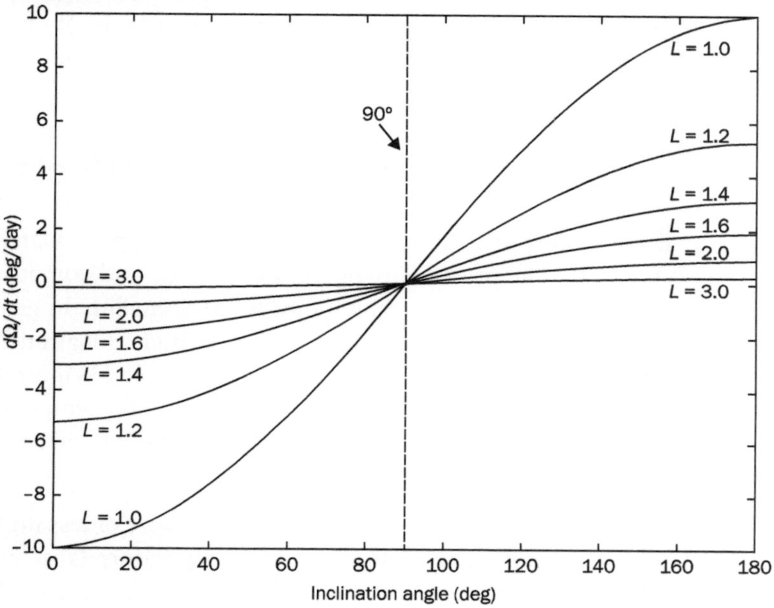

Figure 3.6 Rate of precession of argument of perigee for a circular orbit

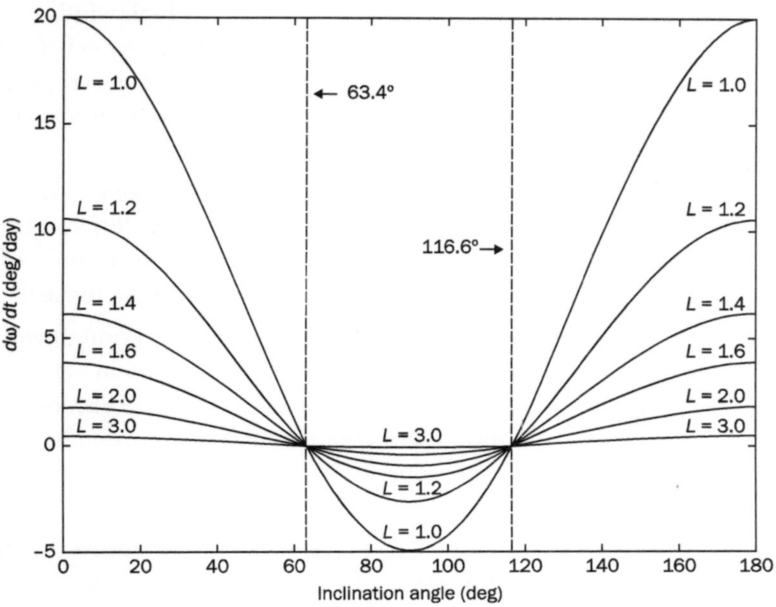

increasing, the precession effects of the J2 perturbation decrease. Once $L = 3.0$, the J2 perturbation is no longer affecting the orbit as compared with $L = 1.0$. The J2 perturbation is an important disturbance if the orbit altitude is lower than 800 km [10].

3.4 The near-Earth atmosphere effects

The near-Earth atmosphere effects happen at altitudes approximately up to 10,000 km [19]. The pressure, temperature, density, and composition of the atmosphere can affect the orbital motion of a satellite. In general, these parameters are a function of the geometry (latitude, longitude, and altitude) of the orbit and the time (Spring, Summer, Fall, and Winter) [20].

The Earth's atmosphere can be classified depending on the temperature. The following are the classification of the regions of the Earth's atmosphere [19] according to temperature:

a. Troposphere – Region nearest the surface having more or less uniform decrease of temperature with altitude. The temperature lapse rate is –6.5 °C/km. This atmosphere contains the weather, has a convective equilibrium with the Sun, and warms the Earth. The troposphere extends to about 6 km in altitude from sea-level.

b. Tropopause – This region occurs between 6 and 18 km of altitude (and is higher over the Equator). This region has high winds and has the highest cirrus clouds. The temperature declines to –60°C at the top of tropopause. This temperature occurs because the troposphere is not being heated by incident energy and only part of the visible light of the spectrum passes through the ionosphere.

c. Stratosphere – The region just above the troposphere having a nominally constant temperature. Thicker over the poles, and thinner to non-existent over the equator. At about 25 km altitude of the stratosphere and in the middle latitudes, ozone is being formed.

d. Mesosphere – The region of the first temperature maximum which is below the major temperature minimum. This variation in temperature is observed at about 80 km altitude within this region. The mesosphere is in irradiative equilibrium between the ultraviolet ozone heating by the upper fringe of the ozone region, and the infrared ozone and carbon dioxide cooling is radiated into Space.

e. Thermosphere – The region of rising temperature above the major temperature minimum beginning at 80 km altitude and has no upper altitude limit. This is the domain of the aurora. The temperature rises at the base of the thermosphere. The thermosphere has frequent molecular collisions to maintain thermodynamic equilibrium. Potentially, the enormous infrared irradiative cooling by carbon dioxide is not actually realized owing to inadequate conditions.

Also, the Earth's atmosphere can be classified by its composition. The composition classification is described in the following list [19]:

a. Homosphere – The region of substantially uniform composition. This region has a constant mean molecular weight that extends to between 80 and 100 km of altitude. The upper part of the homosphere is called the homopause. The composition of the homopause changes due to the dissociation of carbon dioxide. This region is composed of the troposphere, stratopause, and stratosphere.
b. Heterosphere – The region of significantly varying composition above the homosphere and extending significantly out to Space. In this region, the disassociation of nitrogen and the diffuse separation sets in lighter atoms and molecules rising to the top. The heterosphere contains the ionosphere and exosphere. The ionosphere is commonly known because it is the region of sufficiently large electron density that affects radio communication.

Table 3.1 shows the composition of the atmosphere at the Earth's surface in terms of percentage [19]. The total mass of the Earth's atmosphere due

Table 3.1 **Composition of the atmosphere at Earth's surface**

Composition of the atmosphere	Composition percentage
Nitrogen (N_2)	78.09
Oxygen (O_2)	20.95
Argon (Ar)	0.93
Carbon Dioxide (CO_2)	0.03
Neon (Ne)	1.8×10^{-3}
Helium (He)	5.2×10^{-4}
Krypton (Kr)	1.0×10^{-4}
Hydrogen (H_2)	5.0×10^{-5}
Xenon (Xe)	8.0×10^{-6}
Methane (CH_4)	5.0×10^{-6}
Ozone (O_3)	7.0×10^{-6}

to its composition is 29 grams/molecule. It is expressed in terms of mass because it can vary depending on the formulation that is used.

As explained previously, the pressure, temperature, composition, and density are a function of the geometry of the Earth and time. There are five major phenomena that result in variations of the Earth's atmosphere due to time. These are the following:

a. Diurnal Variation – The density of the atmosphere varies in a 24 hour period. The maxima due to solar energy cause the gas molecules to 'boil up' from lower regions of higher density.

b. Semi-annual Variation – The effects due to this cause are not well understood. This variation may be associated with the high altitude winds which move along the lines of longitude as summer and winter poles of the Earth are interchanged. This variation is world-wide at all latitudes, longitude, and high altitudes. The maximum density occurs during mid April and October, and the minimum density occurs during mid July and early January.

c. Annual Variation – The density of the Earth's atmosphere indicates annual change in density at particular latitudes. There is a peak of maximum atmospheric density in the northern latitudes in the summer and in the southern latitudes in the winter. When the Sun is far into the northern hemisphere, an increase in the density of the atmosphere occurs in the northern hemisphere at satellite altitudes.

d. Solar Cycle – There is an increased number of sun spots occurring approximately in 11 year cycles. These variations in solar activity can cause variations in the Earth's atmosphere.

e. Geomagnetic Activity – There are short periods of time in which the Earth's atmospheric density varies by an order of magnitude with magnetic storm activity.

The Earth's atmosphere results in a perturbation known as the aerodynamic drag force. The drag force is a surface force that acts against the motion of the vehicle [21]. For an orbit [22], the aerodynamic drag force is directed along the velocity vector of the satellite and interacts with a contact surface of the satellite. The drag force is expressed as,

$$F_A = \frac{1}{2}\rho(R)v^2 C_D S \tag{3.41}$$

where ρ is the atmospheric density, v is the velocity of the spacecraft in the orbit, C_D is the drag coefficient, and S is the surface area in contact with the Earth's atmosphere. In many satellites, the drag coefficient is between 2.2 and 2.5 [10] [23]. The atmospheric density is described as follows [22],

$$\rho(R) = \rho_0 e^{-\frac{R-r_p}{H}} \tag{3.42}$$

ρ_0 is the atmospheric density at the radius of perigee, and H is the density scale height. The scale height is the distance at some altitude at which the atmospheric density will change by a factor of $1/e$. The density and the scale height can be obtained from different atmospheric models. References [10] and [20] provide tabulated values for the 1976 atmospheric density model. This model is commonly used for the development of control systems. Equation (3.41) is used to specify the aerodynamic force associated with the Earth's atmosphere, but the surface force vector is expressed as follows,

$$\vec{F}_A = \frac{1}{2}\rho(R)v^2 C_D S\hat{V} \tag{3.43a}$$

$\hat{V}$ is the unit vector of the velocity vector of the satellite. To understand the effects of the aerodynamic force on the orbital elements, it is assumed that the drag force has dominant secular effects against the relative motion of the in-plane motion of the satellite [24]. Using this assumption, $\hat{V} = -\hat{\imath}_{\vec{B}}$; then, the Lagrange planetary equations for the effects of the aerodynamic drag force are equal to,

$$\frac{da}{dt} = -\frac{2na^3\sqrt{1-\epsilon^2}}{\mu}\rho(R)v^2 B_C \tag{3.44a}$$

$$\frac{d\epsilon}{dt} = -\frac{na^2\sqrt{1-\epsilon^2}}{\mu}\rho(R)v^2 B_C(\cos(f) + \cos(E)) \tag{3.44b}$$

$$\frac{di}{dt} = \frac{d\Omega}{dt} = 0 \tag{3.44c}$$

$$\frac{d\omega}{dt} = -\frac{na^2\sqrt{1-\epsilon^2}}{\mu\epsilon}\rho(R)v^2 B_C\left(1+\frac{\mathrm{R}}{\mathrm{p}}\right)\sin(f) \tag{3.44d}$$

B_C is the ballistic coefficient and is equal to $C_D S/2m$. m is the mass of the satellite. The semimajor axis and the eccentricity of the orbit keep decaying with respect to time. In addition, there is a variation in the argument of perigee that causes a regression of the semimajor axis. From equation (3.44a) and (3.44b), the orbit decays toward the center of the Earth as time increases; and the decaying orbit forms a spiral pattern

Figure 3.7 **Orbit affected by the atmospheric drag**

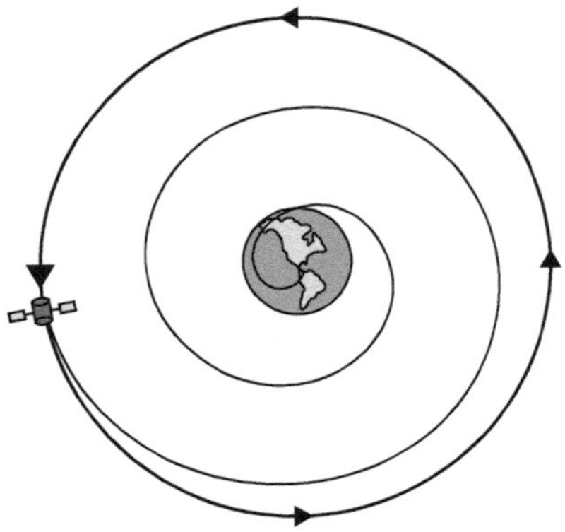

moving toward the center of the Earth. Figure 3.7 shows the spiral pattern of the decaying orbit. The inclination angle and the RAAN are not affected by the atmospheric drag force. On the other hand, the atmospheric drag can be used to perform plane change maneuvers [25] and to stabilize the orientation of the satellite [26].

As shown in Figure 3.7, the atmospheric force acts against the satellite motion so that the orbit size keeps decreasing as time increases. In addition, this force causes a variation in the energy associated with the orbit. The work done by the atmospheric force on the orbit can be described as,

$$dW = -F_A R df \Rightarrow \frac{\Delta W}{\Delta N} = -\oint DR\,df \tag{3.45}$$

where $\Delta W/\Delta N$ is the energy loss per orbit. Assuming a circular orbit, the energy loss per orbit can be written as,

$$\frac{\Delta W}{\Delta N} = -\pi C_D S\rho\mu \tag{3.46}$$

The energy associated with a circular orbit is equal to,

$$W = U = \frac{-m\mu}{2a} = \frac{-m\mu}{2R} \tag{3.47}$$

Equation (3.47) describes the constant orbital energy for a conservative system. The conservation of system energy implies that a change in potential energy equals to the corresponding change in kinetic energy. Using this relation, equation (3.47) can be rewritten as follows,

$$\frac{\Delta W}{\Delta N} = \frac{-m\mu}{2R^2}\frac{\Delta R}{\Delta N} \tag{3.48}$$

$\Delta R/\Delta N$ describes the variation of the radius of the circular orbit per orbit. Equating equation (3.48) and equation (3.46), the change in the radius of the circular orbit per orbit can be written as,

$$\frac{\Delta R}{\Delta N} = \frac{4\pi\rho g_0 R^2}{B_C} \tag{3.49}$$

Because of the decrement in the semimajor axis per orbit due to the atmospheric force, the period of the circular orbit is decreasing per time as follows,

$$\frac{\Delta P}{\Delta N} = \frac{2\pi R^{3/2}}{\mu^{1/2}} \tag{3.50}$$

Dividing equation (3.49) by equation (3.50), the change of the period per change in the radius of the circular orbit can be defined as follows,

$$\frac{\Delta P}{\Delta N} = \frac{2\rho g_0 (\mu R)^{1/2}}{B_C}$$

From which,

$$\Delta P = P_2 - P_1 = \frac{B_C}{2\mu^{1/2}}\int_{R_1}^{R_2} \frac{dR}{\rho R^{1/2}}$$

Using equation (3.42) and knowing that $R = \bar{s} + R_E$, the integral of the previous equation can be written as,

$$P_2 - P_1 \cong \frac{B_C H}{\rho_0 2(\mu R_E)^{1/2}}\left[e^{\bar{s}_1/H} - e^{\bar{s}_2/H}\right] \tag{3.51}$$

where $\bar{S}_1$ and $\bar{S}_2$ are the initial and final orbit altitudes, respectively. Equation (3.51) provides the orbit lifetime for a spacecraft to reenter the Earth's atmosphere; and, Figure 3.8 shows the satellite lifetime. Wolverton [20] shows the solution of more complex formulations for the orbit lifetime of an elliptical orbit. The atmospheric drag force is a dominant perturbing force for a satellite under an altitude of approximately 800 km.

Figure 3.8 Satellite lifetime

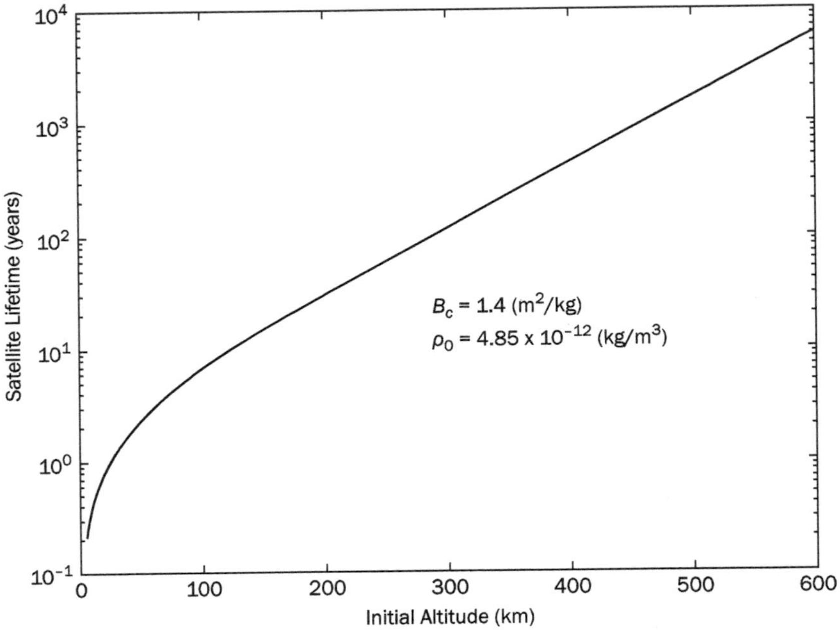

3.5 Solar radiation pressure force

The solar radiation pressure is one of the long term forces that acts on the surface of the satellite. This disturbing force causes variations in the motion of the satellite due to the materials used for the construction of the satellite [27]. The solar radiation force occurs when a light beam falls upon a surface and causes a loss of energy resulting in a force on the surface of the spacecraft. If the surface is part of the spacecraft, there can also be a torque about the center of mass of the satellite through its center of pressure [21].

The solar radiation forces are due to the impact of the light on a surface [28]. This force can be classified as a disturbing force on the satellite. In other cases, the solar force can be used for the attitude control system to maintain the orientation of the satellite [29] and for the orbit maintenance of a formation flying system [27] [30] [31]. The acceleration due to the incidence of light over the surface of a satellite is described as,

$$a = \frac{\bar{h}(1+\varepsilon)}{c}\frac{S}{m}cos^2\alpha = \frac{\bar{h}_0(1+\varepsilon)}{c}\frac{r_0}{R_S}\frac{S}{m}cos^2\alpha \tag{3.52}$$

S is the area of the surface in contact with the sunlight, a is the acceleration of the body, $\bar{h}$ is the power of electromagnetic wave per unit surface, and $\bar{h}_0$ is the solar radiation, which is equal to 1.35×10^6 (erg/cm^2 – sec). ε is the reflectivity coefficient and describes the type of material on the surface of the satellite. If $\varepsilon = 1$, the material can reflect the light. For $\varepsilon = 0$, the material is absorbing the light. This equation only treats the specular effects of the light on the surface of the satellite. r_0 is the mean distance of the Earth to the Sun, R_S is the distance from the center of gravity of the body to the Sun. α describes the angle of the incident light over the surface of the satellite.

Equation (3.52) is a simplified formulation that only takes into account the contact surface affected by the Sun. Karymov [32] developed a more general formulation that depends on the shape of the satellite and on the direction of the solar rays. In general, the satellite can be modeled as a geometrical figure. In practice, the body of the satellite must be divided into sections to actually obtain a complete description of the solar effects [33]. This is the idea behind the formulation developed by Karymov [32]. As part of this formulation, it is necessary to know the number of surfaces in contact with the sunlight because the perturbing forces acting on the satellite can be underestimated.

To understand the interaction of light over an infinitesimal surface, Figure 3.9 can be used to explain how the light interacts with a smooth surface. The momentum of a light (K) beam is described as,

$$K = \frac{\bar{w}}{\bar{c}} = \frac{\text{momentum}}{\text{volume}} \tag{3.53}$$

Figure 3.9 Light interaction with surface

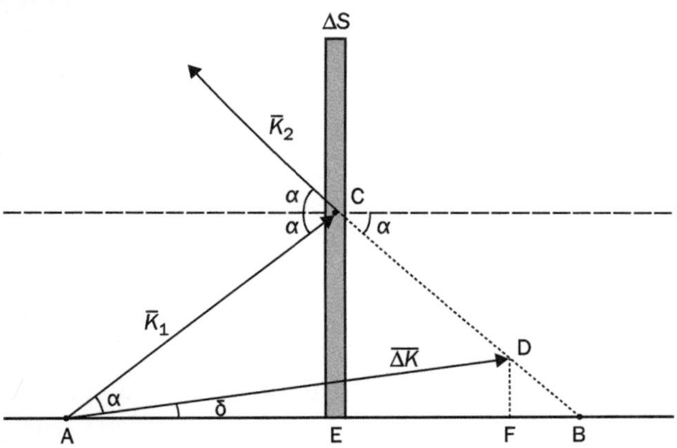

$\bar{w}$ is the light energy per volume, and $\bar{c}$ is the velocity of light in a vacuum. $\bar{K}_1$ is the direction of the incident radiation momentum which interacts the surface at an angle α. $\bar{K}_2$ is the direction of the radiation momentum reflected by the body. Due to the loss of energy, $|\bar{K}_2| < |\bar{K}_1|$. There is a small amount of light that stays within the body which is equal to the difference between the incident and reflected radiation momentum,

$$\overline{\Delta K} = \overline{K_1} - \overline{K_2} = \frac{\bar{f}\Delta t}{v\Delta S} \tag{3.54}$$

$\overline{\Delta K}$ is an impulsive force ($\bar{f}$) applied through the surface in a short period of time (Δt) with a velocity (v). In Figure 3.9, δ is the angle between $\overline{\Delta K}$ (or impulsive force ($\bar{f}$)) and the normal to the surface element (ΔS). $|\overline{K_2}|$ and $|\overline{K_1}|$ are related as follows,

$$|\overline{K_2}| = \varepsilon|\overline{K_1}| \tag{3.55}$$

If $\varepsilon = 1$, all the light is reflected by the surface. In Figure 3.9, $\overline{K_2}$ has a mirror refraction which is shown by the dashed line. Using this refraction of $\overline{K_2}$ and the law of cosines in the triangle ACD, the impulsive force due to the radiation of light on the surface can be written as,

$$\begin{aligned}\overline{\Delta K} &= \sqrt{|\overline{K_1}|^2 + |\overline{K_2}|^2 - 2\,|\overline{K_1}|\,|\overline{K_2}|\cos(\pi - 2\alpha)} \\ &= |\overline{K_1}|\sqrt{1+\varepsilon^2+2\varepsilon\cos(2\alpha)}\end{aligned} \tag{3.56}$$

From the triangle ACE and the triangle ADF,

$$CE = |\overline{K_1}|\sin\alpha \qquad\qquad DF = \overline{\Delta K}\sin\delta$$

The triangles DBF and BCE are similar to each other, and the following relationship can be written,

$$\frac{DF}{CE} = \frac{|\overline{K_1}| - |\overline{K_2}|}{|\overline{\Delta K}|} = 1 - \varepsilon$$

Then,

$$\sin\delta = \frac{(1-\varepsilon)\sin\alpha}{\sqrt{1+\varepsilon^2+2\varepsilon\cos(2\alpha)}} \tag{3.57}$$

Equation (3.57) tells the direction of the impulsive force on the surface. The magnitude, $|\overline{K_1}|$, is equal to the motion of the light flux arriving in the volume of the parallelepiped with base area ΔS and sides $\bar{c}\Delta t$ long as shown in Figure 3.10. The light stream makes an angle with respect to

Figure 3.10 **Light interaction in a parallelepiped**

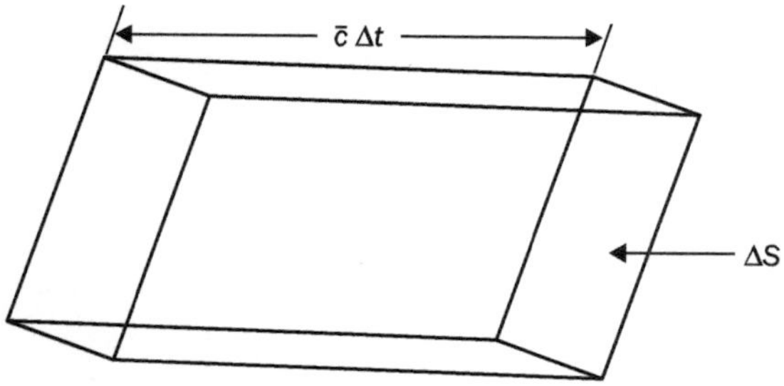

the normal to the surface; then, the volume of the parallelepiped is equal to $\Delta S\bar{c}\Delta t\ cos\ \alpha$. The magnitude of the incident momentum magnitude, $|\overline{K_1}|$, is equal to,

$$|\overline{K_1}| = |\overline{K}| \times (\text{volume of element}) = \frac{\overline{w}}{\overline{c}}\Delta S\bar{c}\Delta t\cos\alpha \tag{3.58}$$

The force acting on the elemental surface dS can be determined from the limit as $\Delta t \to 0$. From equation (3.53), (3.56), and (3.58), the differential force is equal to,

$$df = \lim_{\Delta t\to 0}\frac{|\overline{K}|}{\Delta t} = w\sqrt{1+\varepsilon^2+2\varepsilon\cos(2\alpha)}\cos\alpha\ dS \tag{3.59}$$

Equation (3.59), df, encloses an angle with the normal to the surface given by equation (3.57). Analyzing the limiting cases, the following conclusions can be obtained:

Total reflection ($\varepsilon = 1.0$), $\delta = 0$ and $df = 2\overline{w}cos^2\ \alpha\ dS$. For this case, the radiation force is perpendicular to the surface.

Total absorption ($\varepsilon = 0.0$), $\delta = \alpha$ and $df = \overline{w}\cos\alpha\ dS$. In this case, the radiation force is directed along the incident light stream.

The light energy density, $\overline{w}$, is related to the flow of light energy (E) as,

$$\overline{W} = \frac{E}{\bar{c}} = \frac{E_0}{\bar{c}}\left(\frac{r_0}{R_S}\right)^2 = \bar{b}_0 \tag{3.60}$$

E_0 is the light energy flow arriving at a unit ΔS of the body placed at r_0 distance from the light source. The ratio E_0/c is equal to 4.72×10^{-7} (kg/m^2).

As shown in Figure 3.11, let the exterior normal to the unit vector at every point on the surface of the body be given by,

Figure 3.11 **Light interaction with the surface with respect to the normal**

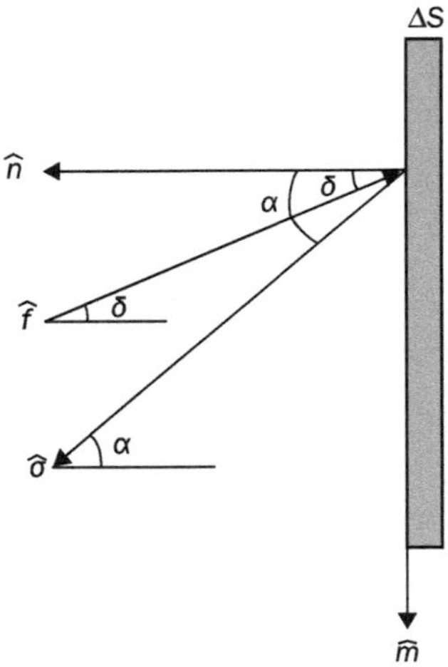

$$\hat{n} = \cos(\hat{n}, \hat{b}_1)\hat{b}_1 + \cos(\hat{n}, \hat{b}_2)\hat{b}_2 + \cos(\hat{n}, \hat{b}_3)\hat{b}_3$$

And the direction opposite to that of the light flux,

$$\hat{\sigma} = a_0\,(t)\hat{b}_1 + b_0\,(t)\hat{b}_2 + c_0(t)\hat{b}_3$$

$\hat{b}_1$,$\hat{b}_2$, and $\hat{b}_3$ are the body coordinate unit vectors. $\hat{\sigma}$ is known as the direction of the incident solar ray on the elemental surface. The elemental force acting on an element of mass is,

$$d\vec{f} = \hat{f}df = (p\hat{b}_1 + q\hat{b}_2 + u\hat{b}_3)df \tag{3.61}$$

$\hat{f}$ has direction cosines p, q, and u in the body coordinate system. In Figure 3.11, the incident ray and reflected ray are assumed to lie in the plane containing the normal to ΔS. The force $\hat{f}$ and the incident solar ray are written as vectors associated with the direction of the normal ($\hat{n}$) and the surface ($\hat{m}$) as follows,

$$\hat{f} = -\cos\delta\,\hat{n} - \sin\delta\,\hat{m} \tag{3.62a}$$

$$\hat{\sigma} = \cos\alpha\,\hat{n} + \sin\alpha\,\hat{m} \tag{3.62b}$$

Equation (3.62a) can be rewritten as,

$$\hat{f} = (-\cos\delta + \sin\delta\cot\alpha)\hat{n} - \frac{\sin\delta}{\sin\alpha}\hat{\sigma} \tag{3.63}$$

From equation (3.57), it can be found that,

$$\frac{\sin\delta}{\sin\alpha} = \frac{(1-\varepsilon)}{\sqrt{1+\varepsilon^2+2\varepsilon\cos(2\alpha)}} \tag{3.64a}$$

Using the trigonometric identity [**13**],

$$\cos\delta = \sqrt{1-\sin^2\delta}$$

And using equation (3.57), the following relationship can be obtained after some trigonometric manipulations,

$$-\cos\delta + \sin\delta\cot\alpha = -\frac{2\varepsilon\cos\alpha}{\sqrt{1+\varepsilon^2+2\varepsilon\cos(2\alpha)}} \tag{3.64b}$$

Substituting equations (3.64) into equation (3.63), the unit force applied to the surface of the body can be described as,

$$\hat{f} = -\frac{(1-\varepsilon)\hat{\sigma}+2\varepsilon\cos\alpha\,\hat{n}}{\sqrt{(1-\varepsilon)^2+4\varepsilon\cos^2\alpha}} \tag{3.65}$$

where $\cos\alpha = \hat{\sigma}\cdot\hat{n}$.

The total force associated with the solar radiation pressure can be written as,

$$\vec{F} = \int_S d\vec{f} \tag{3.66}$$

S is the illuminated part of the spacecraft whose boundary is determined from the condition $\hat{\sigma}\cdot\hat{n} = 0$. The force acting on a body with a surface having an arbitrary reflection coefficient is written as,

$$\vec{F} = (1-\varepsilon)\,\vec{F}^{+} + \varepsilon\vec{F}^{-} \tag{3.67a}$$

where,

$$\vec{F}^{+} = -\bar{h}_0\,\hat{\sigma}\int_S(\hat{\sigma}\cdot\hat{n})\,dS \tag{3.67b}$$

$$\vec{F}^{-} = -2\bar{h}_0\int_S\hat{n}\,(\hat{\sigma}\cdot\hat{n})^2\,dS \tag{3.67c}$$

The illuminated region S has the following condition: $\hat{\sigma}\cdot\hat{n}\geq 0$. $\vec{F}^{+}$ is the solar radiation force due to a total absorbing surface, and $\vec{F}^{-}$ is the force due total reflective surface. $\hat{\sigma}$ describes the direction of the sun light with respect to the body and with condition $a^2_0 + b^2_0 + c^2_0 = 1$. If equations (3.67) are substituted into equations (3.2), the orbital elements can be highly affected by this perturbation. For satellites with altitudes higher than 800 km [**10**], the solar radiation pressure can become a dominant disturbing force.

To demonstrate the use of this formulation, Figure 3.12 shows the schematic of a cylinder. The Cartesian coordinate system (x, y, z) is located at the assumed center of gravity of the satellite. D is the diameter of the cylinder, and l_C is the height of the cylinder. In equations (3.67) and Figure 3.12, the surface integral must be divided into two different integrals containing the lateral surface of the body of the cylinder and the top (or bottom) of the cylinder. For a cylinder, only two surfaces can be illuminated at a time by the Sun. The surface integral in the Cartesian (x, y, z) coordinate system is complicated, but this surface integral can be changed to cylindrical coordinates to perform this calculation. These transformations can be used to calculate the forces for an absorbing and reflective surface for the top and body of the cylinder. For this reason, dS is defined as follows,

$$dS_B = R_c d\phi dz \tag{3.68a}$$

$$dS_T = r_c dr_c d\phi \tag{3.68b}$$

where dS_B is the change over the surface of the body of the cylinder, dS_T is the change over the surface of the top (or bottom) of the cylinder, R_c is given by half of the diameter of the cylinder, and r_c is any given radius of the cylinder defined in the range $0 \leq r_c \leq R_c$. In Figure 3.12, ϕ is the angle measured from the x axis to the arbitrary radius of the cylinder.

Figure 3.12 **Schematic of the cylinder for solar pressure calculation**

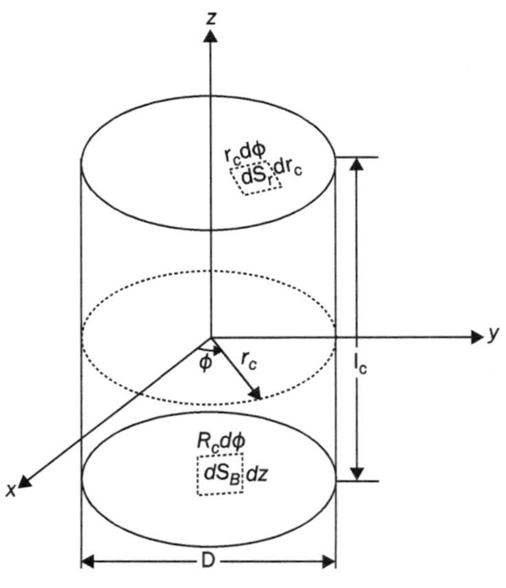

The cylinder is represented by the following equation,

$$x^2 + y^2 = r_c^2,\ z = l_C \tag{3.69}$$

where r_c is an arbitrary radius for the cylinder, and l is the height of the cylinder. The unit normal for the lateral surface of the cylinder is given as,

$$\hat{n} = \frac{\nabla f}{|\nabla f|} = \frac{x}{r_c}\hat{\imath} + \frac{y}{r_c}\hat{\jmath} \tag{3.70}$$

where $f(x, y) = x^2 + y^2 - r^2{}_c$, ∇f is the gradient of the function $f(x, y)$, and $|\ |$ is the norm of a vector. From equations (3.67), the following term can be calculated in the integrals for the total absorbing and total reflective lateral surface,

$$\hat{\sigma}\cdot\hat{n} = \frac{a_0 x}{r_c} + \frac{b_0 y}{r_c} \tag{3.71}$$

From Figure 3.12, the x and y directions can be written in cylindrical coordinates as,

$$x = r_c \cos\phi \qquad y = r_c \sin\phi \tag{3.72}$$

Using equations (3.67b), (3.67c), (3.71), (3.72), and (3.68a), the force due to an absorbing and a reflecting surface over the body of the cylinder is defined as,

$$\vec{F}_B^+ = -4\bar{h}_0\hat{\sigma}\int_0^{\pi}\int_0^{lc/2}(a_0\cos\phi + b_0\sin\phi)R_c\,dz\,d\phi = -4l_c R_c b_0 h_0\hat{\sigma} \tag{3.73a}$$

$$\begin{aligned}\vec{F}_B^- &= -8\bar{h}_0\int_0^{\pi}\int_0^{l_C/2}\left(\cos\phi\,\hat{\imath} + \sin\phi\,\hat{\jmath}\right)(\hat{\sigma}\cdot\hat{n})^2 R_C\,dz\,d\phi\\ \hat{F}_B^- &= -\frac{4}{3}l_C R_C\bar{h}_0\left[4a_0b_0\hat{\imath} - \left(2a_0^2 + 8b_0^3\right)\hat{\jmath}\right]\end{aligned} \tag{3.73b}$$

The normal over the top (or bottom) surface is always along the z direction and is written here as $\hat{n} = \hat{k}$; then, using equations (3.67b), (3.67c), (3.71), (3.72), and (3.68b), the force due to an absorbing and a reflective surface over the top of a cylinder is obtained as,

$$\vec{F}_T^+ = -2\bar{h}_0\int_0^{\pi}\int_0^{RC} C_0 r_c\,dr_c d\phi = -\pi R_c^2 C_0\bar{h}_0\,\hat{\sigma} \tag{3.74a}$$

$$\vec{F}_T^- = -4\bar{h}_0\int_0^{\pi}\int_0^{RC} c_0^2 r_c\hat{k}\,dr_c d\phi = -2c_0^2\bar{h}_0\pi R_c^2\hat{k} \tag{3.74b}$$

The total force over the body of a cylinder can be written,

$$\vec{F}^+ = \vec{F}_B^+ + \vec{F}_T^+ \qquad \vec{F}^- = \vec{F}_B^- + \vec{F}_T^- \tag{3.75}$$

To understand the effects of the reflectivity coefficient, equation (3.75) requires the direction of the light and the ratio between the vectors r_0 and R_S (in equation (3.60)). To simplify the constant $\bar{h}_0$, Ref. [34] assumed that the vectors defining r_0 and R_S are parallel when the solar rays depart from the Sun. This assumption is performed because the relative distance from the satellite to the Sun and the relative distance from the Earth to the Sun does not greatly vary; for this reason, the ratio r_0/R_S is assumed to be equal to one, and

$$\bar{h}_0 = \frac{E_0}{\bar{c}}\left(\frac{r_0}{R_S}\right)^2 \approx 4.72x10^{-7}\ (\text{kg/m}^2) \tag{3.76}$$

Also, the direction of the incident light over the satellite must be established to determine the solar force on the body of the satellite. The direction of the incidence of light can vary depending on the position of the satellite along the orbit and on the inclination angle of the Sun with respect to the Earth. In Reference [35], it is assumed that the direction of the light can be determined with an axis transformation defining the position of the satellites and the inclination angle of the Sun with respect to the Earth. Using this assumption, the direction of the light can be written as:

$$\begin{aligned} a_0 &= \cos f \cos i_S \\ b_0 &= \sin f \cos i_S \\ c_0 &= \sin i_S \end{aligned} \tag{3.77}$$

where i_S is the inclination angle of the Sun with respect to the Earth. Equation (3.77) is normally known as a solar ephemeris model. In practice, this model can provide errors associated with the direction of the light. For conceptual designs and analysis, equation (3.77) can be used. For more detailed analysis, Vallado [36] provides a more complex solar ephemeris model. This solar ephemeris model is shown in Appendix A. Vallado's model can be compared to the Jet Propulsion Laboratory (JPL) solar ephemeris model [37] [38] and has an error of 10^{-4}.

Assume a right circular cylinder satellite with a diameter and height of of 1.016 m and 0.5588 m, respectively. Also, assume that the maximum inclination angle of the Sun with respect to the Earth is 23.5 degrees. Figure 3.13 shows the solution of equation (3.67) for different reflectivity coefficient values when the satellite is located at the Equator (f = 0°). From Figure 3.13, the solar radiation force disturbs the satellite more when the satellite surface is totally reflective. This condition can be considered as the worst-case effect of the solar radiation force. In addition,

Figure 3.13 Solar radiation force vs. reflectivity coefficient

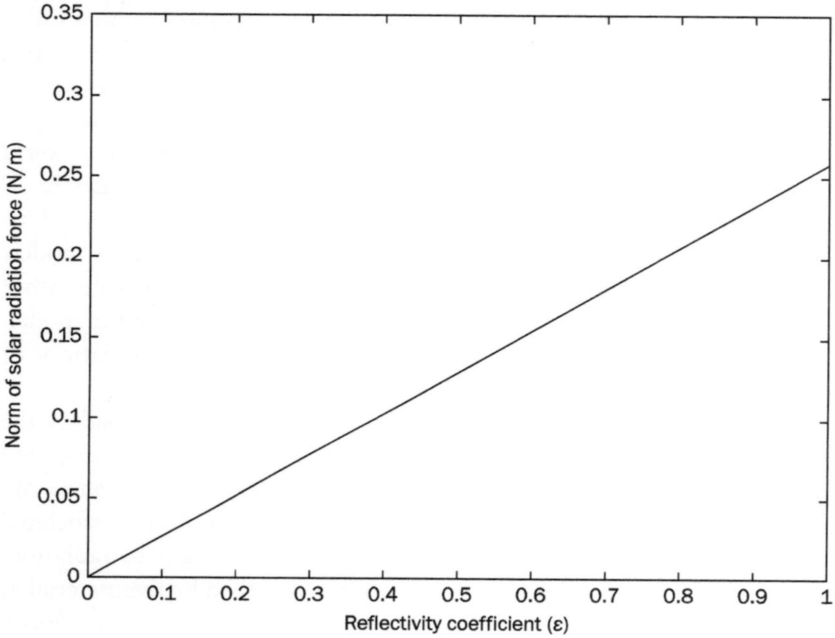

this effect is used in new proposed satellites called solar sails [39]. In solar sailing, a satellite is constructed with a very large area of a reflective surface. When this reflective surface is intersected by the sunlight, the satellite can travel through Space with the solar radiation force. Solar sailing is analogous to the effect of the wind in sailing on the sea.

3.6 Other disturbance effects

The described perturbations in the previous three sections are the common disturbances used in the simulations and calculations. There are higher order terms associated with some of the equations and other terms that are not considered here. The purpose of this section is to provide information about other disturbance effects not mentioned previously.

The perturbation due to the oblateness of the Earth has other higher order terms that describe a more deformed Earth. These higher order terms are derived from the expansion of the Legendre polynomial shown in equation (3.13) and are called higher order zonal harmonics and also the non-zonal harmonics (tesseral harmonics, J_m^n where $n = m$) and other non-zonal harmonics. As an example, J_3 describes an Earth with a bulge below the Equator. In other words, the Earth has a 'pear shaped' form. The J_3 term [10] equals to 2.676×10^{-6}. For conceptual models, the J_2 perturbation effect is enough. For more complex models, the gravitational potential function in equation (3.15) is expanded up to the J_4 term because it provides a more accurate gravitational model of the Earth.

The 1976 US Standard Atmospheric model is commonly used to describe the atmosphere of the Earth. There are more complex models shown in reference [40]. These atmospheric models provide other properties that are not considered in the 1976 US Atmospheric model. The 1976 Atmospheric model can be used in many of the simulations of the satellite motion.

The formulation for the solar radiation pressure only considers the specular effects. Karymov [32] described additional effects associated with the material on the satellite surface. These effects are the emissivity and absorptivity of the satellite. The emissivity of a material is associated with the ability of the surface of the material to emit energy by radiation. The absorptivity is the fraction of light that is absorbed by the material at a specified wavelength. These properties depend on the surface finish, temperature, coating thickness, aging, and contamination, among other factors [41]. For surfaces which are impacted by meteorites and/or other forms of space debris, this may give rise to diffuse specular reflection in addition to the previously discussed reflection. For a complete treatment of this, more complex statistical treatments are required. Also, the solar pressure formulation presented here assumes that the satellite is in full sunlight. In reality, the satellite has periods of time in which the satellite is in the night side of the Earth. The night side of the Earth is called the umbra [10]. Normally, this effect can be modeled such that it can be included in the simulations.

The final disturbance that can be considered is the effects of the Moon. These effects are known as the third body effects [11]. Normally, these third body effects are small in comparison with the disturbance effects of the Earth and Sun. On the other hand, the Moon's disturbance effects should be related to the sphere of influence [8] [36], involving the primary and disturbing planet. The sphere of influence is an imaginary frontier beyond which the Earth–Sun system has a greater effect on the satellites than the Earth–Moon system, or vice versa. According to this concept, certain terms due to the Moon's effect could be maintained or excluded as part of the equations of motion. This is an important relation because it relates to the relative importance of the disturbances of the Moon.

In summary, the disturbance effects mentioned can be considered in studies for the motion of the satellite. In the conceptual designs, these effects are not necessary because the first calculations do not require very detailed information. When the satellite is in the design process, the models should be updated to more complex formulations such that the simulation can be accurate. In practice, the only updated models are the gravitational potential function and the atmospheric models. Normally, the solar radiation pressure is described in terms of the specular effects and the shape of the satellite. For more complete treatments of diffuse reflection, additional complex statistical treatments need to be developed, perhaps based on measurements taken from satellites in orbiting debris fields. It is understood that many current microsatellites are equipped to provide such measurements possibly using optical techniques.

3.7 Suggested problems

Problem 3.1. A sun synchronous orbit is an orbit which moves in such a way that the incidence angle between the sun and the orbit plane is always the same.

a. Plot a curve of the orbital inclination as a function of altitude, $\bar{s}$, for all those circular orbits that are sun synchronous.
b. Find the largest value of $\bar{s}$ that will give a sun synchronous orbit.

Problem 3.2. The observed precession of the orbital planes of the first three Sputnik satellites averaged 2/7 degrees per day, where the initial orbital parameters were:

	Sputnik I	Sputnik II	Sputnik III
Altitude at perigee (miles)	142	140	135
Altitude at apogee (miles)	588	1038	1167
Period (min)	96.2	103.7	106
Inclination (degrees)	64.3	65.4	65.3

Use this data to estimate (approximately) the value of the J_2 coefficient and thus the difference between the Earth's equatorial and polar diameters. You may consider the Earth as an oblate spheroid of revolution formed by rotating an ellipse about its minor axis when relating the radii of gyration to the equatorial and polar diameters. Constants: $M_E = 5.975\times10^{24}$ kg.

Problem 3.3. Develop the general expression for the Legendre polynomial in q of degree n as:

$$P_n(q) = \frac{1\cdot3\cdot5\cdots(2n-1)}{1\cdot2\cdot3\cdots n}\left[q^n - \frac{n(n-1)}{2(2n-1)}q^{n-2} + \frac{n(n-1)(n-2)(n-3)}{2\cdot4\cdot(2n-1)(2n-3)}q^{n-4}\right.$$
$$\left. - \frac{n(n-1)(n-2)(n-3)(n-4)(n-5)}{2\cdot4\cdot6\cdot(2n-1)(2n-3)(2n-5)}q^{n-4} + \cdots\right]$$

Include sufficient number of terms in your binomial series expansion so that the coefficients appearing above may be clearly identified.

Problem 3.4. According to *Sky and Telescope*, March, 1958, the Sputnik I sphere (Russian Earth satellite) which was 22.5 inches in diameter and weighed 185 pounds was observed by radio technique to break up into a number of pieces during its final days. The initial and final orbital period and altitude during this time were recorded as shown below.

Date	Period of orbit (minutes)	Altitude (feet)
Jan. 1, 1958 – 12:ØØ GMT	90.8	1,039,460
Jan. 5, 1958 – 12:ØØ GMT	89.4	790,000

a. Calculate the change in orbital energy, ΔU, from Jan. 1 to Jan. 5. If we assume that the change in energy is equivalent to the work done by the drage force acting over the orbital path, calculate the average drag force. (Assume a circular orbit and use average values of altitude and period in calculating the drag.) In your opinion, based on this information, did drag forces account for Sputnik's break-up? Explain.

b. Over the period under observation, did the satellite slow down or speed up? Explain.

c. If the drag force is derivable from a scalar function of the generalized position coordinates, determine this function. If not, explain.

d. The drag coefficient of a spherical body in the free molecular flow regime is approximately 2.0. Using this value and the average value of the drag force from (a), calculate the average value of the atmosphere density. Compare the calculated density with the value contained in atmospheric tables at the mean altitude [10] [20]. How would you account for any discrepancy?

Problem 3.5. For a spherically shaped satellite of radius, r, calculate $\vec{F}^+$ (the solar pressure force on a totally absorbing surface), $\vec{F}^-$ (the solar pressure force on a totally reflecting surface) and $\vec{F}$, the solar pressure force for a general surface with reflection coefficient, ε, $0 \le \varepsilon \le 1$. Is the total force, $\vec{F}$, dependent on the surface properties? What is the direction of $\vec{F}$?

Hint: Since a sphere has symmetry, without loss of generality, assume direction of solar incidence is along one of the Cartesian coordinate axes.

Problem 3.6. For the case of a satellite in the shape of a right circular cylinder with a totally absorbing surface ($\varepsilon = 0$), calculate the force, $\vec{F}^+$. Assume that the radius of the cylinder is r, the z coordinate of the top surface is $z = a$, and of the bottom surface, $z = -a$.

Hint: Since the cylinder is symmetrical about z, you can assume that $\hat{\sigma} = b_0\hat{J} + c_0\hat{k}$ without loss of generality. Remember that the top and bottom surfaces will not be illuminated at the same time.

$$\text{Answer: } \vec{F}^+ = -\bar{h}_0\pi\hat{\sigma}r^2\left[\left|c_0\right| + \frac{4a}{\pi r}\sqrt{1-c_0^2}\right]$$

Problem 3.7. For the case of a satellite in the shape of a right circular cylinder with a totally reflecting surface ($\varepsilon = 1$), calculate the force $\vec{F}^-$. Assume the cylinder is represented by: $x^2 + y^2 = r^2$ and $|z| \le a$.

Hint: Take advantage of any symmetry and remember that the top and bottom surfaces will not be illuminated at the same time.

$$\text{Answer: } \vec{F}^- = \begin{bmatrix} -\frac{16}{3}\bar{h}_0 a_0 ra\sqrt{1-c_0^2} \\ -\frac{16}{3}\bar{h}_0 b_0 ra\sqrt{1-c_0^2} \\ -2\bar{h}_0\pi r^2 c_0^2\, sign(c_0) \end{bmatrix}$$

4

Frame rotations and quaternions

Abstract: This chapter highlights the use of rotational matrices to transform from one coordinate system to another. In practice, there are different coordinate systems within the spacecraft that specify a desired rotation. Some examples for the use of rotational matrices are the Ares V rocket and the orbital motion of a satellite. In addition, this chapter provides a brief description about quaternions and describes the use of the quaternions in space rotations. The quaternions and the rotational angles are related to each other by the rotational matrix to obtain the transformations between quaternions and Euler angles.

Key words: rotational matrices, rotational sequences, Euler angles, aerospace rotation, Ares V rocket, quaternions, Euler angle and quaternion transformations.

4.1 Introduction to rotations and quaternions

In space flight mechanics, it is necessary to explain the rotation of a space vehicle in free space. Some examples of the use of rotational coordinate frames for space vehicles are the following: 1) for rocket ascent trajectory, 2) for airplane ascending, descending, and in-flight motion, 3) for satellite orientation. In any of these cases, two frames are required: the reference and body frames. The reference frame describes the desired coordinate frame in which the body is traveling. The body frame describes the orientation of the body while it travels in free space. Before defining the attitude motion of the body in Chapter 5, the purpose of this chapter is to explain the rotational sequences and the formulation that explains the orientation of the vehicle.

First, the two-dimensional frame rotations are explained. The two-dimensional frame rotations explain how a vector can be transferred from one coordinate frame to another. Secondly, the three-dimensional frame rotations are explained based on the two-dimensional frame formulations. In three-dimensional rotations, the frame is rotated about one of the axes; and the rotation is expressed in terms of the other two axes. These three dimensional rotations are commonly used to express differences between frames in orbits and aerospace vehicle body frames. Two commonly used sequences are known as the Euler and aerospace sequences. The Euler sequence is used to express the orientation of a satellite in an orbit. The aerospace sequence, as the name clearly states, is used for to explain the orientation of space vehicles. For both cases, it is possible to obtain a relation for the position, velocity, angular velocities, and the forces between the body and reference frames.

Third, the quaternion formulation is defined for rotational sequences. The quaternions are a very useful notation to express rotations of a body [42]. In addition, the quaternions are used to simplify the implementation of algorithms for any space vehicles [43] and orbits. Fourth, the relations between the quaternions and the rotational matrices are explained. This is a very important step because it helps the designer in the analysis of the data.

In summary, this chapter provides a practical explanation of how the orientation of a space vehicle can be described with respect to different coordinate frames. Two examples are developed using rotational matrices and quaternions. The examples provide a good idea of how a rotational sequence can be used to define the orientation of a vehicle.

4.2 Two-dimensional frame rotations

The two-dimensional frame rotations are the simple rotations that can be performed. These rotations refer to the rotation of a plane over a point. Figure 4.1 shows the rotation of the (x, y) Cartesian plane to the (x', y') Cartesian plane. In addition, Figure 4.1 shows a point (Q) in the two coordinate systems, and the position vector ($\vec{r}$) related to this point. The angle θ is the rotation angle of the Cartesian plane (x, y) relative to the Cartesian plane (x', y'). The angle α describes the polar coordinate of the vector $\vec{r}$ at point Q. In the Cartesian plane (x, y), the vector $\vec{r}$ can be written in polar coordinates as,

$$x = |\vec{r}| \cos \alpha \tag{4.1a}$$

Figure 4.1 **Plane rotation**

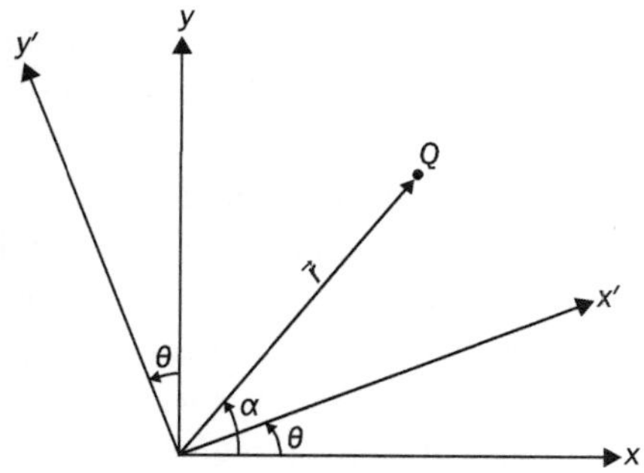

$$y = |\vec{r}|\sin\alpha \tag{4.1b}$$

In the Cartesian plane (x', y'), the vector $\vec{r}$ can be written as,

$$x' = |\vec{r}|\cos(\alpha - \theta) = |\vec{r}|(\cos\alpha\cos\theta + \sin\alpha\sin\theta) \tag{4.2a}$$

$$y' = |\vec{r}|\sin(\alpha - \theta) = |\vec{r}|(\sin\alpha\cos\theta + \cos\alpha\sin\theta) \tag{4.2b}$$

Substituting equations (4.1) into equations (4.2), the transformation from the Cartesian plane (x, y) to the Cartesian plane (x', y') can be written as,

$$\begin{bmatrix} x' \\ y' \end{bmatrix} = \mathbf{R}\begin{bmatrix} x \\ y \end{bmatrix} \tag{4.3a}$$

where,

$$\mathbf{R} = \begin{bmatrix} \cos\theta & \sin\theta \\ -\sin\theta & \cos\theta \end{bmatrix} \tag{4.3b}$$

Equation (4.3b) is known as the rotational matrix and transforms any vector in the (x, y) plane to the (x', y') plane. The point Q is invariant under any rotation [**11**]. The rotational matrix satisfies the orthogonality condition which establishes that $\mathbf{R}^T = \mathbf{R}^{-1}$ and $\mathbf{R}\mathbf{R}^T = \mathbf{R}\mathbf{R}^{-1} = I$. $(\ \ )^T$ refers to the transpose of a matrix, and $(\ \ )^{-1}$ is the inverse of a matrix. **I** is the identity matrix. If it is desired to rotate the frame from the Cartesian plane (x', y') to the Cartesian plane (x, y), equation (4.3b) can be rewritten as follows,

$$\begin{bmatrix} x \\ y \end{bmatrix} = \mathbf{R}^T\begin{bmatrix} x' \\ y' \end{bmatrix}$$

4.3 Three-dimensional frame rotations

The three-dimensional rotations are derived from the plane rotations. The difference is the rotational matrix that is used to establish the rotation from one plane to another in the three-dimensional space. The first condition in three-dimensional frame rotations [42] is that the rotations follow the right hand rule. When the three dimensional frame is rotated, the rotation is performed about one of the axis of the Cartesian system. Figure 4.2 shows the three different rotations that can be performed in a three-dimensional Cartesian system. A rotation about the x axis is described with the following rotational matrix and shown in Figure 4.2,

$$\mathbf{R}_x = \begin{bmatrix} 1 & 0 & 0 \\ 0 & \cos\theta & \sin\theta \\ 0 & -\sin\theta & \cos\theta \end{bmatrix}$$

In Figure 4.2, the rotation about the x axis only rotates the (y, z) plane. In the rotational matrix, the element corresponding to the axis about which the rotation is performed is set to one and is always located on the

Figure 4.2 Rotations about *x* (top left), *y* (top right), and *z* (bottom)

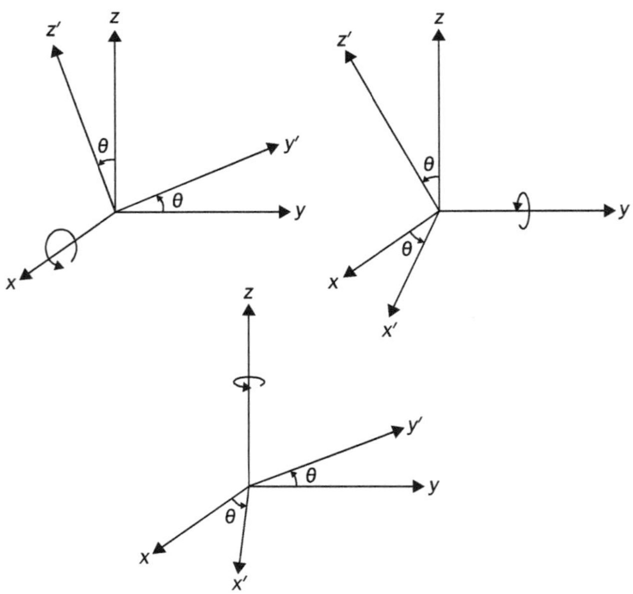

principal diagonal of the matrix. As shown in Figure 4.2, the rotations about the y and z axis have the following rotational matrices:

$$\mathbf{R}_y = \begin{bmatrix} \cos\theta & 0 & -\sin\theta \\ 0 & 1 & 0 \\ \sin\theta & 0 & \cos\theta \end{bmatrix} \qquad \mathbf{R}_z = \begin{bmatrix} \cos\theta & \sin\theta & 0 \\ -\sin\theta & \cos\theta & 0 \\ 0 & 0 & 1 \end{bmatrix}$$

For illustration, the rotational matrices are described by the same angle. In reality, the angles can be different for all the rotational matrices.

The plane rotations are described as a sequence that defines the order of the frame rotations. And, the full rotational matrix can be determined. Assume the following rotation sequence: 1) Rotate about the x axis through an angle θ, 2) Rotate the new frame about the y axis through an angle α, and 3) Rotate the new frame about the z axis through an angle β. The first rotational sequence can be written as follows,

$$\vec{T}_1 = \mathbf{R}_x\,(\theta)\,\vec{T} \tag{4.4a}$$

$\vec{T}$ is the vector in the original frame, $\mathbf{R}_x\,(\theta)$ is the rotational matrix corresponding to a rotation about the x axis through an angle θ, and $\vec{T}_1$ is the new vector in the new coordinate frame. The second rotation about the new axis frame is defined as,

$$\vec{T}_2 = \mathbf{R}_y(\alpha)\vec{T}_1 \tag{4.4b}$$

$\mathbf{R}_y(\alpha)$ is the rotational matrix corresponding to a rotation about the y axis through an angle α, and $\vec{T}_2$ is the new rotated vector. The third rotation is described as,

$$\vec{T}_3 = \mathbf{R}_x(\beta)\vec{T}_2 \tag{4.4c}$$

$\mathbf{R}_x(\beta)$ is the third and final rotation about the x axis through an angle β. Substituting equation (4.4a) and (4.4b) into equation (4.4c), the rotational matrix associated with the rotational sequence x-y-x can be defined as,

$$\vec{T}_3 = \mathbf{R}\vec{T} \tag{4.5a}$$

where,

$$\mathbf{R} = \mathbf{R}_x(\beta)\mathbf{R}_y(\alpha)\mathbf{R}x(\theta) \tag{4.5b}$$

Equation (4.5) defines the order of multiplication for the rotation sequence. Also, the rotational matrix $\mathbf{R}$ satisfies the orthogonality conditions. Because of this condition, the transpose of equation (4.5) can be used to determine the vector in the original frame, i.e. $\vec{T}_1 = \mathbf{R}^T\vec{T}_3$. This rotational sequence theory is used in the examples. Two commonly used frame rotations are explained in the following for the orientation of a rocket and a satellite.

4.4 Example of frame rotations

There are two examples that are commonly used in orbital and space flight mechanics. The first example is used to demonstrate the transformation from the Earth Centered Inertial (ECI) frame to the orbital frame. The second example is to explain the rotation of an aerospace vehicle. Both rotations show different rotational sequences that can produce different information about the motion of a body. There are more rotations that can be performed [**44**]; but these rotations are the commonly used sequences [**11**] [**42**].

4.4.1 *Euler rotation*

The first example is the transformation from the ECI frame to the orbital frame. This rotation sequence is also used to describe the precession and nutation of the Earth. The Earth is not a static body and rotates about its own axis as it moves about the Sun. As shown by Vallado [**36**], the ECI frame can be corrected by the precession and nutation motion. Also, Goldstein [**11**] demonstrates the use of the same rotation sequence to develop the precession and nutation motion of the Earth. To understand the development of a rotation sequence in a problem, the orbital frame rotation of a satellite is introduced here. This rotational sequence is also known as the Euler rotation.

Figure 4.3 shows the ECI frame and the orbital frame. The coordinate system (X, Y, Z) is the ECI frame, and the coordinate system (x, y, z) is the orbital frame. The rotation sequence can be described as follows: 1) First, rotate the ECI frame about the z axis by the angle Ω, 2) Second, rotate about the new x axis by an angle i, 3) Third, rotate about the new z axis by an angle w where $w = f + \omega$. The first rotation locates the new x axis along the line of nodes. The second rotation inclines the coordinate frame to be at the same inclination angle as the orbit. The third rotation provides the location of the perigee point in the orbit. The rotation sequence is Z-X-Z as shown in Figure 4.3. Using the rotation sequence Z-X-Z, the rotational matrix can be written as,

$$\vec{T}_{orbit} = \boldsymbol{R}\vec{T}_{ECI}$$

Figure 4.3 **Earth centered inertial frame to orbital frame rotation**

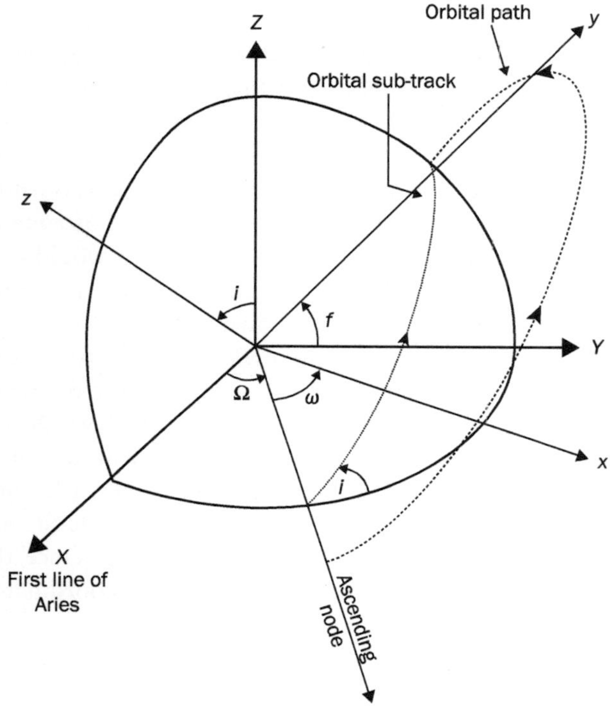

where,

$$\mathbf{R} = \mathbf{R}_z(w)\,\mathbf{R}_x(i)\,\mathbf{R}_z(\Omega) \tag{4.6}$$

Using the rotational matrices shown in Section 4.3 and performing the multiplications, the rotational matrix from the ECI to the orbital frame is written as follows,

$$\mathbf{R} = \begin{bmatrix} \cos\Omega\cos w - \sin\Omega\cos i\sin w & \sin\Omega\cos w + \cos\Omega\cos i\sin w & \sin i\sin w \\ -\cos\Omega\sin w - \sin\Omega\cos i\cos w & -\sin\Omega\cos w + \cos\Omega\cos i\cos w & \sin i\cos w \\ \sin\Omega\sin i & -\cos\Omega\sin i & \cos i \end{bmatrix}$$

Reference [42] demonstrates how to use this rotational matrix to calculate the location of the satellite in terms of the longitude and latitude of the Earth. This information is known as the orbital ephemeris of a satellite.

4.4.2 Aerospace rotation

This second example is to demonstrate the use of the rotational matrix to describe the orientation of a space vehicle that can be a rocket, airplane, helicopter, and satellite. To demonstrate this rotation, the Ares rockets of the Constellation program are used. Ares I is a two stage rocket that was proposed to send the crew exploratory vehicle (CEV) called Orion into a circular orbit about the Earth. Ares V is known as the heavy lifter which was proposed to carry the other modules to put the astronauts into a lunar orbit and descend on the Moon. Ares I and V would rendezvous near the Earth before the motors are ignited to send the astronauts to the Moon. Ares I and Ares V rockets are shown in Figure 4.4 [45].

Figure 4.5 shows a schematic of the Ares V rocket and the coordinate frame. The coordinate frame (*X*, *Y*, *Z*) is known as the reference frame. Other authors also called these frame as the North-East-Down (NED) frame because during flight, the belly of the rocket will be looking down to the Earth. In addition, the East direction is used to describe the location of the platform with respect to the rocket; and the North direction is the direction in which the nose of the rocket is pointing. Once the rocket begins to translate from the reference frame, the new coordinate frame is

Figure 4.4 **Ares I and V rockets***

* Obtained from NASA/courtesy of www.nasaimages.org.

Figure 4.5 **Ares V front (left) and side view (right) of the reference frame**

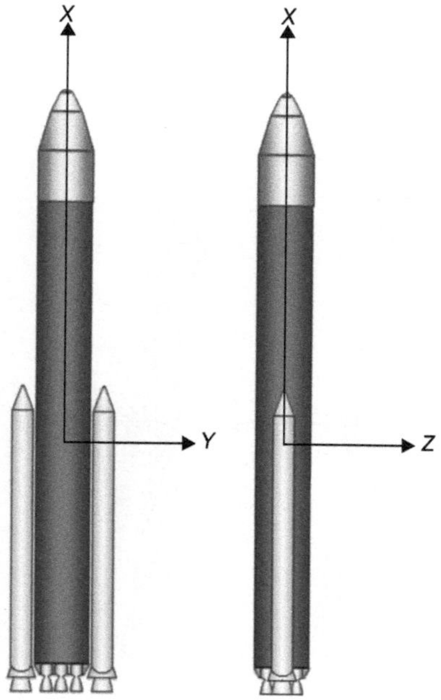

known as the body frame. The body frame (X', Y', Z') in Figure 4.6 describes the orientation of the rocket as it moves in free space. In Figure 4.6, θ is known as the pitch angle and is the angle that rotates about the Y axis; ψ is the yaw angle and describes the rotation about the Z axis; and φ is known as the roll angle and is the angle that explains the rotation about the X axis. The roll angle is not shown in Figure 4.5 but affects the (Y, Z) plane as shown in Figure 4.2. The effects of the roll angle are observed from the nose of the rocket, looking down to the motors.

The rotation sequence is the following: 1) First, rotate the reference frame about the Z axis by the yaw angle; 2) Second, rotate the new frame about the Y axis by the pitch angle; and 3) Third, rotate the new frame to the body frame about the X axis by the roll angle. The rotational matrix for the sequence Z-Y-X can be written as:

$$\vec{T}_{Body} = \mathbf{R}\vec{T}_{Ref} \tag{4.6a}$$

where,

$$\mathbf{R} = \mathbf{R}_x(\varphi)\,\mathbf{R}_y(\theta)\,\mathbf{R}_z(\psi) \tag{4.6b}$$

Figure 4.6 **Ares V front (left) and side (right) view of the body frame**

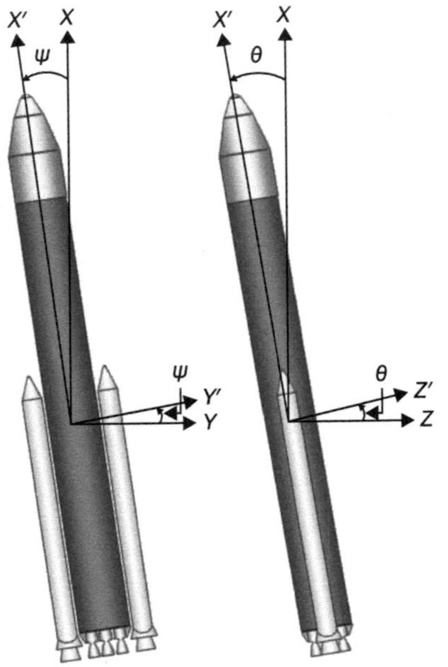

Using the rotational matrices in Section 4.3, equation (4.6b) can be written as,

$$\mathbf{R} = \begin{bmatrix} \cos\psi\cos\theta & \sin\psi\cos\theta & -\sin\theta \\ \cos\psi\sin\theta\sin\varphi - \sin\psi\cos\varphi & \sin\psi\sin\theta\sin\varphi + \cos\psi\cos\varphi & \cos\theta\sin\varphi \\ \cos\psi\sin\theta\cos\varphi + \sin\psi\sin\varphi & \sin\psi\sin\theta\cos\varphi - \cos\psi\sin\varphi & \cos\theta\cos\varphi \end{bmatrix}$$

This matrix is commonly used to describe the orientation of a spacecraft such as the Ares V. It is also used to determine the relation of the angular velocities between the reference and body frame. The angular velocity in the reference frame can be obtained from the rotational matrices used in equation (4.6b). Using Figure 4.5 and the aerospace sequence, the rotational angular velocity in the reference frame can be described as a sum of rotational sequences [43]. First, the roll angle is the last rotation that is performed and is only reflected about the *X'* axis of the body frame. Then, the first angular velocity becomes,

$$\vec{\varpi}_2 = \begin{bmatrix} \dot{\varphi} \\ 0 \\ 0 \end{bmatrix} \tag{4.7a}$$

Second, the pitch rate is the rotational angular velocity of the body frame about the Y axis. To represent that rotation in the body frame, the pitch rate has to be rotated by the roll angle and is equal to,

$$\vec{\varpi}_2 = \mathbf{R}_x(\varphi)\begin{bmatrix} 0 \\ \dot{\theta} \\ 0 \end{bmatrix} = \begin{bmatrix} 0 \\ \dot{\theta}\cos\varphi \\ -\dot{\theta}\sin\varphi \end{bmatrix} \tag{4.7b}$$

Last, the yaw rate is associated with the angular velocity about the Z axis. The yaw rate has to be transformed to the body frame by multiplying the roll and pitch rotational matrix as follows,

$$\vec{\varpi}_3 = \mathbf{R}_x(\varphi)\mathbf{R}_y(\theta)\begin{bmatrix} 0 \\ 0 \\ \dot{\psi} \end{bmatrix} = \begin{bmatrix} -\dot{\psi}\sin\theta \\ \dot{\psi}\sin\varphi\cos\theta \\ \dot{\psi}\cos\varphi\cos\theta \end{bmatrix} \tag{4.7c}$$

Adding equations (4.7), the angular velocity in the reference frame in terms of the angular velocity in the body frame can be written as,

$$\vec{\varpi} = \begin{bmatrix} \dot{\varphi} - \dot{\psi}\sin\theta \\ \dot{\theta}\cos\varphi + \dot{\psi}\sin\varphi\cos\theta \\ -\dot{\theta}\sin\varphi + \dot{\psi}\cos\varphi\cos\theta \end{bmatrix} \tag{4.8}$$

$\vec{\varpi}$ is the angular velocity associated with the reference frame of the rocket. Equation (4.8) can be separated in terms of the angular rates of the body frame.

As shown in Sections 4.4.1 and 4.4.2, the rotational sequences can vary depending on the designer. It is not necessary to perform the same rotation sequences as shown here but it is required to be consistent with the rotations. Also, the relations of the angular velocities between the reference and body frame are required to be consistent with respect to the rotational sequence used.

4.5 Quaternion definition and rotations

In more recent years, the methods used for the attitude determination [46] and controls [47] of spacecraft has been simplified with the use of quaternions. A quaternion is a hypercomplex number [42] that defines a direction and the orientation of a Cartesian frame. The quaternion is written in terms of a vector and a magnitude as follows,

$$q = q_4 + \vec{q} = q_4 + q_1\hat{\imath} + q_2\hat{\jmath} + q_3\hat{k}$$

$\vec{q}$ is the vector defining the direction of the rotation, and q_4 is the magnitude defining the rotation of the vector $\vec{q}$. Figure 4.7 provides the visualization of a quaternion. Appendix B shows all the algebra associated with the quaternion, and Reference [42] provides more definitions and properties of the quaternions. The use of the rotational matrices is not simple because of the linear algebra involved in the problem. Quaternions use a very simple algebra to solve the rotational sequence problem.

The purpose of this section is to present the quaternions from the perspective of a rotational sequence. To express the rotation between two coordinate frames (such as the aerospace and Euler angle sequences), a unit quaternion is used to perform these calculations. The unit quaternion can be described as follows,

$$q = q_4 + \vec{q} = \cos\theta + \hat{q}\sin\theta \tag{4.7}$$

where $\hat{q}$ is a unit vector associated with the orientation of the rotation. Equation (4.7) satisfies the restriction that the norm of the quaternion in equation (4.7) is equal to one. If a rotation about the X axis of the coordinate frame is performed, the quaternion can be written as,

Figure 4.7 Quaternion diagram

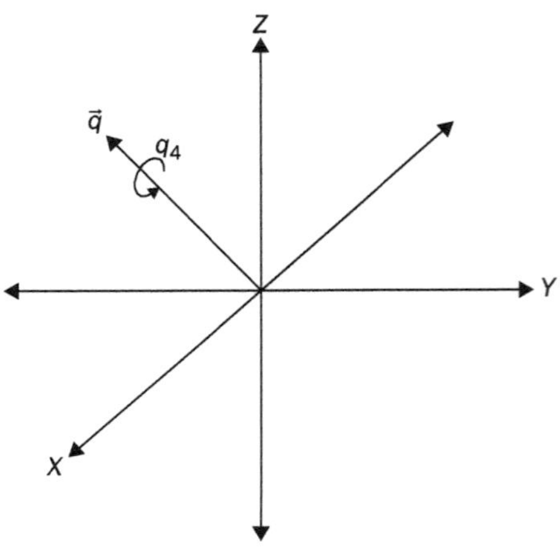

$$q_x(\theta) = \cos\theta + \hat{\imath}\sin\theta$$

For the Y and Z axis rotations, the quaternions can be written as,

$$q_y(\theta) = \cos\theta + \hat{\jmath}\sin\theta \quad q_z(\theta) = \cos\theta + \hat{k}\sin\theta$$

As shown in Appendix B, the quaternions can be written as linear operators that can affect a vector. For the rotational sequences, those linear operators can be used to describe the rotation from one coordinate frame to another. Using the unit quaternion in equation (4.7), the linear operator [48] defining the rotation sequence [42] as in equation (4.5a) can be written as,

$$\vec{w} = L_{q^*}(\vec{v}) = q^*\vec{v}q \tag{4.8a}$$

Equation (4.8a) can be interpreted geometrically as the rotation of the vector $\vec{v}$ through an angle 2θ about $\hat{q}$ as the axis of rotation. In other words; the rotation with equation (4.8a) must be performed by half of the angle of rotation. Equation (4.8a) represents the rotation of a vector to another coordinate frame, but the output of equation (4.8a) is a quaternion. To transform the quaternion into a vector, the following linear operator is used,

$$\vec{v} = L_q(\vec{w}) = q\vec{w}q^* \tag{4.8b}$$

Equation (4.8b) is the analog of taking the transpose of the rotational matrix, and Equation (4.8a) is similar to multiplying the rotational matrix by the vector. The output of equation (4.8b) is a vector.

One question remains: how the rotational matrix used in equation (4.4) can be written in quaternion format? This can be answered with equation (4.8a). Using the example of the rotational sequence X-Y-X in equation (4.4), the quaternions can be written as follows for every rotation,

$$1^{\text{st}}\text{ Rotation, } \vec{T}_1 = L_{q^*}(\vec{T}) = q_x^*\left(\frac{\theta}{2}\right)\vec{T}q_x\left(\frac{\theta}{2}\right) \tag{4.9a}$$

$$2^{\text{nd}}\text{ Rotation, } \vec{T}_2 = L_{s^*}(\vec{T}_1) = s_y^*\left(\frac{\alpha}{2}\right)\vec{T}_1 s_y\left(\frac{\alpha}{2}\right) \tag{4.9b}$$

$$3^{\text{rd}}\text{ Rotation, } \vec{T}_3 = L_{u^*}(\vec{T}_2) = u_z^*\left(\frac{\beta}{2}\right)\vec{T}_2 u_z\left(\frac{\beta}{2}\right) \tag{4.9c}$$

Using equations (4.9a) and (4.9b), equation (4.9c) can be written as,

$$\vec{T}_3 = v^*\vec{T}v \tag{4.10a}$$

where,

$$v = q_x\left(\frac{\theta}{2}\right)s_y\left(\frac{\alpha}{2}\right)u_z\left(\frac{\beta}{2}\right) \tag{4.10b}$$

There is one difference between equation (4.10b) and equation (4.5b). The difference is the order of multiplication between the different rotational sequences. Reference [43] presents a format that can help in the multiplication of the rotational sequences for the matrices. This method is extended here for the quaternions. The rotational sequence X-Y-X in matrix format can be represented as: $\mathbf{R}_x(\beta) \leftarrow \mathbf{R}_y(\alpha) \leftarrow \mathbf{R}_x(\theta)$. The first rotation is located at the right hand side of the sequence. The same rotational sequence in quaternion format can be described as: $q_x\left(\frac{\theta}{2}\right) \rightarrow s_y\left(\frac{\alpha}{2}\right) \rightarrow u_z\left(\frac{\beta}{2}\right)$ in which the first rotation is written at the left-hand side of the sequence.

4.6 Quaternion to Euler angle relations

The reader may think that there is a relation between equation (4.5) and equation (4.10). The answer is no because the output of equation (4.10) is a quaternion, and equation (4.5) is a vector. The only relation between the rotational matrix in equation (4.5) is another rotational matrix in terms of the quaternions and is expressed as [42],

$$\mathbf{R} = \begin{bmatrix} 2q_4^2 + 2q_1^2 - 1 & 2q_1q_2 + 2q_4q_3 & 2q_1q_3 - 2q_4q_2 \\ 2q_1q_2 - 2q_4q_3 & 2q_4^2 + 2q_2^2 - 1 & 2q_2q_3 + 2q_4q_1 \\ 2q_1q_3 + 2q_4q_2 & 2q_2q_3 - 2q_4q_1 & 2q_4^2 + 2q_3^2 - 1 \end{bmatrix} \tag{4.11}$$

Using equation (4.11), a relation between the Euler angles and the quaternions can be established.

In order to demonstrate the use of equation (4.11), the aerospace sequence is used as an example. In quaternion format, the rotational sequence can be described as: $v_z(\alpha) \rightarrow s_y(\beta) \rightarrow u_x(\gamma)$ where $\alpha = \psi/2$, $\beta = \theta/2$, and $\gamma = \varphi/2$. Using the format in equations (4.9) and (4.10), the body frame orientation in quaternion format can be written as,

$$\vec{T}_{Body} = q^*\vec{T}_{Ref}q \tag{4.12a}$$

where,

$$q = q_4 + \vec{q} = q_4 + q_1\hat{\imath} + q_2\hat{\jmath} + q_3\hat{k} \tag{4.12b}$$

and,

$$q_1 = \cos\alpha\cos\beta\sin\gamma - \sin\alpha\sin\beta\cos\gamma$$

$$q_2 = \cos\alpha\sin\beta\cos\gamma + \sin\alpha\cos\beta\sin\gamma$$

$$q_3 = \sin\alpha\cos\beta\cos\gamma - \cos\alpha\sin\beta\sin\gamma$$

$$q_4 = \cos\alpha\cos\beta\cos\gamma + \sin\alpha\sin\beta\sin\gamma$$

If the Euler angles are known, equations (4.12) are substituted into equation (4.11) to determine the rotational matrix. The rotational matrix into equation (4.11) can be used to transform a vector from the reference frame to the body frame.

Equation (4.6b) equals to equation (4.11) and can be used to determine the relation between the aerospace angles and the quaternions. Equating both equations, the result is the following,

$$\mathbf{R} = \begin{bmatrix} \cos\psi\cos\theta & \sin\psi\cos\theta & -\sin\theta \\ \cos\psi\sin\theta\sin\varphi - \sin\psi\cos\varphi & \sin\psi\sin\theta\sin\varphi + \cos\psi\cos\varphi & \cos\theta\sin\varphi \\ \cos\psi\sin\theta\cos\varphi + \sin\psi\sin\varphi & \sin\psi\sin\theta\cos\varphi - \cos\psi\sin\varphi & \cos\theta\cos\varphi \end{bmatrix}$$

$$= \begin{bmatrix} 2q_4^2 + 2q_1^2 - 1 & 2q_1q_2 + 2q_4q_3 & 2q_1q_3 - 2q_4q_2 \\ 2q_1q_2 - 2q_4q_3 & 2q_4^2 + 2q_2^2 - 1 & 2q_2q_3 + 2q_4q_1 \\ 2q_1q_3 + 2q_4q_2 & 2q_2q_3 - 2q_4q_1 & 2q_4^2 + 2q_3^2 - 1 \end{bmatrix}$$

By using this result, the following relations can be established between the quaternions and the roll, pitch, and yaw angles,

$$\sin\theta = 2q_4q_2 - 2q_1q_3 \tag{4.13a}$$

$$\tan\varphi = \frac{2q_2q_3 + 2q_4q_1}{2q_4^2 + 2q_3^2 - 1} \tag{4.13b}$$

$$\tan\psi = \frac{2q_1q_2 + 2q_4q_3}{2q_4^2 + 2q_1^2 - 1} \tag{4.13c}$$

If the quaternions are known, the Euler angles can be determined; and the rotational matrix in equation (4.6b) can be determined. These relations are used in the following chapters to develop different control schemes and to analyze the results.

4.7 Suggested problems

Problem 4.1. Using equation (4.3b), show that $R^{-1} = R^T$ and $RR^T = I$ where I is the identity matrix.

Problem 4.2. Using the rotational matrices shown in Section 4.3 and performing the multiplications, show that the rotational matrix, R, from the ECI to the orbital frame can be developed shown on the bottom of page 81.

Problem 4.3. Using the rotations in **Problem 4.2** show that the angular rates in the ECI frame can be written in terms of the angular rates in the orbital frame as,

$$\vec{\omega} = \begin{bmatrix} \sin w \sin i \dfrac{d\Omega}{dt} + \cos \mathrm{w} \dfrac{di}{dt} \\ \cos w \sin i \dfrac{d\Omega}{dt} - \sin w \dfrac{di}{dt} \\ \cos i \dfrac{d\Omega}{dt} + \dfrac{dw}{dt} \end{bmatrix}$$

Problem 4.4. Using the rotational matrices in Section 4.3 and the rotational sequence indicated in Equations (4.6), develop $\mathbf{R} = \mathbf{R}_x(\varphi)\,\mathbf{R}_y(\theta)\,\mathbf{R}_z(\psi)$.

Problem 4.5. Using Equation (4.11), develop the relationship between the Euler angles and the quaternions and develop both forms of the rotational matrix shown on page 75.

5

Rigid body motion

Abstract: The rotational motion of a body about its center of mass is called attitude dynamics. This chapter highlights the formulation for the development of the attitude dynamics equations of motion. First, the torques associated with the motion of a single particle and a system of particles are explained. Second, a description of the moments of inertia is explained such that it can be used for the development of the rotational equations of motion. Third, the equations of motion are explained in terms of the angular momentum, Lagrange's equations, and the quasi-coordinates system. Also the kinematic equations associated with quaternions are explained. Finally, the translational and rotational equations of motion are determined for the Ares V rocket as an example.

Key words: rate of change of a vector, torques, angular momentum, moments of inertia, principal moments of inertia, kinematics equations for a quaternion, Ares V translational and rotational equations of motion, rotational Lagrange's equations, Lagrange's equations for quasi-coordinates.

5.1 Introduction to attitude dynamics

In Chapter 4, the rotation of a body about a reference axis is explained and developed. As a continuation, the objective of this chapter is to explain the motion of a rotating body when there are surface forces acting on the body. The surface forces acting on a rotational body about a reference point are known as torques. Mathematically, the torques are referred to as the cross product between the distance from a reference point to the applied force. In Section 5.2, the rate of change of a vector is determined and is used to explain the applied torques on an arbitrary body. Then, the time derivative is expanded to define the translational and rotational motion of a body.

When the J2 perturbation is explained in Chapter 3, the moments of inertia are introduced but not defined. In Section 5.3, a formal definition of the moments of inertia is explained in terms of the angular momentum of a body. The moments of inertia express the rotation of a body about its own axis. The moments of inertia are analogous to the constant mass in Newton's Second Law of Motion and can be written in a matrix called the inertia matrix. In Section 5.4, the eigenvalue problem is solved with this matrix. The eigenvalues define the principal moments of inertia. In practice, the principal moments of inertia are commonly used to simplify the equations of motion. On the other hand, it is required to transform any vector associated with translation and rotation motion of the body to the principal axis coordinate system. The principal axis coordinate system is the rotated frame from the body axis frame and is explained by the eigenvectors of the eigenvalue problem solution of the inertia matrix.

After determining the moments of inertia, the rotational motion of a body is described in Section 5.5. This is performed by taking the time derivative of the angular momentum which is equal to the applied torque [**11**]. In addition, the Lagrange's equations and the quasi-coordinates are explained to determine the rotational motion of a body. The Lagrange's equations use an energy approach in which the kinetic and potential energy are expressed in terms of the rotation of the body. The quasi-coordinates are obtained from the Lagrange's equations but are only used to define the kinetic energy of the rotating body. This is a major difference between both methods but can be used to determine the equations of motion.

In Section 5.6, the quaternions are used to express the kinematic equations for the rotation of a body. In Chapter 4, the kinematic equations for the rotation of Ares V are determined for the aerospace sequence. A similar formulation can be performed with the quaternions. The main advantage of using the quaternions is the simplification of the integration process of the equations of motion. In Section 5.7, the equations of motion for the Ares V rocket are determined to describe the attitude motion during the ascent trajectory. Ares V provides a good example to demonstrate the formulation for the attitude motion. In summary, this chapter demonstrates how to write the dynamic equations of a rotating body.

5.2 Rate of change of a vector

In a translational motion, the particle can move in a linear path in a three-dimensional space. If the particle is seen from an observer (marked as 1 in Figure 5.1), the observer 1 sees that the particle moves in a linear path.

For the rotational motion of a body, a non-rotating observer (marked as 2 in Figure 5.1) moves with linear velocity and acceleration of point O. From observer 2, the body appears to be in pure rotation about point O. From observer 1, the particle is translating about the reference point and is rotating about point O. From Newton's Second Law, the motion of the particle can be written as,

$$\vec{F} = m\,(\vec{a}_{0/1} + \vec{a}_{2/0}) \tag{5.1}$$

where $\vec{a}_{0/1}$ is inertial acceleration of the particle with respect to the reference point (observer 1), and $\vec{a}_{2/0}$ is the acceleration of the particle about the point O as seen from the moving observer (observer 2). Taking the cross product of $\vec{r}_2$ with respect to $\vec{F}$, the torques associated with the motion of the particle can be described as,

$$\vec{N} = \vec{r}_2 \times \vec{F} = m\,(\vec{r}_2 \times \vec{a}_{0/1} + \vec{r}_2 \times \vec{a}_{2/0}) \tag{5.2}$$

where $\vec{N}$ are the applied torques, and $\vec{r}_2$ is the distance from the point O to the particle. Using the following vector identity [13],

$$\frac{d}{dt}(\vec{r} \times \vec{v}) = \vec{r} \times \frac{d\vec{v}}{dt}$$

Figure 5.1 Rotating motion of a particle

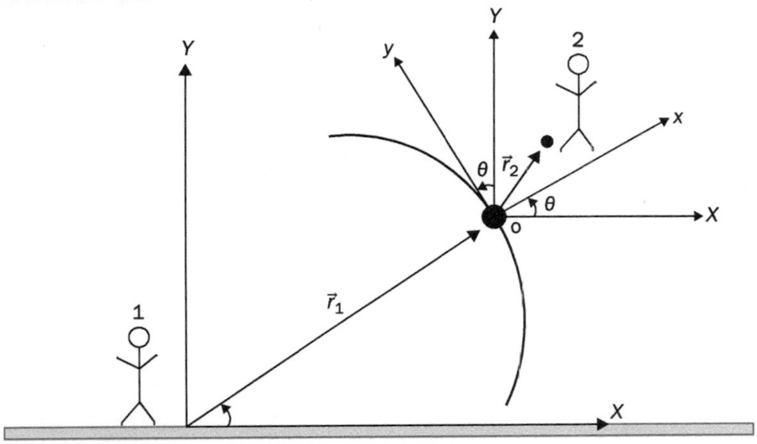

Equation (5.2) can be written as,

$$\vec{N} = \vec{r}_2 \times \vec{F} = m\frac{d}{dt}(\vec{r}_2 \times \vec{v}) + m\vec{r}_2 \times \vec{a}_{0/1}$$

If there is a system of particles, the torque is equal to,

$$\vec{N} = \vec{r}_2 \times \vec{F} = m\frac{d}{dt}(\vec{r}_2 \times \vec{v}) + \Sigma_k^N m_k \vec{r}_{k/a} \times \ddot{\vec{r}}_a \tag{5.3}$$

In equation (5.3), the last term could vanish for any of the following three reasons: 1) if the point O in Figure 5.1 is the center of mass, 2) if the point O is an unaccelerated point in the body, and 3) if the acceleration of point O has no moment about the center of mass. An application of this equation is observed in Reference [49]. Equation (5.3) can be written as follows,

$$\vec{N} = \frac{d\vec{H}}{dt} + \vec{s} \times \vec{a} \tag{5.4a}$$

where $\vec{s} = \Sigma_k^N m_k \vec{r}_{k/a}$, and $\vec{a}$ is the inertial acceleration of point O. Also,

$$\vec{H} = \vec{r} \times \vec{p} = \vec{r} \times m\vec{v} \tag{5.4b}$$

$\vec{H}$ is known as the angular momentum of the particle and describes the rotation of the particle about point O. The angular momentum is normal to the distance and the velocity of the particle when the right hand rule is applied [50].

The reference point in Figure 5.1 is used for the development of the equations of motion [51]. In Figure 5.1, the observer 1 sees a linear motion of the particle, but the observer 2 sees the rotation of the particle about the point O. The distance from observer 1 to observer 2 can be written as,

$$\vec{r} = x\hat{i} + y\hat{j} = \begin{bmatrix} x \\ y \end{bmatrix} \tag{5.5}$$

The motion of the particle can be described by the two-dimensional rotational matrix in equation (4.3b). Taking the time derivative of the position vector seen from observer 2, the motion of the particle can be described as,

$$\left.\frac{d\vec{r}}{dt}\right|_2 = \frac{d}{dt}(\boldsymbol{R}\vec{r}|_1) = \boldsymbol{R}\frac{d}{dt}(\vec{r}|_1) + \frac{d}{dt}(\boldsymbol{R})\vec{r}|_1 \tag{5.6}$$

Taking small angle assumptions, the velocity of an assumed body about point O can be written as,

$$\left.\frac{d\vec{r}}{dt}\right|_2 = \left.\frac{d\vec{r}}{dt}\right|_1 + x\frac{d\theta}{dt}\hat{j} - y\frac{d\theta}{dt}\hat{\imath} \tag{5.7}$$

If $\vec{\varpi} = \dot{\theta}\hat{k}$, then,

$$x\frac{d\theta}{dt}\hat{j} - y\frac{d\theta}{dt}\hat{\imath} = \vec{\varpi} \times \vec{r}$$

And,

$$\vec{v}|_2 = \vec{v}|_1 + \vec{\varpi} \times \vec{r} \tag{5.8}$$

Generalizing equation (5.8), the rate of change of a vector [11] seen from observer 2 can be written as,

$$\frac{d}{dt}(\;)\Big|_2 = \frac{d}{dt}(\;)\Big|_1 + \vec{\varpi} \times (\;) \tag{5.9}$$

The calculation in equation (5.8) can be performed one more time to determine the acceleration of the particle. The acceleration of the particle is performed at the end of the chapter with an example to explain other properties of the equation.

5.3 Moment of inertia

The velocity of a particle when it is rotating about a reference point is described by the angular velocity. An example is the motion of a satellite about the Earth. In this case, the angular velocity is described by the mean motion of the satellite. Taking the derivative of the position vector and assuming that the particle is not moving in a linear motion, the linear velocity is written as,

$$\vec{v} = \vec{\varpi} \times \vec{r} \tag{5.11}$$

where $\vec{\varpi}$ is the angular velocity of the particle. Substituting equation (5.11) into equation (5.4b), the angular momentum of a body with constant mass can be described as,

$$\vec{H} = m\vec{r} \times (\vec{\varpi} \times \vec{r})$$

Expanding the triple vector product [13],

$$\vec{H} = m(\vec{\varpi}|\vec{r}|^2 - \vec{r}(\vec{r} \cdot \vec{\varpi})) \tag{5.12}$$

Equation (5.12) can be written as,

$$H_x = m\,\varpi_x(|\vec{r}|^2 - x^2) - \varpi_y mxy - \varpi_z mxz \tag{5.13a}$$

$$H_y = m\,\varpi_y(|\vec{r}|^2 - y^2) - \varpi_x mxy - \varpi_z myz \tag{5.13b}$$

$$H_z = m\,\varpi_z(|\vec{r}|^2 - z^2) - \varpi_x mxz - \varpi_y myz \tag{5.13c}$$

but [52],

$$m = \int_{\mathbb{V}} \rho\, d\mathbb{V}$$

where $\mathbb{V}$ is the volume of a body. Using the mass integral, the following definition for the moments of inertia is obtained from equation (5.13a),

$$H_x = I_{xx}\varpi_x - I_{xy}\varpi_y - I_{xz}\varpi_z \tag{5.14}$$

where,

$$I_{xx} = \int_{\mathbb{V}} \rho\, (|\vec{r}|^2 - x^2)\, d\mathbb{V} \qquad I_{xy} = \int_{\mathbb{V}} \rho\, xy\, d\mathbb{V} \qquad I_{xz} = \int_{\mathbb{V}} \rho\, xz\, d\mathbb{V}$$

The moments of inertia are analogous to the constant mass in Newton's Second Law of Motion but describe the rotation of the mass about an axis of rotation. The same mass integral is used for equations (5.13b) and (5.13c), and the angular momentum can be written as,

$$\vec{H} = \hat{\boldsymbol{I}}\vec{\varpi} = \begin{bmatrix} I_{xx} & I_{xy} & I_{xz} \\ I_{xy} & I_{yy} & I_{yz} \\ I_{xz} & I_{yz} & I_{zz} \end{bmatrix} \begin{bmatrix} \varpi_x \\ \varpi_y \\ \varpi_z \end{bmatrix} \tag{5.15}$$

I_{xx}, I_{yy}, and I_{zz} are the moments of inertia about the axes of the coordinate system. The principal moments of inertia describe the rotation of the mass about one of the axes of the reference coordinate system. I_{xy}, I_{xz}, and I_{yz} are the products (of moment) of inertia. The products of inertia are the inertia associated with the rotation of the mass about two of the axes of the coordinate system. $\hat{\boldsymbol{I}}$ is known as the inertia matrix and is a symmetric matrix. The moments of inertia can be written in general format with the indicial notation [52] as,

$$I_{ij} = \int_{\mathbb{V}} \rho\, (r^2\delta_{ij} - x_i x_j)\, d\mathbb{V} \tag{5.16}$$

where $i, j = 1, 2, 3$ and $1 \rightarrow x$, $2 \rightarrow y$, $3 \rightarrow z$. Also, $x_1 \rightarrow x$, $x_2 \rightarrow y$, and $x_3 \rightarrow z$. δ_{ij} is known as the kronecker delta. If $i = j$, $\delta_{ij} = 1$; otherwise, $\delta_{ij} = 0$ for $i \neq j$.

The moments of inertia can be determined at different points in the body. To transfer the moment of inertia from one point to another, the parallelogram theorem is used [11]. The parallelogram theorem is written as,

$$I_2 = I_1 + M|\vec{d}|^2 \tag{5.17}$$

I_1 is the moment of inertia in the first coordinate system, I_2 is the moment of inertia in the second coordinate system that is parallel with the first coordinate system, M is the mass of the body, and $\vec{d}$ is the distance vector from one coordinate system to another.

Figure 5.2 shows a simple example to determine the moments of inertia for a right circular cylinder. The Cartesian coordinate system (X,Y,Z) is known as the body coordinate system because it is located at the center of gravity of the body. The Cartesian coordinate system (x,y,z) is called the structural coordinate system and is displaced from the center of gravity. The moment of inertia of the right circular cylinder is first calculated about the body coordinate system. The moment of inertia about the X axis is equal to,

$$I_{XX} = \int_{\mathbb{V}} \rho(r_c^2 - x^2)\, d\,\mathbb{V} = \int_{\mathbb{V}} \rho(y^2 + z^2)\, d\mathbb{V}$$

$$= 4\rho \int_0^{l_c/2} \int_0^{\pi} \int_0^{R_c} (r_c^2 \sin^2\gamma + z^2) r_c\, dr_c\, d\gamma dz = \frac{\pi\rho l_c R_c^2}{12}(3R_c^3 + l_c^2) \tag{5.18}$$

In equation (5.18), the cylindrical coordinate system is used to solve the problem. In the cylindrical coordinate system, $x = r_c \cos\gamma$ and $y = r_c \sin\gamma$. Also, $\mathbb{V} = \pi R_c^2 l_c$ which is the volume of the cylinder. Using the definition

Figure 5.2 **Right circular cylinder for inertia calculation**

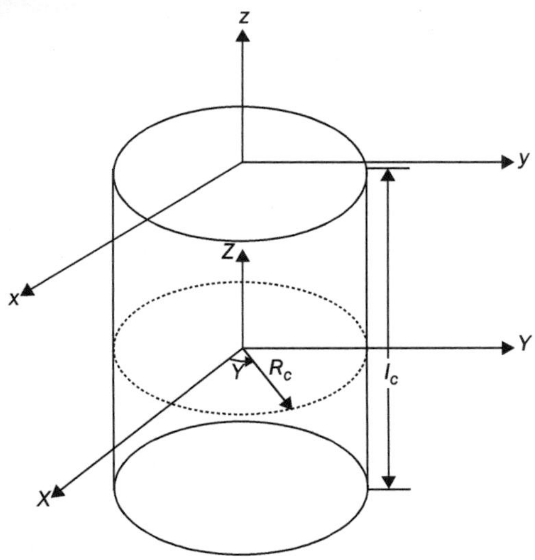

for the volume of the cylinder, the moment of inertia about the X axis becomes as,

$$I_{XX} = \frac{m}{12}(3R_c^3 + l_c^2) \tag{5.19}$$

I_{YY} can also be calculated with equation (5.16). Because the right circular cylinder is axisymmetric, $I_{YY} = I_{XX}$. Using equation (5.16) for the Z axis, the moment of inertia is equal to,

$$\begin{aligned} I_{ZZ} &= \int_{\mathbb{V}} \rho(r_c^2 - z^2)\, d\mathbb{V} = \int_{\mathbb{V}} \rho(x^2 + y^2)\, d\mathbb{V} \\ &= 4\rho \int_0^{l_c/2} \int_0^{\pi} \int_0^{R_c} r_c^2 \; dr_c\, d\gamma\, dz = \frac{\pi\rho l_c R_c^4}{2}\; I_{ZZ} = \frac{mR_c^2}{2} \end{aligned} \tag{5.20}$$

The products of inertia can be calculated for every axis but are equal to zero using equation (5.16). Hence, the inertia matrix in the body frame can be written as,

$$\hat{I} = \begin{bmatrix} \frac{m}{12}(3R_c^3 + l_c^2) & 0 & 0 \\ 0 & \frac{m}{12}(3R_c^3 + l_c^2) & 0 \\ 0 & 0 & \frac{mR_c^2}{2} \end{bmatrix}$$

To determine the moments of inertia with respect to the structural coordinate system (x,y,z) shown in Figure 5.2, the parallelogram theorem is used. The distance vector between the body and structural coordinate system is $\vec{d} = \frac{l}{2}\hat{k}$. From Figure 5.2, the only moments of inertia parallel to the structural frame are about the x and y axes. Then, the moment of inertia about the X axis becomes as,

$$I_{xx} = I_{XX} + m\frac{l^2}{4} = \frac{m}{12}(3R_c^3 + 4l_c^2) = I_{yy}$$

And the inertia matrix for the structural frame is,

$$\bar{I}_S = \begin{bmatrix} \frac{m}{12}(3R_c^3 + 4l_c^2) & 0 & 0 \\ 0 & \frac{m}{12}(3R_c^3 + 4l_c^2) & 0 \\ 0 & 0 & \frac{mR_c^2}{2} \end{bmatrix}$$

5.4 Principal moments of inertia

The previous example allows the reader to understand the calculations of the moments of inertia in a body. In practice, there are products of inertia that can be calculated in the body. This is due to the masses associated with sensors, actuators, payloads, and any other mass that the vehicle is carrying. Because of these mass variations, the inertia matrix is generally fully populated. It is possible to simplify the inertia matrix in terms of three specific values in a different coordinate system. These three specific values are known as the principal moments of inertia and are determined from the eigenvalue problem [12] as follows,

$$|\hat{\boldsymbol{I}} - \lambda \boldsymbol{I}| = 0 \tag{5.21}$$

λ are the eigenvalues and represent the principal moments of inertia. The eigenvectors associated to these eigenvalues are the direction cosines expressed in the following rotational matrix,

$$\boldsymbol{R}_J = \begin{bmatrix} c_{11} & c_{12} & c_{13} \\ c_{21} & c_{22} & c_{23} \\ c_{31} & c_{32} & c_{33} \end{bmatrix} \tag{5.22}$$

c_{ij} are the direction cosines where the columns of equation (5.22) are the eigenvectors. To visualize the use of the rotational matrix in equation (5.22), Figure 5.3 represents the body and principal coordinate systems.

Figure 5.3 Principal and body axes

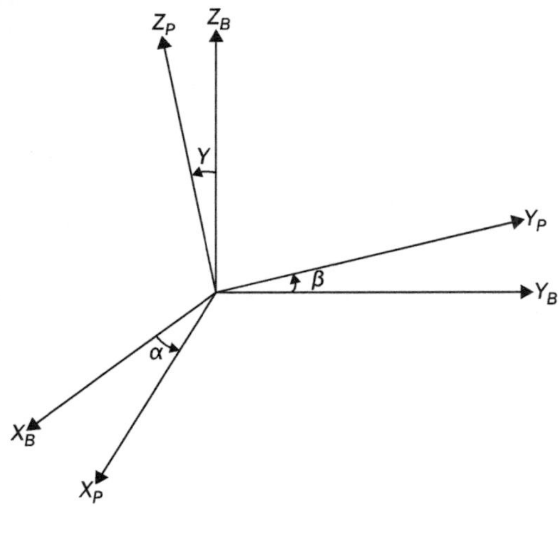

The angles α, β, and γ are the direction cosines describing the rotation of the principal coordinate system with respect to the body coordinate system. In many applications, the inertia matrix is expressed in terms of the principal coordinate system to simplify the control problem and the equations of motion. The only constraint is that any vector represented in the body frame must be transformed to the principal axis coordinate frame. If the principal coordinate system is the same as the body axis coordinate system, the direction cosine matrix is identity. This is observed in the inertia matrix calculated for the example of the right circular cylinder. The principal moment of inertia matrix is represented as,

$$\boldsymbol{J} = \begin{bmatrix} J_1 & 0 & 0 \\ 0 & J_2 & 0 \\ 0 & 0 & J_3 \end{bmatrix} \tag{5.23}$$

Also, the inertia matrix can be obtained from the following relation,

$$\hat{\boldsymbol{I}} = \boldsymbol{R}_J^T \boldsymbol{J} \boldsymbol{R}_J = \boldsymbol{R}_J \boldsymbol{J} \boldsymbol{R}_J^T \tag{5.24}$$

5.5 Energy formulation

The rotation of a particle is described in terms of the angular momentum and its motion about a single point as shown in Figure 5.1. The next step is to relate these rotations to the actual equations of motion. In Chapter 2, the motion of a particle is explained for the linear motion with the Lagrange's equations of motion. The equations of motion for the rotation of a body can be described by the angular momentum, the Lagrange's equation, and the Lagrange's equation for a quasi-coordinate system. These three forms provide different means to determine the equations of motion for a rotating body.

5.5.1 Angular momentum equations of motion

The angular momentum describes the rotation of a body about a reference point as shown in Figure 5.1 and is described in equation (5.15). Taking the time derivative of the angular momentum, the equations of motion related to the rotation of a vehicle about the center of mass can be described as,

$$\frac{d}{dt}(\vec{H})\Big|_2 = \frac{d}{dt}(\vec{H})\Big|_1 + \vec{\varpi} \times \vec{H} \tag{5.25}$$

Substituting the definition for the angular momentum in terms of the moments of inertia, equation (5.25) can be expanded to,

$$\hat{I}\dot{\vec{\varpi}} + \vec{\varpi} \times \hat{I}\vec{\varpi} = \vec{N} \tag{5.26}$$

Equation (5.26) describes the rotation of the body about its center of mass. This equation is known as the Euler equations of motion for the rotation of a body. When Newton's Second Law of motion is expanded for rotational systems, the second term on the left hand side of equation (5.26) is set to zero for small angle rotations. This second term is known as the gyroscopic term because it is associated with the combination of the three angular velocities. The cross product of the second term can be represented in matrix format as,

$$\vec{\varpi} \times \hat{I}\vec{\varpi} = \Omega\hat{I}\vec{\varpi} \tag{5.27a}$$

where,

$$\mathbf{\Omega} = \begin{bmatrix} 0 & -\varpi_3 & \varpi_2 \\ \varpi_3 & 0 & -\varpi_1 \\ -\varpi_2 & -\varpi_1 & 0 \end{bmatrix} \tag{5.27b}$$

Ω is known as the skew symmetric or anti-symmetric matrix. This matrix can be used to simplify the cross product between two vectors.

In practice, the equations of motion are explained in terms of the principal axis system. Because of the principal moments of inertia, the angular velocities must be transformed from the body axis to the principal axis. Using equation (5.23), equation (5.26) can be described as,

$$J_1\varpi_1 + \varpi_2\varpi_3 (J_3 - J_2) = N_1 \tag{5.28a}$$

$$J_2\varpi_2 + \varpi_1\varpi_3 (J_1 - J_3) = N_2 \tag{5.28b}$$

$$J_3\varpi_3 + \varpi_1\varpi_2 (J_2 - J_1) = N_3 \tag{5.28c}$$

These equations are very useful when the attitude control schemes are developed.

5.5.2 Lagrange's equation for rotational motion

The Lagrange's equation can be used to obtain the equations of motion for a body rotating about its center of mass. To write the Lagrange's equation, it is required to specify the kinetic and potential energy associated to the rotation of a body.

The kinetic energy of a body can be described as,

$$T_K = \frac{1}{2} m |\vec{v}|^2 = \frac{1}{2} m \vec{v} \cdot \vec{v}$$

Using the mass integral and equation (5.11), the kinetic energy for a rotating body can be written as,

$$T_K = \frac{1}{2} \vec{\varpi} \cdot \int_{\mathbb{V}} \rho(\vec{r} \times \vec{v})\, d\mathbb{V} = \frac{1}{2} \vec{\varpi} \cdot \vec{H} = \frac{1}{2} \vec{\varpi} \cdot \hat{\boldsymbol{I}} \vec{\varpi} \tag{5.29}$$

In equation (5.29), the integral term can be recognized as the angular momentum of the body.

The potential energy describing the rotation of a body is written differently as in the linear motion of a body. Figure 5.4 shows the center of the Earth and an arbitrary mass rotating about the Earth. $\vec{R}_{CM}$ is the distance from the center of the Earth to the center of mass, $\vec{R}$ is the distance from the center of mass to a differential mass in the body, and β is the angle associated with the vectors $\vec{R}$ and $\vec{R}_{CM}$. The potential energy associated with the rotation of a body about the Earth can be written as,

$$U = -\frac{\mu m}{|\vec{r}|} = -\int_{\mathbb{V}} \frac{\mu \rho}{|\vec{r}|} d\mathbb{V}$$

Figure 5.4 Arbitrary body potential function about the Earth

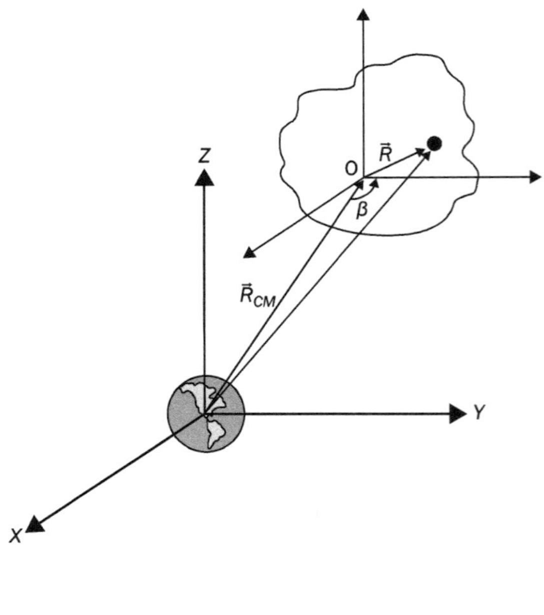

Using the law of cosines, the denominator of the potential function can be defined as follows,

$$|\vec{r}|^2 = |\vec{R}|^2 + |\vec{R}_{CM}|^2 - 2|\vec{R}|\,|\vec{R}_{CM}| \cos \beta$$

Then, the potential function can be written as,

$$U = -\int_{\mathbb{V}} \frac{\mu\rho}{\left|\vec{R}\right|^2 + \left|\vec{R}_{CM}\right|^2 - 2\left|\vec{R}\right|\left|\vec{R}_{CM}\right|\left|\cos\beta\right|} d\mathbb{V}$$

$$= -\frac{\mu}{\left|\vec{R}_{CM}\right|}\int_{\mathbb{V}} \rho\left(1+\frac{\left|\vec{R}\right|^2}{\left|\vec{R}_{CM}\right|^2} - 2\frac{\left|\vec{R}\right|}{\left|\vec{R}_{CM}\right|}\cos\beta\right)^{-1/2} d\mathbb{V}$$

As shown in equation (3.12), the binomial function can be used to expand the integrand term in the potential function. The integrand can be approximated with the binomial expansion as,

$$\left(1+\frac{\left|\vec{R}\right|^2}{\left|\vec{R}_{CM}\right|^2} - 2\frac{\left|\vec{R}\right|\left|\vec{R}_{CM}\right|}{\left|\vec{R}_{CM}\right|^2}\cos\beta\right)^{-1/2} \approx 1 - \frac{\left|\vec{R}\right|}{\left|\vec{R}_{CM}\right|}\cos\beta + \frac{1}{2}\frac{\left|\vec{R}\right|^2}{\left|\vec{R}_{CM}\right|^2}(3\cos^2\beta - 1)$$

Then,

$$U = -U_0 + U_1 - U_2 \tag{5.30a}$$

where,

$$U_0 = \int_{\mathbb{V}} \frac{\mu\rho}{\left|\vec{R}_{CM}\right|} d\mathbb{V} \tag{5.30b}$$

$$U_1 = \frac{\mu}{\left|\vec{R}_{CM}\right|}\int_{\mathbb{V}} \rho\left|\vec{R}\right| \cos\beta \, d\mathbb{V} \tag{5.30c}$$

$$U_2 = \frac{\mu}{\left|\vec{R}_{CM}\right|^2}\int_{\mathbb{V}} \frac{\rho}{2}\left|\vec{R}\right|^2 (3\cos^2\beta - 1)\, d\mathbb{V} \tag{5.30d}$$

Integrating equation (5.30b), this term expresses the potential function associated with the location of the center of mass of the body. Equation (5.30b) is written as,

$$U_0 = \int_{\mathbb{V}} \frac{\mu\rho}{\left|\vec{R}_{CM}\right|} d\mathbb{V} = \frac{\mu m}{\left|\vec{R}_{CM}\right|} \tag{5.31}$$

Equation (5.30c) yields to zero because of the definition of the center of mass [11], $U_1 = 0$. Equation (5.30d) can be expanded in the following form,

$$U_2 = \frac{\mu}{|\vec{R}_{CM}|^2}\int_{\mathbb{V}} \rho|\vec{R}|^2 \, d\mathbb{V} - \frac{3\mu}{2|\vec{R}_{CM}|^2}\int_{\mathbb{V}} \rho|\vec{R}|^2 \cos^2\beta \, d\mathbb{V} \tag{5.32}$$

The first term in the RHS of equation (5.32) is related to the principal diagonal of the inertia matrix as,

$$\begin{aligned}\frac{\mu}{|\vec{R}_{CM}|^2}\int_{\mathbb{V}} \rho|\vec{R}|^2 \, d\mathbb{V} &= \frac{\mu}{2|\vec{R}_{CM}|^2}\int_{\mathbb{V}} \rho[(y^2+z^2)+(z^2+x^2)+(x^2+y^2)]\, d\mathbb{V} \\ &= \frac{\mu}{2|\vec{R}_{CM}|^2} tr(\hat{\boldsymbol{I}})\end{aligned} \tag{5.33a}$$

where $tr(\ )$ is the trace of a matrix. The second term of equation (5.32) can be determined from the ellipsoid of inertia shown in Chapter 3 for the calculation of the potential function associated with the Earth. Taking similar steps, the second term on the right hand side of equation (5.32) can be written as,

$$\begin{aligned}\frac{3\mu}{2|\vec{R}_{CM}|^2}\int_{\mathbb{V}} \rho|\vec{R}|^2 \cos^2\beta \, d\mathbb{V} &= \frac{3\mu}{2|\vec{R}_{CM}|^2}\int_{\mathbb{V}} \rho\vec{R}_{CM}\cdot\vec{R}_{CM} \, d\mathbb{V} \\ &= \frac{3\mu}{2|\vec{R}_{CM}|^2}\hat{R}_{CM}\cdot\hat{\boldsymbol{I}}\cdot\hat{R}_{CM}\end{aligned} \tag{5.33b}$$

$\hat{R}_{CM}$ is the unit vector associated with the location of the center of mass. Substituting equations (5.33) into equation (5.32), equation (5.30d) is equal to,

$$U_2 = \frac{\mu}{2|\vec{R}_{CM}|^2}(tr(\hat{\boldsymbol{I}}) - 3\hat{R}_{CM}\cdot\hat{\boldsymbol{I}}\cdot\hat{R}_{CM})$$

Then, the potential energy for the rotation of a body in equation (5.30a) is equal to,

$$U = -\frac{\mu m}{|\vec{R}_{CM}|} + \frac{\mu m}{2|\vec{R}_{CM}|^2}(3\hat{R}_{CM}\cdot\hat{\boldsymbol{I}}\cdot\hat{R}_{CM} - tr(\hat{\boldsymbol{I}})) \tag{5.34}$$

The Lagrange's equation of motion for a rotating body is solved with the following equation,

$$\frac{d}{dt}\left(\frac{\partial L}{\partial \dot{q}_j}\right) - \frac{\partial L}{\partial q_j} = 0 \tag{5.35a}$$

where,

$$L = T_K - U \tag{5.35b}$$

The generalized coordinates q_j describe the rotational angles of the body about the center of mass. Reference [53] shows the use of equations (5.35) to determine the equations of motion for a rotating body.

5.5.3 Lagrange's equations for a quasi-coordinate system

The Lagrange's equations can be used to determine the equations of motion for a rotating body in a general format, but the Lagrange's equations for a quasi-coordinate system do not require the determination of the potential function. This means that the Lagrange's equations are only expressed in terms of the kinetic energy, angular velocity of the body, and torques associated with the surface forces. Meirovitch [54] explains the use of the Lagrange's equation expressed in quasi-coordinates which can be written as,

$$\frac{d}{dt}\left(\frac{\partial T_k}{\partial \varpi_1}\right) - \varpi_3 \frac{\partial T_K}{\partial \varpi_2} + \varpi_2 \frac{\partial T_K}{\partial \varpi_3} = N_1 \tag{5.36a}$$

$$\frac{d}{dt}\left(\frac{\partial T_k}{\partial \varpi_2}\right) - \varpi_1 \frac{\partial T_K}{\partial \varpi_3} + \varpi_3 \frac{\partial T_K}{\partial \varpi_1} = N_2 \tag{5.36b}$$

$$\frac{d}{dt}\left(\frac{\partial T_k}{\partial \varpi_3}\right) - \varpi_2 \frac{\partial T_K}{\partial \varpi_1} + \varpi_1 \frac{\partial T_K}{\partial \varpi_2} = N_3 \tag{5.36c}$$

These equations can be used to obtain the same solution as in equations (5.28). The kinetic energy for the principal axis system can be written as,

$$T_K = \frac{1}{2}\vec{\varpi} \cdot \boldsymbol{J}\vec{\varpi} = \frac{1}{2}J_1\varpi_1^2 + \frac{1}{2}J_2\varpi_2^2 + \frac{1}{2}J_3\varpi_3^2 \tag{5.37}$$

Substituting equation (5.37) into equations (5.36), the solution is the Euler equations of motion in the principal coordinate system shown in equations (5.28). These equations can be used to simplify the formulation of the equations of motion. The Lagrange equations for a quasi-coordinate system are useful when the satellite has actuators that are related to the position and angular velocity. References [55], [56], and [57] show some problems in which the Lagrange's equations for quasi-coordinate system are used.

5.6 Rate of change of a quaternion

In Chapter 4, the kinematic equations for the aerospace sequence are determined. The kinematic equations are related to the angular velocity between the body and reference frames. In a similar manner, there are kinematic equations for the quaternions as shown in References [43] and [47]. The derivation is presented here for the kinematic equation with quaternions [42].

As shown in Chapter 4, the multiplication of two unit quaternions can be related to another intermediate quaternion. The relation for two quaternions in the present and future time can be written as,

$$q(t + \Delta t) = q(t)\Delta p(t) \tag{5.38}$$

where $q(t)$ is the quaternion in the present time, $q(t + \Delta t)$ is the quaternion in the future time, and $\Delta p(t)$ is an incremental unit quaternion which can be described as,

$$\Delta p(t) = \cos \Delta\theta + \vec{u}(t) \sin \Delta\theta$$

where $\vec{u}(t)$ is the unit quaternion. Using small angle assumptions, the incremental unit quaternion can be written as,

$$\Delta p(t) = 1 + \vec{u}(t)\Delta\theta \tag{5.39}$$

Substituting equation (5.39) into (5.38), the quaternion in the future time can be written as,

$$q(t + \Delta t) - q(t) = q(t)\vec{u}(t)\Delta\theta$$

Dividing by Δt at both side of the previous equation and taking the limit as $\Delta t \to 0$, the time derivative of a quaternion is equal to,

$$\frac{dq}{dt} = \lim_{\Delta t \to 0} \frac{q(t + \Delta t) - q(t)}{\Delta t} = \lim_{\Delta t \to 0} \frac{q(t)\vec{u}(t)\Delta\theta}{\Delta t} = q(t)\vec{\varpi}(t) \tag{5.40}$$

In matrix format, equation (5.40) can be written as,

$$\frac{dq}{dt} = \begin{bmatrix} 0 & -\varpi_1 & -\varpi_2 & -\varpi_3 \\ \varpi_1 & 0 & \varpi_3 & -\varpi_2 \\ \varpi_2 & -\varpi_3 & 0 & \varpi_1 \\ \varpi_3 & \varpi_2 & \varpi_1 & 0 \end{bmatrix} \begin{bmatrix} q_1 \\ q_2 \\ q_3 \\ q_4 \end{bmatrix} \tag{5.41}$$

Or, equation (5.40) can be written per element as follows,

$$\dot{\vec{q}} = \frac{1}{2}\Omega\vec{q} + \frac{1}{2}q_4\vec{\varpi} \tag{5.42a}$$

$$\dot{q}_4 = -\frac{1}{2}\vec{\varpi}\cdot\vec{q} \tag{5.42b}$$

The kinematic equations for the quaternion simplify the algorithm implementation in comparison to the kinematic equations shown in equation (4.8). This is the main advantage of using quaternions instead of the angles expressing the rotation of a body.

5.7 Ares V equations of motion

The next step is to understand how to develop the equations of motion for a rotating frame. One of the most interesting examples is the attitude motion of a launch vehicle during the ascent trajectory. The Ares V rocket is used to demonstrate the development of the equations of motion. The Ares V could represent the next generation of launch vehicles that takes cargo into Space. This rocket is also known as the 'heavy-lifter' because it can carry approximately 414,000 pounds of equipment [58] to low-Earth orbit. The Ares V is a vertically stacked vehicle composed of two stages. A two-stage vehicle refers to how many components are used during the ascending flight.

For the first initial ascent, two five stage and a half segment solid rocket boosters (SRB) and six RS-68b motors are used. The two SRBs are derived for the four segment SRB for the shuttle launcher. The six RS-68b motors are derived from the motors used in the Saturn V and use a mixture of liquid hydrogen and oxygen. The SRBs are located at the side of the Ares V rocket, but the RS-68bs are located at the center aft of the vehicle forming a circle. The SRBs are reusable motors, but the RS-68bs are expendable motors. The five segment SRB motors produce more than 3,467,300 pounds of force. The RS-68b motors produce more than 663,000 pounds of force. These motors constitute the first stage (FS) of the Ares V rocket.

During the trajectory toward the orbit insertion, there is a second engine that is fired. This engine is called the J-2X. The J-2X is derived from the J-2 engine in the Saturn V rocket, and it produces about 294,000 pounds of force. Since the Ares V would be designed to carry the Altair (lunar lander), the J-2X engine could also be used for the Earth-departure to the Moon. The J-2X engine is powered by liquid hydrogen and oxygen and is located at the upper stage (US) of the Ares V rocket.

During the initial launch to the upper atmosphere of the Earth, the rocket is affected by the aerodynamic forces. The aerodynamic forces and the wind affect the trajectory of the rocket [59]. In the dynamic equations presented here, the wind is not described because it is a complex model to explain. Once the rocket has reached the upper atmosphere, the aerodynamic forces are not affecting the rocket. The main reason is that the rocket has reached a high velocity that depends on the mission and moves toward the orbit insertion [60].

Before developing the equations of motion, there are a few steps that should be described for formulating the equations of motion for a rotating body. The steps are the following:

1. Select the rotation sequence for the reference and body frame.
2. Understand the direction and location of the forces acting on the vehicle.
3. Write the forces acting on the vehicle in terms of the body frame.
4. Write the translational equations of motion for a rotating body expressed in the body frame.
5. Write the kinematic equations describing the relation between the reference and body frame.
6. Write the attitude dynamic equations describing the rotation of the vehicle.

These steps allow the reader to understand how to describe the translational motion of a vehicle. Now, the equations of motion for the Ares V can be written. Figure 5.5 shows the side view of the Ares V rocket during ascension, and Figure 5.6 shows a view of the body.

In step 1, the rotation sequence must be selected. As shown in Chapter 4, the rotation sequence for the Ares V is described by the aerospace rotation. In Figures 5.5 and 5.6, the Cartesian coordinate system (X_B, Y_B, Z_B) is the body frame, and the Cartesian coordinate system (X_R, Y_R, Z_R) is the reference frame. These coordinate systems are centered at the center of gravity (CG) of the vehicle. The aerospace rotation is described by the following sequence: $\mathbf{R}_x(\varphi) \leftarrow \mathbf{R}_y(\theta) \leftarrow \mathbf{R}_z(\psi)$. The rotational matrix due to the aerospace rotation is described as follows,

$$\mathbf{R} = \mathbf{R}_x(\varphi)\,\mathbf{R}_y(\theta)\,\mathbf{R}_z(\psi) \tag{5.43}$$

The rotational matrix for the aerospace rotation, **R**, from Section 4.3 is,

$$\mathbf{R} = \begin{bmatrix} \cos\psi\cos\theta & \sin\psi\cos\theta & -\sin\theta \\ \cos\psi\sin\theta\sin\varphi - \sin\psi\cos\varphi & \sin\psi\sin\theta\sin\varphi + \cos\psi\cos\varphi & \cos\theta\sin\varphi \\ \cos\psi\sin\theta\cos\varphi + \sin\psi\sin\varphi & \sin\psi\sin\theta\cos\varphi - \cos\psi\sin\varphi & \cos\theta\cos\varphi \end{bmatrix}$$

Figure 5.5 Diagram for the ascent trajectory for Ares V in the pitch plane

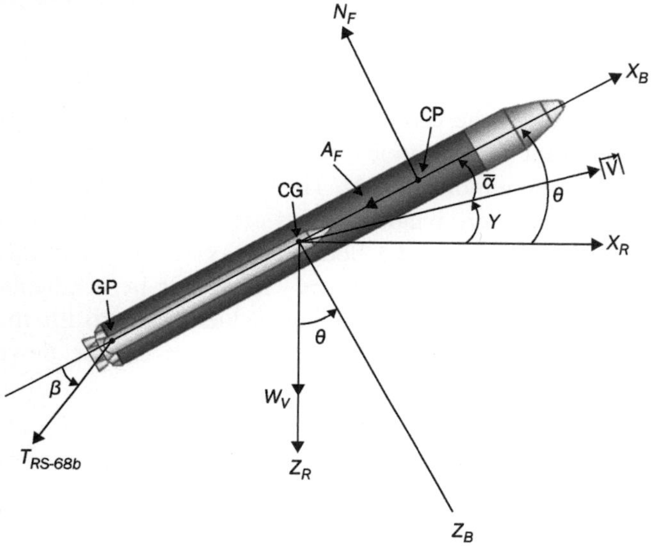

Figure 5.6 Diagram for the ascent trajectory for Ares V about yaw

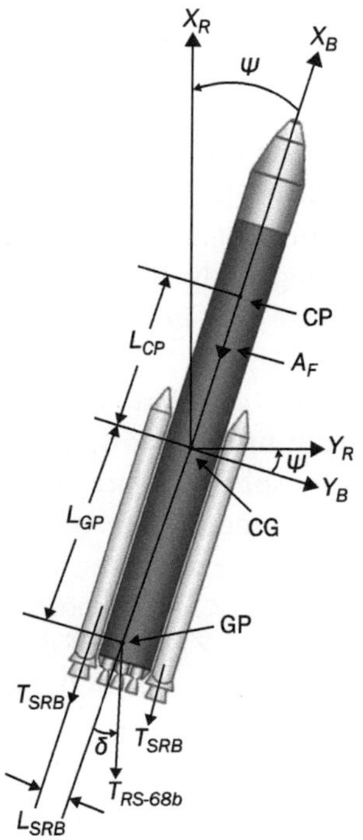

In step 2, the forces acting on the vehicle must be understood. In Figure 5.5, the X_R direction in the reference frame is the horizontal line associated with the motion of the vehicle. The weight (W_V) of the vehicle is along the Z_R direction in the reference fame and is in the direction pointing downward toward the Earth. From the CG, there is another point called the center of pressure (CP). The CP is the point where the resultant aerodynamic forces are assumed to act on the vehicle, causing a force and a moment [21]. The CP is separated from the CG by a distance L_{CP}. The aerodynamic forces are described by the axial (A_F) and normal (N_F) force on the body. The axial force acts against the motion of the vehicle, and the normal force acts perpendicular to the axial force (in the nominal pitch plane) which tends to lift the vehicle from the ground. The normal and axial forces can be described by the following equations,

$$N_F = \frac{1}{2}\rho\left|\vec{V}\right|^2 C_N S = QC_N S \tag{5.44a}$$

$$A_F = \frac{1}{2}\rho\left|\vec{V}\right|^2 C_A S = QC_A S \tag{5.44b}$$

ρ is the atmospheric density, S is the contact surface area with the atmosphere, C_N is the normal coefficient force, C_A is the axial coefficient force, and Q is known as the dynamic pressure which is equal to $\frac{1}{2}\rho\left|\vec{V}\right|^2$. $\vec{V}$ is the velocity vector of the rocket. The angle from the horizontal line to the velocity vector is known as the flight path angle, γ. This angle describes the flight path of the vehicle [43]. The angle measured from the velocity vector to the X_B axis is the angle of attack, $\bar{\alpha}$. The angle of attack describes the orientation of the lifting body through a fluid while the body is moving. The sum of the angle of attack and the flight path angle is equal to the pitch angle, θ.

There are four more forces acting at different points on the vehicle when the motors are fired during the lift off, within the Earth's atmosphere and during the upper atmosphere flight. During the lift off and first stage (FS) flight, the RS-68b motors are turned on and can rotate the vehicle about the pitch and yaw axes frame as shown in Figures 5.5 and 5.6. To rotate the vehicle, the RS-68b motors are rotated about the gimbal point (GP) on the Y_B and Z_B axes of the body frame by an angle β and δ, respectively. The GP for the RS-68b motors are separated by a distance $L_{RS\text{-}68b}$ measured from the center of gravity (CG). During the lift off and part of the FS flight, the two solid rocket boosters (SRB) will be fired to provide enough thrust force for the Ares V rocket to move against the gravity acceleration. Both SRB motors cannot be gimbaled and have an applied force in the direction of the X_B axis. The SRB motors are offset by a distance L_{SRB} from the X_B axis to the center line of the SRBs. Once the vehicle is in the upper atmosphere, the RS-68b and SRB motors are detached from the vehicle, and the J-2X engine will be fired to impulse the upper stage (US) of the vehicle to the orbit insertion. During this upper atmosphere flight, there is a GP found at the US of the vehicle for the J-2X motor and is located a distance $L_{J\text{-}2X}$ from the CG. During flight, the CG keeps changing because of the varying mass inside the rocket; for this reason, the FS and US flight have different composite CG points and measurements from the CG to the GP. The thrust force for the SRBs, RS-68b, and J-2X motors can be written as [43],

$$T_{SRB} = -gI_{SP}^{SRB}\dot{m}_{SRB} \tag{5.45a}$$

$$T_{J\text{-}2X} = -gI_{SP}^{J\text{-}2X}\dot{m}_{J\text{-}2X} \tag{5.45b}$$

$$T_{RS\text{-}68b} = -gI_{SP}^{RS\text{-}68b}\dot{m}_{RS\text{-}68b} \tag{5.45c}$$

For the three type of motors, g is the sea level gravitational acceleration constant of the Earth and is equal to 9.81 m/sec^2, I_{SP} is the specific impulse of the vehicle, and $\dot{m}$ is the mass flow rate of the gases coming from the exit nozzle of the motors. The specific impulse describes the efficiency of the motors in a rocket [43].

The third step is to write the forces described in the previous two paragraphs in terms of the body frame. The weight of the vehicle in the body frame can be written as,

$$F_G = \mathbf{R}\begin{bmatrix} -W_V \\ 0 \\ 0 \end{bmatrix} = \begin{bmatrix} -\cos\psi\cos\theta \\ -\cos\psi\sin\theta\sin\varphi + \sin\psi\cos\varphi \\ -\cos\psi\sin\theta\cos\varphi - \sin\psi\sin\varphi \end{bmatrix} W_V \tag{5.46a}$$

The axial force acts against the motion of the vehicle in the $-X_R$ axis, and the normal force acts against the motion of the vehicle in the $-Z_R$. The aerodynamic forces can be written as,

$$F_A = \mathbf{R}\begin{bmatrix} -A_F \\ 0 \\ -N_F \end{bmatrix} = -\begin{bmatrix} \cos\psi\cos\theta & -\sin\theta \\ \cos\psi\sin\theta\sin\varphi - \sin\psi\cos\varphi & \cos\theta\sin\varphi \\ \cos\psi\sin\theta\cos\varphi + \sin\psi\sin\varphi & \cos\theta\cos\varphi \end{bmatrix} \begin{bmatrix} A_F \\ N_F \end{bmatrix} \tag{5.46b}$$

The force due to the SRB motors can be defined as,

$$F_{SRB} = \mathbf{R}\begin{bmatrix} -2T_{SRB} \\ 0 \\ 0 \end{bmatrix} = -2\begin{bmatrix} \cos\psi\cos\theta \\ \cos\psi\sin\theta\sin\varphi - \sin\psi\cos\theta \\ \cos\psi\sin\theta\cos\varphi + \sin\psi\sin\theta \end{bmatrix} T_{SRB} \quad (5.46c)$$

The thrust of the J-2X and the RS-68b engines acts on the body axis frame of the rocket. Both engines can be rotated about the Y and Z axis. The rotation sequence for these engines is the following: $R_z(\delta) \leftarrow R_y(\beta)$. The force due to the RS-68b and J-2X engine can be written as,

$$F_{RS\text{-}68b} = \mathbf{R}_z(\delta)\mathbf{R}_y(\beta)\begin{bmatrix} -6T_{RS\text{-}68b} \\ 0 \\ 0 \end{bmatrix} = -6\begin{bmatrix} \cos\delta\cos\beta \\ -\cos\beta\sin\delta \\ \sin\beta \end{bmatrix} T_{RS\text{-}68b} \quad (5.46d)$$

$$F_{J\text{-}2X} = \mathbf{R}_z(\delta)\mathbf{R}_y(\beta)\begin{bmatrix} -2T_{J\text{-}2X} \\ 0 \\ 0 \end{bmatrix} = -\begin{bmatrix} \cos\delta\cos\beta \\ -\cos\beta\sin\delta \\ \sin\beta \end{bmatrix} T_{J\text{-}2X} \quad (5.46e)$$

The total forces acting on the body frame are equal to the sum of equations (5.46).

In step 4, the translational equations are written in the body frame. When the rocket is launched from the platform, the rocket sees the platform as a stationary object. Once the rocket begins to gain altitude, the platform begins to move away from the launch trajectory because of the rotation of the Earth. Using the ECI frame, the acceleration of the rocket in the rotating frame can be described by taking the time derivative of equation (5.8) as follows,

$$\vec{a} + \dot{\vec{\Omega}}_E \times \vec{r} + 2\vec{\Omega}_E \times \vec{v} + \vec{\Omega}_E \times \vec{\Omega}_E \times \vec{r} = \frac{1}{m}(F_g + F_A + F_{SRB} + F_{RS\text{-}68b}) \quad (5.47)$$

Equation (5.47) is the acceleration of the vehicle in the rotating frame. The third term on the left hand side of equation (5.47) is known as the Coriolis Effect and describes the rotation of the Earth. Ω_E is the angular rotation of the Earth which is equal to 7.292×10^{-5} rad/sec and rotates about the Z axis of the ECI frame (Figure 5.4). F_{SRB} and $F_{RS\text{-}68b}$ are the forces during the lift off and FS flight of the rocket; this means that the Ares V rocket is within the Earth's atmosphere. For this reason, the forces described in equations (5.46c) and (5.46d) are included in equation (5.47). Once the rocket is in the upper atmosphere or the US flight, equation (5.47) is rewritten as,

$$\vec{a}+\dot{\vec{\Omega}}_E\times\vec{r}+2\vec{\Omega}_E\times\vec{v}+\vec{\Omega}_E\times\vec{\Omega}_E\times\vec{r}=\frac{1}{m}(F_g+F_{J\text{-}2X}) \tag{5.48}$$

At this point of the launching sequence, the FS, the RS-68b motors, and the SRB motors have been detached from the vehicle.

In step 5, the kinematic equations relating the rotation of the body about the reference frame is developed. As shown in Section 4.4.2, equation (4.8) describes the angular velocity for the reference frame in terms of the body frame. Equation (4.8) can be written in terms of the angular velocities of the body frame as,

$$\begin{bmatrix}\dot{\varphi}\\ \dot{\theta}\\ \dot{\psi}\end{bmatrix}=\begin{bmatrix}1 & \tan\theta\sin\varphi & \tan\theta\cos\varphi\\ 0 & \cos\varphi & -\sin\varphi\\ 0 & \sec\theta\sin\varphi & \sec\theta\cos\varphi\end{bmatrix}\begin{bmatrix}\varpi_x\\ \varpi_y\\ \varpi_z\end{bmatrix} \tag{5.49}$$

The final step is to write the equations of motion associated with the rotation of the body. Euler's equations can be written either in terms of the body or reference frame. When the problem is fully solved, the equations are written in terms of the reference frame to simplify the integration process in the algorithm. For the control analysis, the equations are linearized and expressed in terms of the body frame. In this step, the equations of motion are written in terms of the body and reference frame. The Ares V rocket is axisymmetric about the X_B axis; for this reason, the Euler's equation for the principal axis system is used. The torques acting on the body are determined from the CG of the rocket. The torques affecting the rotation of the body and the location of the forces from the CG are defined as,

$$\vec{r}_A=L_{CP}\hat{\imath}\quad N_A=L_{CP}\begin{bmatrix}0 & 0\\ -\cos\psi\sin\theta\cos\varphi-\sin\psi\sin\varphi & -\cos\theta\cos\varphi\\ \cos\psi\sin\theta\sin\varphi-\sin\psi\cos\varphi & \cos\theta\sin\varphi\end{bmatrix}\begin{bmatrix}A_F\\ N_F\end{bmatrix} \tag{5.50a}$$

$$\vec{r}_{SRB}=-L_{GP}\hat{\imath}\pm L_{SRB}\hat{\jmath}\quad N_{SRB}=-2T_{SRB}L_{GP}\begin{bmatrix}0\\ \cos\psi\sin\theta\cos\varphi+\sin\psi\sin\varphi\\ \cos\psi\sin\theta\sin\varphi-\sin\psi\cos\varphi\end{bmatrix} \tag{5.50b}$$

$$\vec{r}_{RS\text{-}68b}=-L_{GP}\hat{\imath}\quad N_{RS\text{-}68b}=-6T_{RS\text{-}68b}L_{GP}\begin{bmatrix}0\\ \sin\beta\\ \cos\beta\sin\delta\end{bmatrix} \tag{5.50c}$$

$$\vec{r}_{J\text{-}2X} = -L_{GP}\hat{\imath} \quad N_{J\text{-}2X} = -T_{J\text{-}2X}L_{GP}\begin{bmatrix} 0 \\ \sin\beta \\ \cos\beta\sin\delta \end{bmatrix} \tag{5.50d}$$

From equations (5.50a) to (5.50d), the torques about the X_B axis are equal to zero. The only way to rotate the vehicle is by adding additional thrusters along the cylindrical body of the Ares V rocket. These thrusters are known as the roll control system (RoCS) and are used along the ascent trajectory of the rocket. Adding the RoCS to the Ares V, the torque about the X_B axis can be written as,

$$\vec{r}_{RoCS} = R_{AresV}\hat{\jmath} \qquad \vec{N}_{RoCS} = n_T T_{RoCS} R_{AresV}\hat{\imath} \tag{5.50e}$$

n_T are the number of thrusters used along the body of the Ares V rocket, R_{AresV} is the radius of the body of the Ares V, and T_{RoCS} is the thrust associated with the RoCS. The thrust is described as,

$$T_{RoCS} = -gI_{SP}^{RoCS}\dot{m}_{RoCS} \tag{5.51}$$

In the reference frame, the Euler's equation for the rotation of the body can be written as,

$$J\dot{\vec{\varpi}} + \vec{\varpi}\times J\vec{\varpi} = \vec{N}_A + \vec{N}_{SRB} + \vec{N}_{RS\text{-}68b} + \vec{N}_{RoCS} \tag{5.52a}$$

$$J\dot{\vec{\varpi}} + \vec{\varpi}\times J\vec{\varpi} = \vec{N}_{J\text{-}2X} + \vec{N}_{RoCS} \tag{5.52b}$$

Equation (5.52a) describes the rotation of the body within the Earth's atmosphere, and Equation (5.52b) accounts for the rotation of the body above the upper atmosphere. To write equations (5.52) in terms of the body frame, equation (4.8) is used and is shown here,

$$\vec{\varpi} = \begin{bmatrix} \dot{\varphi} - \dot{\psi}\sin\theta \\ \dot{\theta}\cos\varphi + \dot{\psi}\sin\varphi\cos\theta \\ -\dot{\theta}\sin\varphi + \dot{\psi}\cos\varphi\cos\theta \end{bmatrix} \tag{5.53}$$

Equations (5.53) are substituted into equations (5.52) to determine the equations of motion in the body frame. This process is not shown here but is left as an exercise for the reader.

The equations of motion are written in terms of the rotational sequences. This process can also be performed with quaternions. The only difference in the calculation process is the rotational matrix which is described in terms of the quaternions. Also, the solution of step 5 is written with equations (5.40). With these substitutions, the equations of motion are not required to be expressed in terms of the body rotations. In addition, the integration process is simplified with the equations described by the quaternions. This is the main advantage of using the rotational equations in terms of quaternions. In Chapter 7, the control systems are described and use this example as the basis to explain the control schemes.

5.8 Suggested problems

Problem 5.1. Consider the symmetrical cylindrical satellite as shown in Figure 5.7 whose principal moments of inertia are: $J_3 = 4ml^2$ and $J_1 = J_2 = 2ml^2$.

a. If an additional mass, m, is placed on the upper surface at the point $(l, 0, l)$ calculate the moment of inertia tensor referenced to the (x, y, z) coordinate system. Then, determine the new principal moments of inertia.

b. Calculate the eigenvectors associated with the new principal moments and determine the orientation of the new principal axes system relative to the (x, y, z) coordinate system.

Problem 5.2. Develop the expression for the force acting on an arbitrarily shaped mass in an inverse square gravitational field. Note that:

Figure 5.7 Problem 5.1

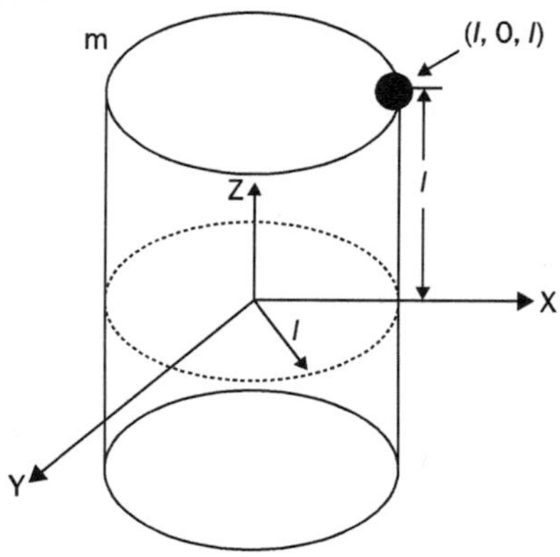

a. $\int |\vec{R}|^2 dm = \int (x^2 + y^2 + z^2) dm = \frac{1}{2} \int [(y^2 + z^2) + (x^2 + z^2) + (x^2 + y^2)] dm$

$$= \frac{1}{2}(I_{xx} + I_{yy} + I_{zz}) = \frac{1}{2} trace\,[\boldsymbol{I}]$$

b. $|\vec{R}_{CM}|^2 I_{RCM} = x_{CM}^2 I_{xx} + y_{CM}^2 I_{yy} + z_{CM}^2 I_{zz}$ where I_{RCM} is the moment of inertia of m about $\vec{R}_{CM}$.

Answer:

$$\vec{F} = \frac{\mu M m}{|\vec{R}_{CM}|^2}\left[\frac{\vec{R}_{CM}}{|\vec{R}_{CM}|} - \frac{3}{2}\frac{\vec{R}_{CM}}{|\vec{R}_{CM}|}\left\{\frac{5I_{R_{CM}} - (I_{xx} + I_{yy} + I_{zz})}{m|\vec{R}_{CM}|^2}\right\}\right.$$
$$\left. + \frac{3\vec{R}_{CM}}{m|\vec{R}_{CM}|^2}\left\{\frac{x_{CM}}{|\vec{R}_{CM}|} I_{xx}\hat{\imath} + \frac{y_{CM}}{|\vec{R}_{CM}|} I_{yy}\hat{\jmath} + \frac{z_{CM}}{|\vec{R}_{CM}|} I_{zz}\hat{k}\right\}\right]$$

Problem 5.3. Derive the equations of motion for a dual-spin spacecraft three different ways. Assume the (x, y, z) axes of the main part which has an inertial angular velocity vector,

$$\vec{\varpi} = \varpi_x \hat{e}_x + \varpi_y \hat{e}_y + \varpi_z \hat{e}_z$$

The rotor is constrained to rotate about the common y axis fixed in both bodies at a relative angular motion of $\bar{s}$ with respect to the main part. Further, assume the rotor is symmetrical about the y axis such that $I_{Rx} = I_{Rz} \neq I_{Ry}$. The principal moment of the main part given by J_1, J_2, and J_3 where $\hat{e}_x$, $\hat{e}_y$, and $\hat{e}_z$ are unit vectors along the principal axes system.

a. Give an expression for the total angular momentum of this system. From this expression derive the three Euler rotational equations of motion. (Equations (5.28))

b. Give expressions for the angular momentum of the main part and the angular momentum of the rotor. Develop separate Euler equations for the main part and rotor, respectively. These may then be combined by noting Newton's Third Law to yield a final set of three equations.

c. Develop an expression for the total rotational kinetic energy of the system. Then, using the Lagrangian equations for quasi-coordinates for the motion of a rigid body about a point, develop the rotational equations of motion.

d. Compare the final results in parts a – c. Would you expect them to be the same or not? Explain.

Problem 5.4. Expand the acceleration term in equations (5.47) and (5.48) to show the acceleration of the rocket relative to the rotation of the Earth.

6

Environmental and actuator torques

Abstract: The purpose of this chapter is to explain in detail the environmental torques that can affect the rotation of a satellite in Space. In a similar manner, the actuator torques are defined and described such that their formulation can be used to compensate the environmental torques to maintain the attitude motion of the satellite. In this chapter, the actuator torques are explained in a more practical formulation such that the reader can understand how to implement the physical actuator to the satellite model. In summary, this chapter provides the first practical concepts of how to use actuators to compensate for the motion of the satellite; in addition, the chapter shows how to explain the disturbing torques due to the Earth and the Sun.

Key words: gravity-gradient torque, atmospheric torque, solar pressure torque, reaction wheels, thrusters, magnetic torques, Earth's magnetic field modeling.

6.1 Introduction to torque formulation

In Chapter 3, the forces acting on the satellite due to the perturbing forces are explained and developed. In a similar manner, the same perturbing forces can cause a disturbing rotation to the satellite. On the other hand, there are torques that can be used to counteract the effects of these disturbing forces. The purpose of this chapter is to explain the environmental torques affecting the attitude dynamics of the satellite and the actuator torques that can be used to compensate for these perturbing forces.

The chapter is divided into two main sections. The first section is dedicated to the environmental torques which include the effects due to the Earth and Sun. The effects of the Earth are the gravity-gradient, atmospheric, and magnetic torques. The Earth's magnetic field is not explained in Chapter 3 because the electric system within the satellite does not create a force to translate the satellite; but, the magnetic field around the satellite can cause a rotation that can create an attitude pointing error. In addition, the different models used to define the Earth's magnetic field are explained in this section. As in the solar pressure force formulation, there are solar pressure torques that can cause the satellite to rotate. The general [10] and Karymov's [32] solar pressure torque formulations are also described in this chapter.

The second section explains the different actuators that can be used to compensate for these disturbing forces. The actuators are the magneto-torques, reaction wheels, and thrusters. In the same way that the induced electric field within the satellite can create a disturbance, a magnetic field can be created with magnetic torque rods to cause a rotation to correct the attitude motion of the satellite. The magnetic torque rod formulation is also explained in detail. The thrusters (not explained in Chapter 3) create a force that can translate the satellite into another position in the orbit. In this chapter, the thruster formulation is explained to demonstrate its capability to correct the orientation of the satellite. Finally, the reaction wheels are explained such that the process of how a reaction wheel works can be understood; in this way, a reaction wheel system can be selected depending on the torque and voltage applied for the correction of the attitude motion of the satellite.

6.2 Environmental torques

In Chapter 2, the disturbance forces are discussed to understand how the satellite motion is affected. In similar manner, there are torques affecting the attitude motion of the satellite as it moves about the Earth. This section presents the torques due to the Earth and Sun. The torques due to the Earth are the gravity-gradient, atmospheric, and magnetic torques. The Sun disturbance torque can be defined with a general [10] and Karymov's formulation [32]. Karymov's formulation [32] has a similar calculation as for the forces, but the integral is presented differently. In many cases, these torques can be used to compensate for the attitude motion of a satellite or can be used to describe a disturbance in the formulation. In the following sections, the disturbance torques are discussed and studied to determine the best approach in the formulation.

6.2.1 Gravity-gradient torques

In Section 5.5.2, a potential function is developed to express the angular rotation of the body due to the gravity of the Earth. Solving Lagrange's equation with the kinetic and potential function, the gravity-gradient torques affecting the motion of the satellite can be determined. The gravity-gradient torques can be used to compensate for the attitude motion of the satellite; in this way, the satellite always tends to pointing toward the center of the Earth or is maintaining an orientation relative to the local vertical/local horizontal (LVLH) frame. The LVLH frame is the same as the North-East-Down (NED) frame which is described by the reference frame of Ares V. In Figure 6.1, the LVLH frame describes the orientation of the satellite in orbit.

Figure 6.1 shows an arbitrary body in an orbit. The coordinate frame in the body is known as the LVLH frame and has the following direction: 1) x is positive along the velocity vector, 2) z is positive toward the center of the Earth, and 3) y is positive when the right handed system is completed. In practice, the reference frame can have any desired direction chosen by the designer, but the LVLH frame is commonly used.

Figure 6.1 ECI and LVLH frame for gravity-gradient torque formulation

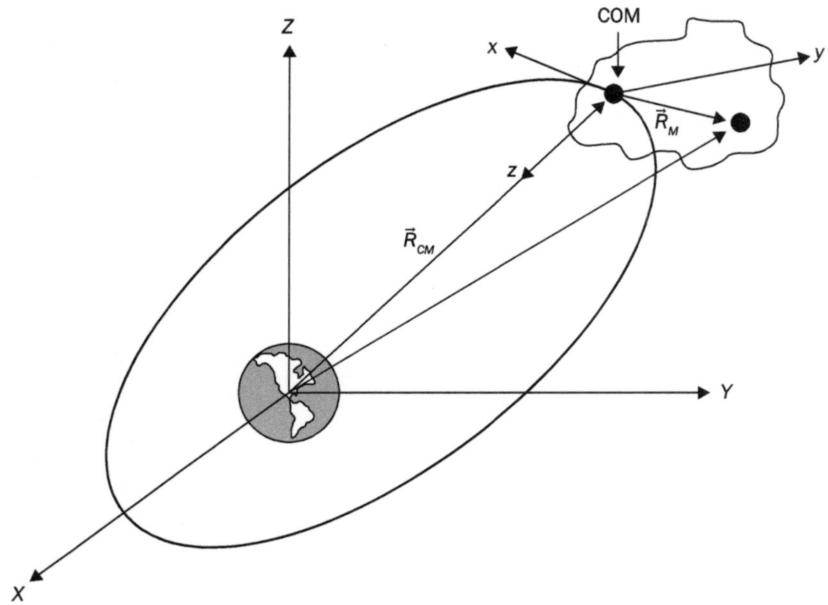

This section develops the gravitational and angular rate torques that can be used to align the body frame of the satellite in the same direction as the radial vector ($\vec{R}$). In Figure 6.1, $\vec{R}_{CM}$ is the vector from the center of the Earth to the center of mass of the satellite, $\vec{R}_m$ is the vector from the center of mass (COM) of the satellite to a point in the body of the satellite, and dm is an elementary mass in the body of the satellite.

In an environment without perturbations, the force acting over the body of the satellite can be represented as [50],

$$\vec{F} = \int \frac{-\mu\left(\vec{R}_m + \vec{R}_{CM}\right)}{\left|\vec{R}_m + \vec{R}_{CM}\right|^3} dm \tag{6.1}$$

where $\mu = GM_E$. The torques acting on the satellite due to gravitational forces can be expressed as,

$$\vec{N}_E = -\mu \int \vec{R}_m \times \frac{\left(\vec{R}_m + \vec{R}_{CM}\right)}{\left|\vec{R}_m + \vec{R}_{CM}\right|^3} dm \tag{6.2}$$

From the binomial theorem [13], it is known that,

$$\left(1 + \frac{y}{x}\right)^{-n} = 1 - n\frac{y}{x} + \frac{n(n-1)}{2!}\left(\frac{y}{x}\right)^2 + \ldots \tag{6.3}$$

This function can be approximated with the first two terms of the expansion. For a linearized system of equations, the division in equation (6.2) can be approximated within the linear range as,

$$\left|\vec{R}_m + \vec{R}_{CM}\right|^{-3} = \frac{1}{\left|\vec{R}_{CM}\right|^3}\left\{1 - \frac{3}{2}\left(\frac{\left|\vec{R}_m\right|^2}{\left|\vec{R}_{CM}\right|^2} + 2\frac{\vec{R}_m \cdot \vec{R}_{CM}}{\left|\vec{R}_{CM}\right|^2}\right)\right\} \tag{6.4}$$

Equation (6.4) is substituted into equation (6.2) to approximate the torques due to gravitational forces as follows,

$$\vec{N}_E = \frac{-\mu}{\left|\vec{R}_{CM}\right|^3} \int \vec{R}_m \times \left\{\left(\vec{R}_m + \vec{R}_{CM}\right)\left[1 - \frac{3}{2}\left(\left(\frac{\left|\vec{R}_m\right|^2}{\left|\vec{R}_{CM}\right|^2} + 2\frac{\vec{R}_m \cdot \vec{R}_{CM}}{\left|\vec{R}_{CM}\right|^2}\right)\right)\right]\right\} dm \tag{6.5}$$

Since $\vec{R}_m \ll \vec{R}_{CM}$, equation (6.5) is reduced to,

$$\vec{N}_E = \frac{-\mu}{\left|\vec{R}_{CM}\right|^3} \int \vec{R}_m \times \left(\vec{R}_m + \vec{R}_{CM}\right) \left(1 - 3\frac{\vec{R}_m \cdot \vec{R}_{CM}}{\left|\vec{R}_{CM}\right|^2}\right) dm \tag{6.6}$$

Performing the following vector calculations,

$$(\vec{R}_m + \vec{R}_{CM})(\vec{R}_m \cdot \vec{R}_{CM}) = \vec{R}_m(\vec{R}_m \cdot \vec{R}_{CM}) + \vec{R}_{CM}\,(\vec{R}_m \cdot \vec{R}_{CM})$$

$$(\vec{R}_m \times (\vec{R}_m + \vec{R}_{CM})\,(\vec{R}_m \cdot \vec{R}_{CM}) = (\vec{R}_m \times \vec{R}_{CM})(\vec{R}_m \cdot \vec{R}_{CM})$$

Equation (6.6) becomes,

$$\vec{N}_E = \frac{-\mu}{\left|\vec{R}_{CM}\right|^3} \int \left(\vec{R}_m \times \vec{R}_{CM} - 3\frac{\left(\vec{R}_m \times \vec{R}_{CM}\right)\left(\vec{R}_m \cdot \vec{R}_{CM}\right)}{\left|\vec{R}_{CM}\right|^2}\right) dm \tag{6.7}$$

$\vec{R}_m$ and $\vec{R}_{CM}$ can be written as,

$$\vec{R}_m = X_m\hat{\imath} + Y_m\hat{\jmath} + Z_m\hat{k} \tag{6.8a}$$

$$\vec{R}_{CM} = X_{CM}\hat{\imath} + Y_{CM}\hat{\jmath} + Z_{CM}\hat{k} \tag{6.8b}$$

Expanding the second term on the right-hand side of equation (6.7),

$$\begin{aligned}(\vec{R}_m \times \vec{R}_{CM})(\vec{R}_m \cdot \vec{R}_{CM}) &= [X_mX_{CM}\,(Y_mZ_{CM} - Z_mY_{CM}) + Y_mZ_m(Z^2_{CM}\\ &- Y^2_{CM}) + Y_{CM}Z_{CM}\,(Y^2_m - Z^2_m)]\hat{\imath} + [Y_mY_{CM}\,(Z_mX_{CM} - X_mZ_{CM})\\ &+ X_mZ_m(X^2_{CM} - Z^2_{CM}) + X_{CM}Z_{CM}\,(Z^2_m - X^2_m)]\,\hat{\jmath} + [Z_mZ_{CM}\,(X_mY_{CM}\\ &- Y_mX_{CM}) + X_mY_m\,(Y^2_{CM} - X^2_{CM}) + X_{CM}Y_{CM}(X^2_m - Y^2_m)]\hat{k}\end{aligned} \tag{6.9}$$

Assuming that the principal axes are centered along the radial direction of $\vec{R}_{CM}$ and substituting equation (6.9) into (6.7), the torques due to the gravitational forces can be defined as,

$$\begin{aligned}\vec{N}_E = \frac{3\mu}{\left|\vec{R}_{CM}\right|^5} \int \big[&Y_{CM}Z_{CM}(Y_m^2 - Z_m^2)\,\hat{\imath} + X_{CM}Z_{CM}(Z_m^2 - X_m^2)\hat{\jmath}\\ &+X_{CM}Y_{CM}(X_m^2 - Y_m^2)\hat{k}\big]\,dm\end{aligned} \tag{6.10}$$

By the definition of the center of mass [11], the first term on the right-hand side of equation (6.7) is equal to zero; also, the products of moment of inertia in equation (6.9) are equal to zero. Expanding equation (6.10), the different torques due to the gravitational forces can be represented in terms of the principal moments of inertia as follows,

$$\vec{N}_E = \frac{3\mu}{\left|\vec{R}_{CM}\right|^5}\int\left[Y_{CM}Z_{CM}\left\{\left(Y_m^2+X_m^2\right)-\left(Z_m^2+X_m^2\right)\right\}\hat{\imath}\right.$$
$$+X_{CM}Z_{CM}\left\{\left(Z_m^2+Y_m^2\right)-\left(Y_m^2+X_m^2\right)\right\}\hat{\jmath} \tag{6.11}$$
$$\left.+X_{CM}Y_{CM}\left\{\left(Z_m^2+X_m^2\right)-\left(Z_m^2+Y_m^2\right)\right\}\hat{k}\right]dm$$
$$\vec{N}_E = N_{E,x}\hat{\imath}+N_{E,y}\hat{\jmath}+N_{E,Z}\hat{k}$$

where,

$$N_{E,x} = \frac{3\mu}{\left|\vec{R}_{CM}\right|^5}(J_3-J_2)yz \tag{6.12a}$$

$$N_{E,y} = \frac{3\mu}{\left|\vec{R}_{CM}\right|^5}(J_1-J_3)xz \tag{6.12b}$$

$$N_{E,z} = \frac{3\mu}{\left|\vec{R}_{CM}\right|^5}(J_2-J_1)xy \tag{6.12c}$$

In equations (6.12), the subscript CM is removed because it is known that the equations define the position of the center of mass of the satellite along the orbit. Equations (6.12) can be reduced as follows,

$$N_{E,x} = \frac{3\mu}{\left|\vec{R}_{CM}\right|^3}(J_3-J_2)\hat{R}_2\hat{R}_3 \tag{6.13a}$$

$$N_{E,y} = \frac{3\mu}{\left|\vec{R}_{CM}\right|^3}(J_1-J_3)\hat{R}_1\hat{R}_3 \tag{6.13b}$$

$$N_{E,z} = \frac{3\mu}{\left|\vec{R}_{CM}\right|^3}(J_2-J_1)\hat{R}_1\hat{R}_2 \tag{6.13c}$$

In equations (6.13), $\hat{\hat{R}} = \hat{R}_1\hat{\imath} + \hat{R}_2\hat{\jmath} + \hat{R}_3\hat{k}$ and is the unit vector associated with the radius measured from the center of the Earth to the location of the satellite in the orbit. From equations (6.13), it can be easily observed that equations (6.11) can be represented as follows,

$$\vec{N}_E = \frac{3\mu}{\left|\vec{R}_{CM}\right|^3}\hat{\hat{R}}\times \boldsymbol{J}\hat{\hat{R}} \tag{6.14}$$

This equation is used to determine the torque on a gravity-gradient stabilized spacecraft such that the satellite is pointing towards the center of the Earth. Equation (6.14) is commonly used for the gravity-gradient stabilization of the attitude motion of a satellite in a circular orbit. On the other hand, equations (6.12) can be used to explain the attitude motion in highly elliptical orbits which can be more complicated as shown in references [44] and [61].

The angular rate can also be included in the formulation to compensate the torques due to the gravitational forces [50]. The angular momentum is the cross product of the radial direction vector and the linear momentum of the vehicle. For a rotating body, the angular momentum is related to the vector from the center of mass to a point in the satellite. This relation for the angular momentum can be written as,

$$\vec{H} = \int \vec{R}_m \times (\vec{\Omega} \times \vec{R}_m)\, dm$$

where $\vec{\Omega} = \dot{f}\hat{k}$. The torques due to the angular momentum for a rotating frame can be written in the following form [50],

$$\vec{N}_{AR} = \frac{d\vec{H}}{dt} = \int\left[\vec{R}_m^2\dot{\vec{\Omega}} - (\vec{R}_m \cdot \dot{\vec{\Omega}})\vec{R}_m\right] dm + \int (\vec{R}_m \cdot \vec{\Omega})(\vec{R}_m \times \vec{\Omega})\, dm \qquad (6.15)$$

Using equation (6.8a) and the definition for $\vec{\Omega}$, equation (6.15) can be expanded and represented in terms of the moments and products of inertia as follows,

$$\vec{N}_{AR} = \int [\ddot{f}(X_m^2 + Y_m^2)\hat{k} - (\ddot{f}X_mZ_m\hat{\imath} + \ddot{f}Y_mZ_m\hat{\jmath}) + \dot{f}^2Y_mZ_m\hat{\imath} - \dot{f}^2X_mZ_m\hat{\jmath}\,]\, dm$$

$$\vec{N}_{AR} = N_{AR,x}\hat{\imath} - N_{AR,y}\hat{\jmath} + N_{AR,z}\hat{k} \qquad (6.16)$$

where,

$$N_{AR,x} = \dot{f}^2 I_{yz} - \ddot{f} I_{xz} \qquad (6.17a)$$

$$N_{AR,y} = \dot{f}^2 I_{xz} + \ddot{f} I_{yz} \qquad (6.17b)$$

$$N_{AR,z} = \ddot{f} I_{zz} \qquad (6.17c)$$

Using equations (6.14) and (6.16), the torques due to the gravitational forces and the angular rate can be used to determine the correct orientation for a single satellite. In a circular orbit, $\dot{f}$ equals to the mean motion of the satellite. The gravity-gradient torques can be used up to an altitude of a geostationary satellite because for higher altitudes, the gravity-gradient torques become relatively small.

6.2.2 Atmospheric torques

As presented in Chapter 3, the atmospheric forces can cause variations in the orbital motion of the satellite. In addition, this disturbance can cause additional rotation about the center of pressure of the satellite. The atmospheric torques can be described as follows [62],

$$\vec{N}_A = \vec{r}_{CP} \times \vec{F}_A \tag{6.18}$$

where

$$\vec{F}_A = \frac{1}{2}\rho(R)v^2 C_D S\hat{V}$$

and $\vec{r}_{CP}$ is the location of the center of pressure measured from the center of gravity of the satellite. This disturbance affects the satellite motion up to an altitude of 800 km.

6.2.3 Solar pressure torques

The solar pressure force can also cause disturbance torques that affect the attitude motion of the satellite. As shown in Chapter 3, there are different formulations to represent the solar pressure force. The general formulation for the solar pressure only uses the surface area in contact with the Sun, and Karymov's formulation [32] expresses the solar pressure effects in terms of the composition of the surfaces on the satellite. In both formulations, the effects due to the reflectivity coefficient are used.

The general formulation for the solar radiation pressure torque is represented as follows [10] [28],

$$\vec{N}_{SP} = h_s S\left(\hat{u}_s\,(\epsilon_a + \epsilon_d) + \hat{u}_n\left[2\varepsilon + \frac{2}{3}\epsilon_d\right]\right) \times \vec{r}_s \tag{6.19}$$

where h_s is the solar pressure constant equal to 4.644×10^{-6} (N/m²), S is the surface area in contact with the Sun, $\hat{u}_s$ is the unit vector normal toward the Sun, $\hat{u}_n$ is the unit vector normal to the surface area, and $\vec{r}_s$ is the vector from the spacecraft center of mass to the surface area. ϵ_a is the surface absorptivity coefficient, ϵ_d is the surface diffuse reflectance coefficient, and ε is the reflectivity coefficient. $\hat{u}_s$ is determined from the solar ephemeris model in Appendix A. Also, $\varepsilon_a + \varepsilon_d + \varepsilon = 1$.

Karymov [32] showed a formulation for the torques due to the solar pressure on the body of an assumed (smooth) opaque satellite in the absence of diffuse reflection. These torques are also defined in terms of the reflectivity coefficient that specifies an absorbing ($\varepsilon = 0$) and reflecting ($\varepsilon = 1$) surface. Karymov [32] established that the body has a center of pressure due to the incidence of the solar rays over the body. This center of pressure is not necessarily at the center of mass of the satellite and can be located at other points in the body. The torques due to the solar pressure for an absorbing and reflective surface, respectively, are described as,

$$\vec{N}^+ = \bar{h}_0 \hat{\sigma} \times \int \vec{r}_{CP} (\hat{\sigma} \cdot \hat{n})\, dS \tag{6.20a}$$

$$\vec{N}^- = 2\bar{h}_0 \int \hat{n} \times \vec{r}_{CP} (\hat{\sigma} \cdot \hat{n}) dS \tag{6.20b}$$

where,

$$\hat{\sigma} = a_0 \hat{\imath} + b_0 \hat{\jmath} + c_0 \hat{k} \tag{6.21a}$$

$$\bar{h}_0 \approx \frac{E_0}{\bar{c}} = 4.72 \times 10^{-7} (kg/m^2) \tag{6.21b}$$

$$\vec{r}_{CP} = x_{CP} \hat{\imath} + y_{CP} \hat{\jmath} + z_{CP} \hat{k} \tag{6.21c}$$

$\hat{\sigma}$ is a unit vector in the direction of the incidence of light over the body of the satellite. $\hat{n}$ is the unit normal to the surfaces of the body of the satellite. The constant $\bar{h}_0$ is determined from E_0, which is the magnitude of the flux of light energy arriving at a unit surface of the body, and $\bar{c}$ is the velocity of light in the vacuum. In Chapter 3, it is assumed that the distance from the Sun to the Earth and the distance from the Sun to the satellite are approximately the same because the solar rays travel in parallel through Space as explained in Reference [34]. $\vec{r}_{CP}$ is the position vector of the center of pressure measured from the center of mass of the satellite body. The general formulation for the solar pressure torques can be written as,

$$\vec{N}_{SP} = (1 - \varepsilon)\vec{N}^+ + \varepsilon\vec{N}^- \tag{6.22}$$

$\vec{N}^+$ is the torque due to a total absorbing surface, and $\vec{N}^-$ is the torque due to a total reflective surface. With both formulations, the solar pressure torque can be determined for the satellite. With equation (6.19), the solar torque can be easily calculated. Instead, Karymov's formulation provides a more comprehensive solution for the solar torques depending on the surface(s) of the satellite in contact with the Sun.

6.2.4 Earth's magnetic field

The Earth's magnetic field can be considered either as a disturbance or as an actuator. In this section, the Earth's magnetic field is introduced and explained as a disturbance. The magnetic torques on a spacecraft may result from a) the interaction of any magnetic dipole moment within the spacecraft with the planet's magnetic field intensity, and b) the electromagnetic induction resulting from the rotation of a current conducting through a body in a magnetic field. The interaction of a dipole moment with the planet's magnetic field can be used as an actuator to maintain the orientation of the satellite. The electromagnetic induction is due to the spin decay which is a result from the energy dissipation as heat from the induced current within the spacecraft.

There are different representations of the Earth's magnetic field. The Earth's magnetic field is described by the 'dynamo theory' which is the result of large circulating currents below the Earth's crust. Figure 6.2 shows the coordinate frame used to represent the orientation of the Earth's magnetic field. The (X, Y, Z) coordinates refer to the ECI frame,

Figure 6.2 Earth magnetic coordinate frame

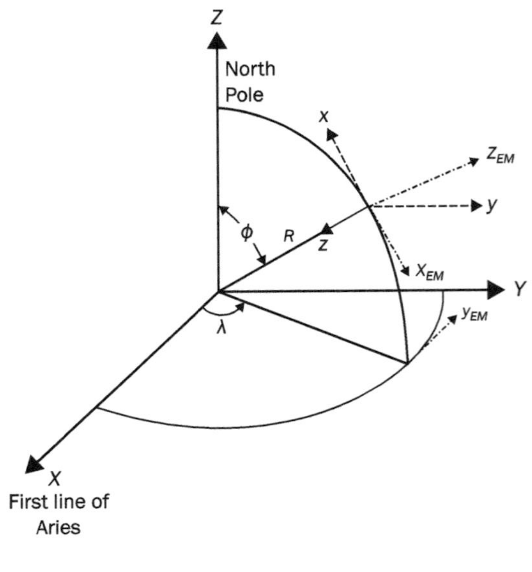

and (x, y, z) coordinates is the LVLH frame in the satellite. The magnetic field lines of the Earth [10] [63] travel from the south pole to the north pole and are represented by the $(x_{EM}, y_{EM}, z_{EM},)$ frame. Hence, the magnetic field direction can be represented as,

$$\vec{B}_m = x + y + z = -x_{EM}\hat{\imath} + y_{EM}\hat{\jmath} - z_{EM}\hat{k} \tag{6.23}$$

The total magnetic field strength is equal to $\left|\vec{B}_m\right| = \sqrt{(x_{EM})^2 + (y_{EM})^2 + (z_{EM})^2}$. The horizontal component of the magnetic field (tangent plane) is described by $\vec{B}_h = x + y = -x_{EM}\hat{\imath} + y_{EM}\hat{\jmath}$. The magnetic inclination (dip) angle, I, describes the deviation of $\vec{B}_m$ from the horizontal plane and is written as follows,

$$\tan I = \frac{z}{\left|\vec{B}_h\right|} \tag{6.24}$$

If $I > 0$, the dip angle is downward. The magnetic declination angle, D, defines how much the horizontal component deviates from the true north. The magnetic declination is described as,

$$\tan D = \frac{y_{EM}}{x_{EM}} \tag{6.25}$$

If $D > 0$, the magnetic field has an eastward declination.

The simplest model for the Earth's magnetic field is when the magnet bar is located at the center of the Earth. This is known as the spin-axis oriented magnetic dipole and is shown in Figure 6.3. The magnetic flux line has a symmetrical contour which resembles a pumpkin. In this model, the Earth's field is represented by a relative short magnetic dipole aligned along the Earth's spin axis. Since the flux lines are always in a longitudinal plane that has no eastward component, the Earth's magnetic field can be represented as,

$$x_{EM} = \frac{D_E \sin\phi}{R^3}, \quad y_{EM} = 0, \quad z_{EM} = \frac{2D_E \cos\phi}{R^3} \tag{6.26}$$

where D_E is the Earth's magnetic dipole moment equal to 7.9×10^{25} pole – cm, and ϕ is the co-latitude angle. A unit pole is a measurement in the centimeter-gram-second (cgs) unit measurement system. The unit conversion to the metric system is shown in Appendix C.

Figure 6.4 shows another representation of the Earth's magnetic model which is the tilted centered magnetic dipole model. In this model, the bar is rotated about the X direction to have the same (magnetic) inclination

Figure 6.3 Earth magnetic dipole model

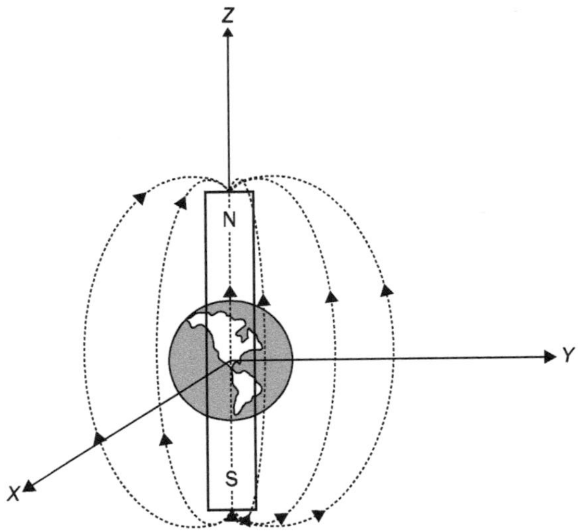

Figure 6.4 Earth magnetic tilted dipole model

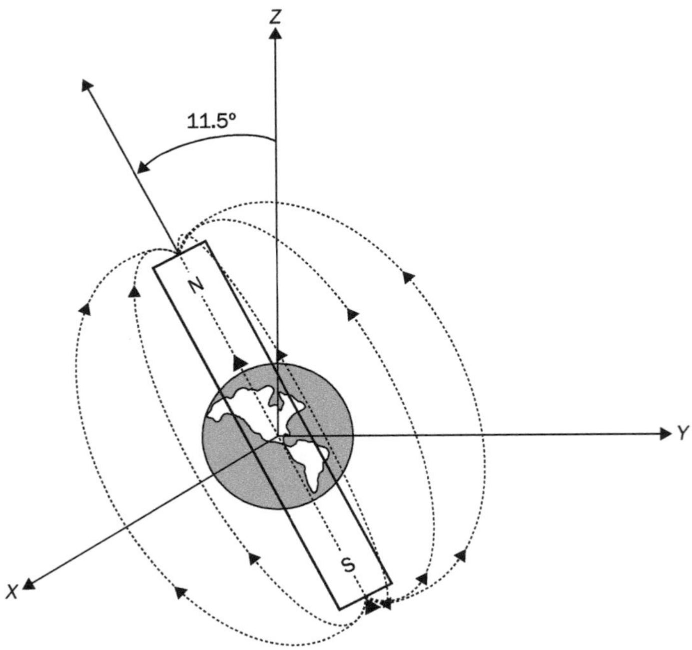

as the Earth. The magnetic north pole is located 28.5°*N*, 69.7°*W*, and the magnetic south pole is located 78.5°*S*, 110.3°*E*. The magnetic bar is inclined 11.5° from the *Z* axis in the ECI frame. The magnetic field in this model is equal to,

$$x_{EM} = \left(\frac{10R_E}{R}\right)^3 (a \sin\phi + b \cos\phi \cos\lambda + c \cos\phi \sin\lambda) \tag{6.27a}$$

$$y_{EM} = \left(\frac{10R_E}{R}\right)^3 (b \sin\lambda - c \cos\lambda) \tag{6.27b}$$

$$z_{EM} = \left(\frac{10R_E}{R}\right)^3 (a \cos\phi + b \sin\phi \cos\lambda + c \sin\phi \sin\lambda) \tag{6.27c}$$

where $a = -30.426$ (gamma), $b = -2.174$ (gamma), and $c = 5.761$ (gamma). The gamma is another measurement of the electromagnetic field (See Appendix C for the unit conversion). For the tilted centered dipole, there is a finite Y (east component). The magnitude of the Earth's magnetic field for the tilted dipole model can be approximated to 8.03×10^{25} (pole – cm).

There are more complex models for the Earth's magnetic field in which the dipole is displaced from the center of the Earth. In addition, there are some models that include anomalies at certain locations of the Earth. One of the more complex models is the IGRF-10 Earth's magnetic model **[64]**. IGRF-10 refers to the 10th generation international geomagnetic reference field. This mathematical model takes into account the effects of the Earth's deep interior, crust, and atmosphere. It is a complex model because it uses an expansion of a Legendre polynomial to describe the Earth's magnetic field. In Reference **[65]**, the IGRF-10 can be downloaded for use in simulations; also, the coefficients of the IGRF-11 have now been released to update the previous model.

The different Earth's magnetic field models can be used in simulations of the attitude motion of the satellite. If there is a residual current within the satellite, a magnetic field can be determined for this current **[63]**. The residual current creates a disturbance torque that causes the satellite to rotate, and the torque is written as,

$$\vec{N}_m = \vec{D} \times \vec{B}_m \tag{6.28}$$

where $\vec{D}$ is known as the magnetic dipole moment **[63]** and is measured in amperes per meter square (A/m^2). $\vec{B}_m$ is the ambient magnetic field strength. For a satellite weighting approximately 133 kg, a magnetic dipole moment of 0.03 (A/m^2) can cause approximately 1.58 degrees of attitude error for the orientation of the satellite. Because of this residual dipole, the satellite should be insulated from the electronic equipment such that the current does not interact with the electromagnetic field.

6.3 Actuator (or control) torques

The disturbance torques can cause variations in the attitude and/or translational motion of a spacecraft; but, there are torques that can be used to correct the attitude and/or translational motion of the satellite. These systems are known as actuators that produce a force and/or a torque against the disturbance to correct the translational and/or attitude motion of the satellite. The correction of the orientation of the satellite has been achieved with the use of thruster systems [66], reaction or momentum wheels [56] [57] [67], magnetic torquers [53] [68], and internal passive mass movement techniques [57] [69].

The thrusters can be used to control the spinning of the satellite and to damp the tip-off rates. The location of the thrusters can cause rotations in the spacecraft as well as translational motion. In addition, the thrusters can be misaligned to perform corrections into other axes. The momentum wheels [50] are used to maintain bias momentum for stiffness. There are other techniques that use gyros and reaction wheels. The control moment gyros are used to apply control torques when it is required to maintain a specific orientation. With these control moment gyros, the base is rotated about some axis to cause an additional rotation of the satellite. The reaction wheels can be used to maintain zero momentum but can be used to change the angular speed or to apply torques about their spinning axis. Three momentum wheels can be located about the principal body axes. For redundancy, a fourth reaction wheel can be located about a skew symmetric axis in case that one of the other three reaction wheels fails to work [67].

As the Earth's magnetic field can create disturbances, a magnetic dipole can be created within the satellite to rotate the satellite into a different orientation. Active control schemes can be developed with magnetic dipoles to either maintain or reorient the satellite into a different attitude pointing [68]. Also, the dipole moments can be used to manage the momentum associated with the rotation of the satellite [70]. The torques can be highly coupled, but the equations can be decoupled depending on how many magneto-torquers are used to correct the satellite orientation [71].

In reference [69], a passive control procedure based on moving masses along the axial or radial direction is used to maintain the rotation of the satellite. These moving masses [69] produce intentional misalignments of the principal axes of rotation in the satellite. The authors used this passive technique to control the spin-stabilization of a satellite about a principal axis. There are other passive techniques that can be used to stabilize the satellite such as the gravity [72]. These techniques can be used to maintain the orientation of the satellite.

In summary, these systems are the main actuators or control torques that can be used to maintain the attitude motion and rotational speed of the satellite. In the following subtopics, the control torques are defined and explained to be as control systems in the following chapters.

6.3.1 Reaction wheels

The reaction wheel is a mechanical system that provides additional angular momentum to the vehicle by spinning a mass about an axis. The reaction wheels are used when the satellite is required to have an attitude pointing error of less than one degree [10]. The satellites can carry up to four reaction wheels in different formations. In practice, three reaction wheels are located on the body axis (Figure 6.5), and one reaction wheel is located on a skew-symmetric axis. There are other formations in which the reaction wheels are located on a pyramid formation [73] as shown in Figure 6.6. The objective of four reaction wheels is to have an additional

Figure 6.5 Reaction wheels in body and skew-symmetric axis

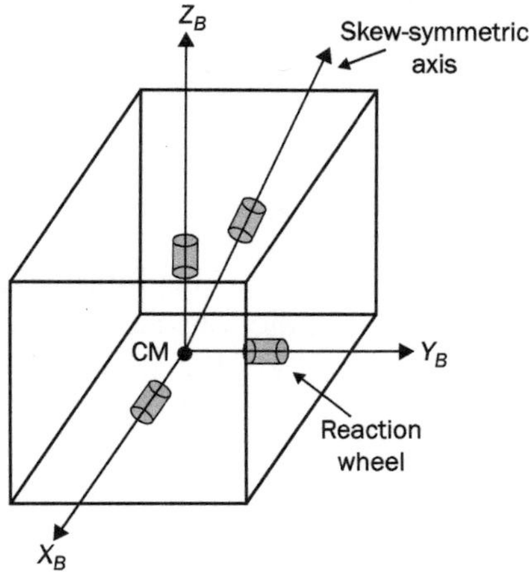

Figure 6.6 **Reaction wheel pyramid formation**

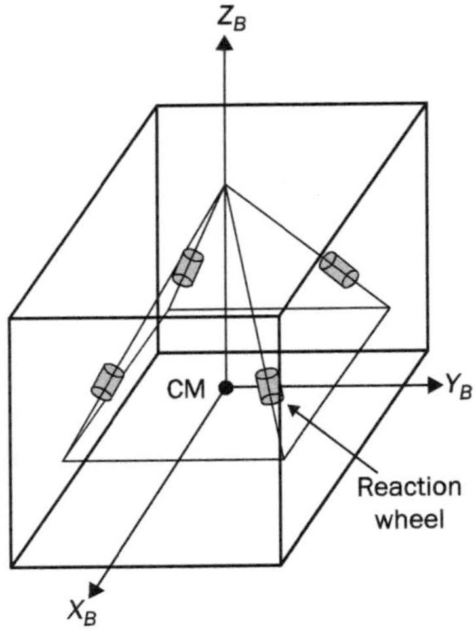

(or redundant) wheel that can be used in case that one of the principal wheels fails to work. The purpose of this topic is to explain in detail how the reaction wheels work such that a controller can be developed based on the characteristics of the reaction wheel.

First, the equations of motion for the rotation of a satellite are defined. Figure 6.5 shows three reaction wheels rotating about the body axes of the satellite. The total angular momentum associated with the motion of the satellite can be written as,

$$\vec{H}_T = \vec{H} + \vec{H}_{RW} \tag{6.29}$$

where $\vec{H}$ is the angular momentum of the spacecraft, and $\vec{H}_{RW}$ is the angular momentum for the reaction wheels. Taking the rate of change of a vector in equation (6.29), the equations of motion for the rotation of a spacecraft is written as,

$$\dot{\vec{H}} + \dot{\vec{H}}_{RW} + \vec{\varpi} \times \vec{H} + \vec{\varpi} \times \vec{H}_{RW} = \vec{N}_p \tag{6.30}$$

where $\vec{N}_p$ is the torque due to the disturbing forces. Equations (6.30) can be separated into the equations of motion associated with the rotation of the spacecraft and the reaction wheels. Equation (6.30) can be written as follows,

$$\dot{\vec{H}} + \vec{\varpi} \times \vec{H} = \vec{N}_p - \vec{N}_{RW} \tag{6.31a}$$

$$\dot{\vec{H}}_{RW} + \vec{\varpi} \times \vec{H}_{RW} = \vec{N}_{RW} \tag{6.31b}$$

Equations (6.31) can be further expanded knowing that $\vec{H} = \hat{I}\vec{\omega}$; then,

$$\hat{I}\dot{\vec{\omega}} + \vec{\omega} \times \hat{I}\vec{\omega} = \vec{N}_p - \vec{N}_{RW} \tag{6.32}$$

This equation explains the rotation of the spacecraft with respect to the reaction wheel.

Equation (6.31b) can be expanded by understanding how the reaction wheel is constructed. A reaction wheel is developed from a brushless DC motor [74]. DC stands for direct current. A brushless DC motor produces a higher velocity with a low voltage by magnetizing the poles in the motor [75]. Figure 6.7 shows the circuit for a DC motor. A stationary magnetic field is developed with the magnets in the motor and is called the fixed field. There is a stationary circuit connected to the motor which is called the armature circuit. The armature circuit contains a resistance and an inductance. In reality, the resistance can be considered as a system that releases energy, and the inductance is a system that saves energy. The armature current that passes through this circuit causes the motor to rotate, resulting in a torque. This torque moves the rotating member that contains a mass.

Because the armature current is rotating in a magnetic field, the voltage is proportional to the angular speed of the rotor. Then, the voltage across the rotor can be written as [76],

$$E_b(t) = K_b \upsilon(t) \tag{6.33}$$

where K_b is a constant of proportionality called the back emf (electromotive force) constant. υ is the angular velocity of only one reaction wheel, and $E_a(t)$ is the voltage applied to the rotor called the back emf. It is called the back emf because the polarity of the induced coil is dropped off from equation (6.33). Using the closed-loop within the electric circuit, the Kirchhoff's voltage law [13] can be written as,

$$L_a \dot{I}_a(t) + R_a I_a(t) + E_b(t) = E_a(t) \tag{6.34}$$

Figure 6.7 **DC motor circuit**

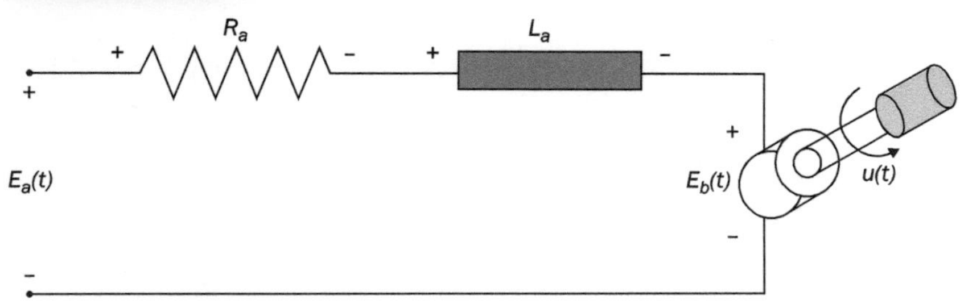

The torque developed by the motor is proportional to the amount of current applied to the motor; then,

$$N_{RW} = K_t I_a(t) \rightarrow I_a(t) = \frac{N_{RW}}{K_t} \tag{6.35}$$

K_t is a constant of proportionally called the motor torque constant. This constant depends on the motor and the magnetic field characteristics. Using equations (6.33) and (6.35), equation (6.34) is equal to,

$$\frac{L_a}{K_t}\dot{N}_{RW}(t) + \frac{R_a}{K_t}N_{RW}(t) + K_b \upsilon(t) = E_a(t) \tag{6.36}$$

Usually for DC motors, the armature resistance is much larger than the inductance armature. Neglecting the armature inductance, equation (6.36) is written as,

$$N_{RW}(t) = -\frac{K_b K_t}{R_a}\upsilon(t) + \frac{K_t}{R_a}E_a(t) \tag{6.37}$$

Equation (6.37) can be analyzed as a linear equation for different input voltages. Figure 6.8 shows a diagram for this equation and some key points in the graphs. This plot is known as a torque-speed curve that

Figure 6.8 Torque-speed curve

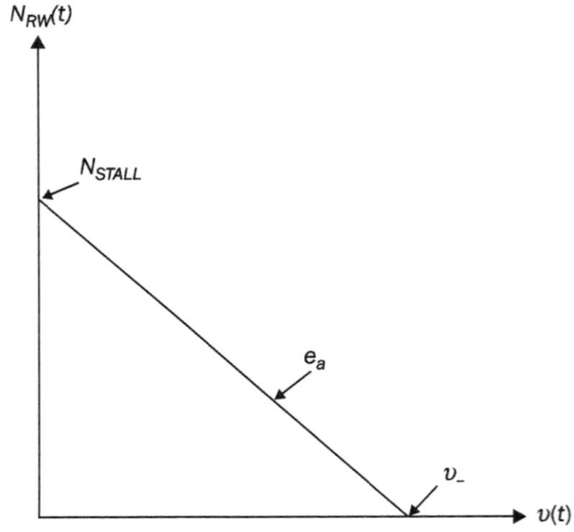

allows the engineer to determine the type of motor that can be used [76]. When the velocity is equal to zero, the torque equals to,

$$N_{STALL} = \frac{K_t}{R_a} e_a \tag{6.38a}$$

When the motor is fully stopped, it is required to apply a large voltage to force the motor to move. The required torque to move the motor is known as the stall torque, N_{STALL}. e_a is a constant voltage that determines the amount of required voltage to force the motor to rotate. If the torque equals to zero, the angular velocity of the motor is equal to,

$$\upsilon_{-} = \frac{1}{K_b} e_a \tag{6.38b}$$

When the motor is running, the torque keeps reducing with respect to time. When the torque is zero, υ_{-} is known as the no-load angular velocity; and e_a specifies the voltage at which the motor rotates without any applied torque. These parameters can be determined from the dynamometer [75]. Normally, these values can be found in the datasheet for the reaction wheels.

In some DC motors, a damper winding can be installed on top of the rotor. This damper winding can be used to control the speed of the motor and to develop a high torque [75]. This system added to the rotor of the motor can be considered as a damping mechanism. The torque due to the damper winding can be written as,

$$N_{RW,Damp} = -d_{RW}\upsilon(t) \tag{6.39}$$

d_{RW} is the damping of the reaction wheel. Then, the total torque associated with the rotation of a DC motor can be written as,

$$N_{RW}(t) = -\frac{K_b K_t}{R_a}\upsilon(t) + \frac{K_t}{R_a} E_a(t) - d_{RW}\upsilon(t) \tag{6.40}$$

Writing equation (6.40) in vector format, the reaction wheel torque can be written as,

$$\vec{N}_{RW}(t) = -\left(d_{RW} + \frac{K_b K_t}{R_a} \right)\vec{\upsilon}(t) + \frac{K_t}{R_a}\vec{E}_a(t) \tag{6.41}$$

Equation (6.41) assumes that the motors have the same constants unless it is stated differently by the designer.

The angular momentum due to the reaction wheels are written as,

$$\vec{H}_{RW} = J_{RW}\,(\vec{\varpi} + \vec{\upsilon}) \tag{6.42}$$

Substituting the time derivative of equations (6.42) into the vector equation (6.41), the equations of motion for the reaction wheel can be defined as follows,

$$\boldsymbol{J}_{RW}(\dot{\vec{\varpi}} + \dot{\vec{\upsilon}}) + \vec{\varpi} \times \boldsymbol{J}_{RW}(\vec{\varpi} + \vec{\upsilon}) + \left(d_{RW} + \frac{K_b K_t}{R_a} \right) \vec{\upsilon}(t) = \frac{K_t}{R_a} \vec{E}_a(t) \tag{6.43a}$$

And the equation of motion for the rotation of the satellite is equal to,

$$\hat{\boldsymbol{I}}\dot{\vec{\varpi}} + \vec{\varpi} \times \hat{\boldsymbol{I}}\vec{\varpi} = \vec{N}_p + \left(d_{RW} + \frac{K_b K_t}{R_a} \right) \vec{\upsilon}(t) - \frac{K_t}{R_a} \vec{E}_a(t) \tag{6.43b}$$

Equations (6.43) can be used to describe the rotation of a spacecraft based on the reaction wheel DC motor. On the other hand, a simple analysis can be performed on equations (6.31) that only takes into account the angular momentum of the reaction wheel. The applied voltage, $\vec{E}_a(t)$, is written as a vector but refers to the number of applied voltages associated with any reaction wheel.

6.3.2 Thrusters

The thrusters are a system used to correct in a short period of time the position and the orientation of the satellite. Some examples of the use of thrusters are the correction of a satellite decaying orbit, the separation distance between two satellites in a constellation, the damping of the tip-off rates of a satellite coming out from a launch vehicle, among other things. The thrusters can be located along the body axis of the satellite to produce a translation of the satellite or can be located along the faces of the satellite to cause a rotation of the spacecraft. The thruster is a system that controls the mass flow rate coming out of an exhaust nozzle. The mass flow rate exiting the nozzle can be made of different propellants such as solids, liquids, and electrostatic forces (composed of inert gases such as Xenon) [10]. As shown in Chapter 5, the equation associated with the thruster force in the motors is expressed as follows,

$$T = -gI_{SP}\dot{m} \tag{6.44}$$

g is the Earth's gravitational acceleration at sea level. I_{SP} is the specific impulse which identifies the type of motor that is used to translate or rotate the vehicle. If the specific impulse is higher, the motor produces a higher thruster force to translate the space vehicle. On the other hand, the propellant mass increases with relatively large specific impulses. $\dot{m}$ is the mass flow rate which describes the amount of fluid mass coming out of the nozzle in a time unit (usually second). Normally, the mass flow rate is constant, and the change in propellant mass can be described as,

$$m(t) = \dot{m}t + m_0 \tag{6.45}$$

where m_0 is the initial propellant mass, and t is the time since the motors are fired.

Figure 6.9 shows the different configurations that can be used to locate the thrusters. Figure 6.9a shows that the thrusters are located along the body axes of the satellite; in this way, the applied force causes a translational motion of the satellite into another position. Figure 6.9b shows that the thrusters are located along the surfaces of the satellite. In Figure 6.9b, the thrusters cause a rotation of the body of the vehicle such that the satellite attitude motion can be affected. Assuming a cube satellite with dimensions a_c in Figure 6.9b, the torques associated with the rotation of the vehicle about the center of mass can be written as,

$$\vec{N}_T = \frac{a_c}{2}T\hat{\imath} + \frac{a_c}{2}T\hat{\jmath} + \frac{a_c}{2}T\hat{k} \tag{6.46}$$

Figure 6.9 Thruster for (a) translational and (b) attitude motion displacement

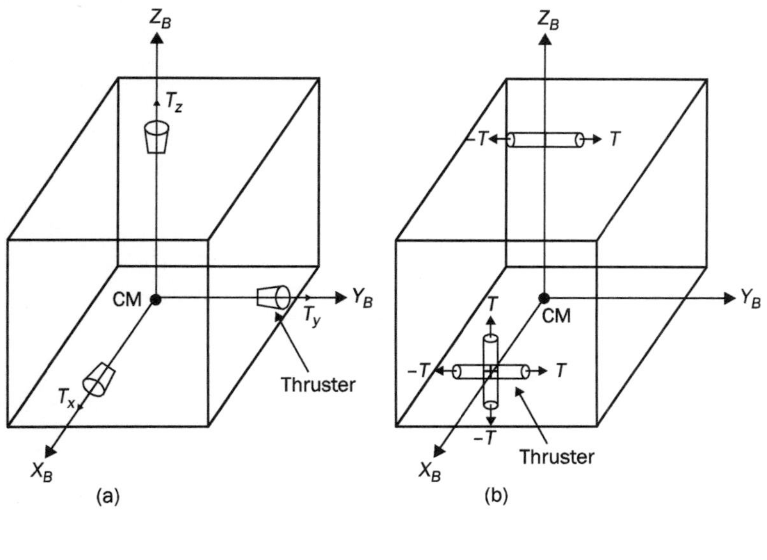

This equation is substituted into Euler's equations of motion which is written as follows,

$$\hat{I}\dot{\vec{\varpi}} + \vec{\varpi} \times \hat{I}\vec{\varpi} = \vec{N}_p + \vec{N}_T \qquad (6.47)$$

There are other configurations that can be used. Figure 6.10 shows a misalignment of the thrusters [66] with respect to the X_B axis. In this case, the thruster can cause a rotation and a translation of the vehicle. Because of this configuration, the equations of motion can be written as,

$$\vec{a} + \dot{\vec{\Omega}}_E \times \vec{r} + 2\vec{\Omega}_E \times \vec{v} + \vec{\Omega}_E \times \vec{\Omega}_E \times \vec{r} = \frac{\vec{F}_T}{m} \qquad (6.48a)$$

$$\hat{I}\dot{\vec{\varpi}} + \vec{\varpi} \times \hat{I}\vec{\varpi} = \vec{N}_p + \vec{N}_T \qquad (6.48b)$$

where,

$$\vec{F}_T = T\cos\beta\,\hat{\imath} - T\sin\beta\,\hat{\jmath}$$

$$\vec{N}_T = aT\sin\beta\,\hat{k}$$

These equations should be solved together in order to obtain the correct amount of control effort. This means that the satellite will perform a rotational motion as well as a translational motion such that the propellant mass can be used efficiently for the complete mission. In practice, this is achieved by understanding the mass flow rate equation.

Figure 6.10 Misaligned thruster

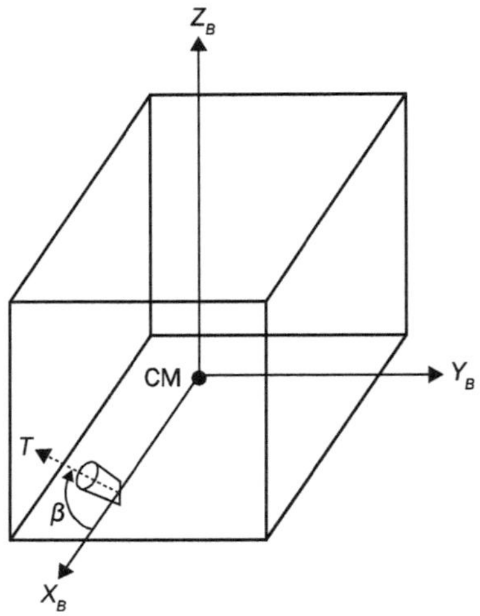

The mass flow rate can be described as follows,

$$\dot{m} = \rho_{Prop} A_{VO} V_{Prop} \quad (6.49)$$

ρ_{Prop} is the density of the propellant, A_{VO} is the area of the valve opening, and V_{Prop} is the velocity of the propellant. As shown in equation (6.49), the area of the valve opening can be controlled such that the satellite can apply the desired amount of thrust; in this way, the amount of thrust applied for the correction of the satellite can be controlled. In summary, the thrusters are a useful actuator for a satellite, but the propellant mass can affect the dimensions and size of the satellite.

6.3.3 Magnetic torquers (or Magneto-torquers)

In Section 6.2.4, the Earth's magnetic field has been explained as a disturbance torque effect that causes a rotation about one of the axis of the satellite. The objective of this section is to provide a comprehensive look at the magnetic actuator system for a satellite to maintain the attitude pointing. In addition, the magnetic control attitude system is explained in terms of the location of the satellite. At certain locations in the orbit, the satellite tends to rotate against or along the Earth's magnetic flux lines.

The magnetic torquers are explained in terms of the magnetic dipoles. A magnetic dipole is defined by the following equation [**63**],

$$\vec{D} = n_t I_s S \vec{n} \quad (6.50)$$

where I_s is the current passing through a metal piece, S is the cross sectional area, n_t is the number of turns of the coil, and $\vec{n}$ is a unit vector normal to the plane of the induced magnetic field and its direction is determined by the right-hand rule [**63**]. The magnetic torquer can be visualized as a long steel rod in which a copper coil is wrapped around it. Figure 6.11 shows the circuit connection of a magnetic torquer. In Figure 6.11, a current source is used to activate the magnetic torquers to produce an induced magnetic field around it. If the magnetic torquers are mounted on the chassis of the satellite, the magnetic torque rods can cause a rotation of the satellite once a current is passed through the copper coil. To cause a rotation to any of the three axes of the satellite, a triad of magnetic torquers can be placed in the satellite. These magnetic torque rods are placed perpendicular to each other and are shown in Figure 6.12. Using equation (6.28), the induced magnetic field in the magneto-torquers can be written as,

Figure 6.11 Magnetic torquer circuit

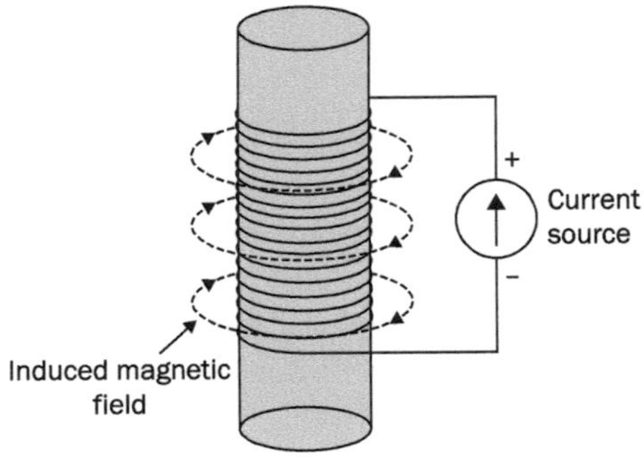

Figure 6.12 Magnetic torquer triad mounted in the satellite

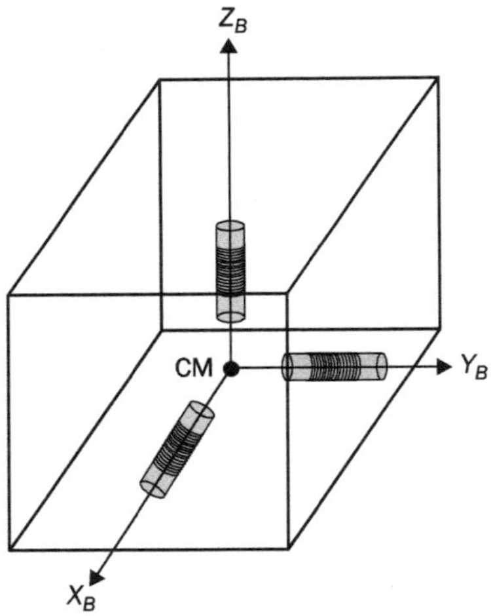

$$\vec{N}_m = -\vec{B}_m \times \vec{D} = \begin{bmatrix} 0 & B_{m,3} & -B_{m,2} \\ -B_{m,3} & 0 & B_{m,1} \\ B_{m,2} & -B_{m,1} & 0 \end{bmatrix} \begin{bmatrix} m_1 \\ m_2 \\ m_3 \end{bmatrix} \tag{6.51}$$

Figure 6.13 shows a satellite with one magnetic torquer placed on the Z_B axis. The torque due to the induced magnetic field of the satellite is defined as [13],

$$\vec{N}_m = -\vec{B}_m \times \vec{D} = DB_0 \sin \varphi \tag{6.52}$$

Figure 6.13 **Satellite rotation with magnetic torque rod**

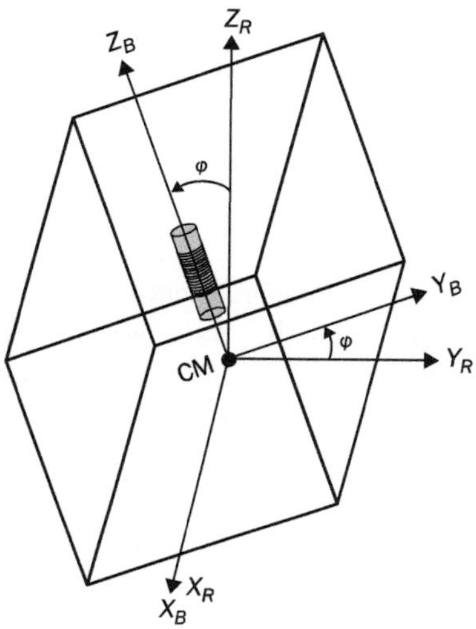

where D is the magnitude of the satellite's magnetic dipole moment, B_0 is the Earth's magnetic filed intensity written as $B_0 = \sqrt{\vec{B}_m \cdot \vec{B}_m}$, and φ is the angle between $\vec{D}$ and $\vec{B}_m$. Letting J_1 to be the principal moment of inertia about the X_B axis and perpendicular to $\vec{B}_m$, the attitude equation of motion without damping and other external torques is equal to,

$$\frac{d^2\varphi}{d^2t} = -\frac{DB_0 \sin\varphi}{J_1} \tag{6.53}$$

The magnetic torque tends to reduce the angle φ in which the dipole mounted in the chassis of the satellite is aligned with the Earth's magnetic field. For small angle approximations of equation (6.53), the solution is a simple harmonic oscillator [**13**] and is described as follows,

$$\varphi_n(t) = \varphi_0 \cos(2\pi f_M t + \varphi_i) \tag{6.54}$$

where f_M is the frequency and equal to,

$$f_M = \frac{1}{2\pi}\sqrt{\frac{DB_0}{J_1}}(\text{Hz})$$

φ_i is the initial phase angle, and φ_0 is the amplitude. For large φ angles, equation (6.53) must be solved in terms of elliptical integrals [**13**], but the equation (6.54) is still a good expression for natural period errors less than 8% and for an approximate maximum φ angle of 60 degrees.

The potential energy of the magnetic dipole moment is found by integrating the torque over the angular displacement. This happens when the magnetic dipole is displaced by an angle φ from the local direction of $\vec{B}_m$; then, the potential energy is equal to,

$$U_m(\varphi) = \int_0^{\varphi} |\vec{N}_m| \, d\varphi = \int_0^{\varphi} DB_0 \sin\varphi \, d\varphi = DB_0 (1 - \cos\varphi) \tag{6.55}$$

From Reference [77], only those magnetic torque rods perpendicular to $\vec{B}_m$ and perpendicular to the axis about which the spacecraft is oscillating will be effective in damping the magnetic oscillation. In the presence of damping, the angular motion of the spacecraft can be written as,

$$\varphi(t) = \varphi_m(t)\varphi_n(t) \tag{6.56}$$

where $\varphi_m(t)$ is the damping function resulting from the energy dissipation, and $\varphi_n(t)$ is the amplitude function associated with the simple harmonic oscillator.

For a single permeable rod perpendicular to $\vec{B}_m$ and perpendicular to the axis of oscillation, the hysteresis loss per cycle of oscillation is equal to [77],

$$U_L = \frac{\mathbb{V}}{4\pi} \oint B_0 dD_m \left(\frac{\text{ergs}}{\text{cycle}}\right) \tag{6.57}$$

where $\mathbb{V}$ is the volume of the rod in cm^3, $\oint B_0 \, dD_m$ is the area of the hysteresis loop, B_0 is the external magnetic field intensity in oersted, and D_m is the flux density in the rod in gauss (See Appendix C for unit conversions). If there are N bars each perpendicular to $\vec{B}_m$ and perpendicular to the axis of oscillation, the change in time of the hysteresis loss per cycle of oscillation is equal to,

$$\frac{dU_L}{dt} = -\frac{N\mathbb{V}}{4\pi} f_M \oint B_0 dD_m \left(\frac{\text{ergs}}{\text{sec}}\right) \tag{6.58}$$

The negative sign in equation (6.58) indicates a decreasing energy. Within the torque rods, there are swirling paths of induced currents when the flux density is increasing within the material. These swirling paths in the torque rods resemble the eddy currents of air or water. The swirling paths are also called eddy currents [75]. Although there is energy loss due to the eddy currents in the rods, for the low oscillatory rates achievable with the magnetic orientation of a spacecraft, the eddy current loss is less than the hysteresis loss. From experimental measurements, it can be deduced (Steinmetz's Law) that [77],

$$\oint B_0 \, dD_m = \lambda_c B_m^3 \text{ (gauss – oersted)} \tag{6.59}$$

where λ_c is a constant of proportionality with units gauss/oersted^2. Equation (6.59) applies for small values of B_m where B_m is the peak value of the magnetic field.

Because of these relations, a hysteresis loop can be obtained and is shown in Figure 6.14. Moving around the loop in Figure 6.14, there are different magnetic torque levels that can be reached (both retarding and accelerating). Because of the hysteresis, these result in a net retarding or de-spin torque on the rod which is reflected on the spacecraft. The peak magnetizing field along the permeable rods is defined as,

$$B_m = B_0 \sin \varphi_m \tag{6.60}$$

where φ_m is the maximum angular displacement from equilibrium during particle oscillation, and B_0 is the total value of Earth's magnetic field intensity in oersted; then,

$$\oint B_0 \, dD_m = \lambda_c B_0^3 \sin^3 \varphi_m \text{ (gauss – oersted)} \tag{6.61}$$

Figure 6.14 **Ideal hysteresis curve**

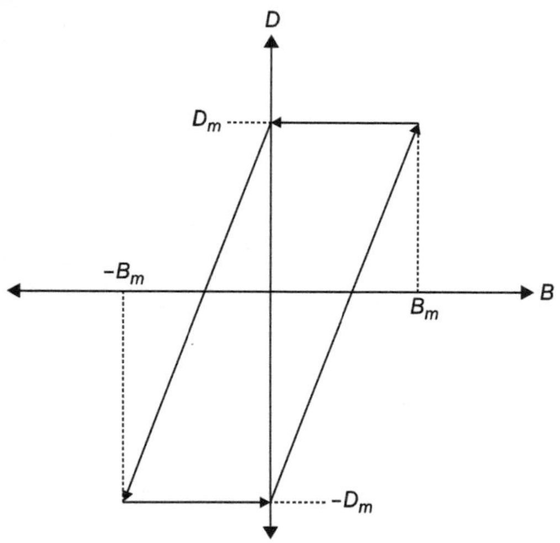

And from equation (6.58),

$$\frac{dU_L}{dt} = -k \sin^3\varphi_m \left(\frac{\text{ergs}}{\text{sec}}\right) \tag{6.62a}$$

where,

$$k = \lambda_c \frac{NV}{8\pi^2}\sqrt{\frac{D}{J_1}}B_0^{7/2} \tag{6.62b}$$

Equation (6.55) can be written in terms of the maximum angular displacement. Taking the time derivative of equation (6.55), the derivative of the potential energy equation in terms of the maximum angular displacement equals to,

$$\frac{dU_L}{dt}(\varphi_m) = DB_0 \sin\varphi_m \frac{d\phi_m}{dt} \tag{6.63}$$

Thus, equation (6.62) equals to equation (6.63) as follows,

$$DB_0 \sin\varphi_m \frac{d\varphi_m}{dt} + k \sin^3\varphi_m = 0 \tag{6.64}$$

Equation (6.64) can be integrated by using separation of variables [**12**] and looking at a table of integrals for the integral of the cosecant function [**13**]. Using the initial condition φ_m $(t = 0) = \varphi_0$, the solution of equation (6.64) is equal to,

$$\varphi_m = arc\ \cot\left(\frac{kt}{DB_0} + \cot\varphi_0\right) \tag{6.65}$$

For reasonable small values of φ, the complete solution of the angular motion of the spacecraft can be written as,

$$\frac{\varphi(t)}{\varphi_0} = arc\ \cot\left(\frac{kt}{DB_0} + \cot\varphi_0\right)\cos(2\pi f_M + \varphi_i) \tag{6.66}$$

Since equation (6.59) comes from experimental data, different powers for B_m can be assumed to resolve the differential equation according to the experimental data. The time required to reach final alignment of D with respect to B_0 is defined as,

$$t_{\varphi\to\varphi_f} = \frac{DB_0}{k}(\cot\varphi_f - \cot\varphi_0) \tag{6.67}$$

If there is no precession and $t_f > t$, the spacecraft is not spinning about the magnetic dipole axis. Consider a simple example to understand how long a satellite with several torque rods can achieve a desired attitude. Assume a satellite with three torque rods in which the torque rods have the following characteristics: $\lambda_C = 25.8x10^3$ gauss/oersted^2, $\mathbb{V} = 6.2$ cm^3, and $D = 7x10^4$ pole – cm. Assuming a moment of inertia equal to 1.1×10^8 gm – cm^2 and $B_0 = 0.3$ oersted, the constant in equation (6.62b) equals, $k = 2.27$. For $\varphi_0 = \pi/2$ and $\varphi_f = 2°$, the time that the satellite takes to achieve the desired attitude pointing equals to,

$$t_{\varphi \to \varphi_f} = 264{,}917.31 \text{ sec} = 3.07 \text{ days}$$

Multiple magnetic rods can be added to damp out the attitude motion of the satellite, but the damping function is also a function of the length and diameter ratio of the magnetic torque rods. The magnetic torque rods can be placed in parallel to increase the damping. If the magnetic torque rods are too close to each other, the flux density varies along the magnetic torque rod length; in addition, the magnetic flux density varies with the length to diameter ratio as shown in the calculation of the dipole in equation (6.50). Placing the magnetic torque rods too close to each other, results in an additional dipole created by the coupling effects between the torque rods. If the magnetic torque rods are used frequently, there is also another dipole that remains within the torque rods. This dipole is known as the magnetic wandering dipole. The wandering dipole can cause an additional degree of error in the attitude pointing of the satellite.

Therefore, for a magnetic inclination angle of zero degrees where the magnitude of the magnetic field intensity of the Earth is constant, the spacecraft aligns itself along the direction of the Earth's magnetic dipole. The oscillation of the spacecraft is damped in a reasonable short period of time. When the spacecraft is placed in a magnetic polar orbit with a single torque rod, the spacecraft can follow the field lines in the polar orbit as shown in Figure 6.15. Above the North Pole, the satellite is pointing toward the Earth; above the South Pole, the satellite is pointing away from the Earth. Near the Equator, the satellite is pointing in the direction of the flux lines. Using these characteristics, different controllers can be developed to maintain the attitude pointing of the satellite. Because of the rotation of the satellite with a single magnetic torquer, the satellite does one revolution in half of an orbit. Once the satellite has damped out the satellite motion, the angular velocity achieved by the satellite is approximately two revolutions per orbit.

The attitude motion of a satellite can be written in terms of the orbital latitude in an assumed circular magnetic polar orbit. β is the angle between the spacecraft magnitude axis and the Earth's dipole axis. The

Figure 6.15 **Satellite angular motion with a single dipole in a polar orbit**

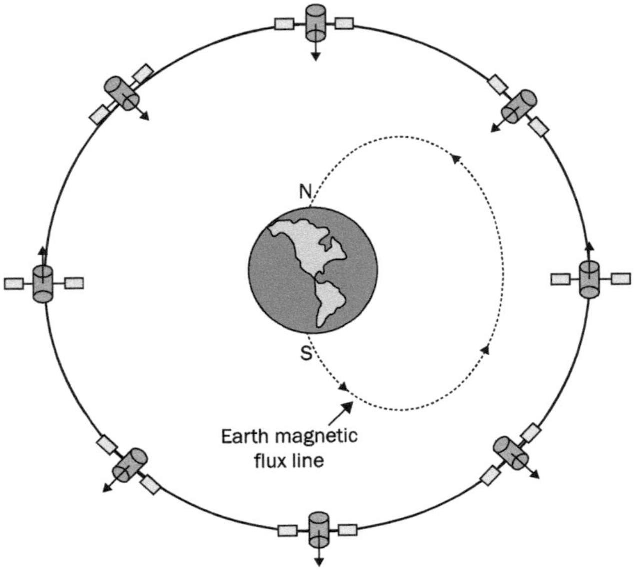

function describing the displacement of the angle β with respect to time can be written as,

$$\beta = \arctan\left(\frac{3\tan f}{1-2\tan^2 f}\right) \tag{6.68a}$$

where $f = nt$, and n is the mean motion (or orbital rate) of the satellite. When $f = 0$, $\beta = 0°$. If $f = \pi/4$, $\beta = 60°$. Taking the limit as $f \rightarrow \pi/2$, $\beta = 0°$. This calculation can be performed for every position of the satellite in the orbit to determine the angular displacement of the satellite. Taking the time derivative of equation (6.68), the satellite angular velocity and acceleration is written as,

$$\frac{d\beta}{dt} = n\left(1+\frac{2}{1+3\sin^2 f}\right)(\text{rad/sec}) \tag{6.68b}$$

$$\frac{d^2\beta}{d^2t} = -n^2\left(\frac{6\sin 2f}{\left(1+3\sin^2 f\right)^2}\right)(\text{rad/sec}) \tag{6.68c}$$

Figure 6.16 shows a plot of equations (6.68). When $f = 0$, $\dot{\beta} = 3n$ (rad/sec). At the poles, $\dot{\beta} = 3n/2$ (rad/sec). Also, the maximum angular

Figure 6.16 **Attitude motion of the satellite with a single magnetic torquer, circular magnetic polar orbit**

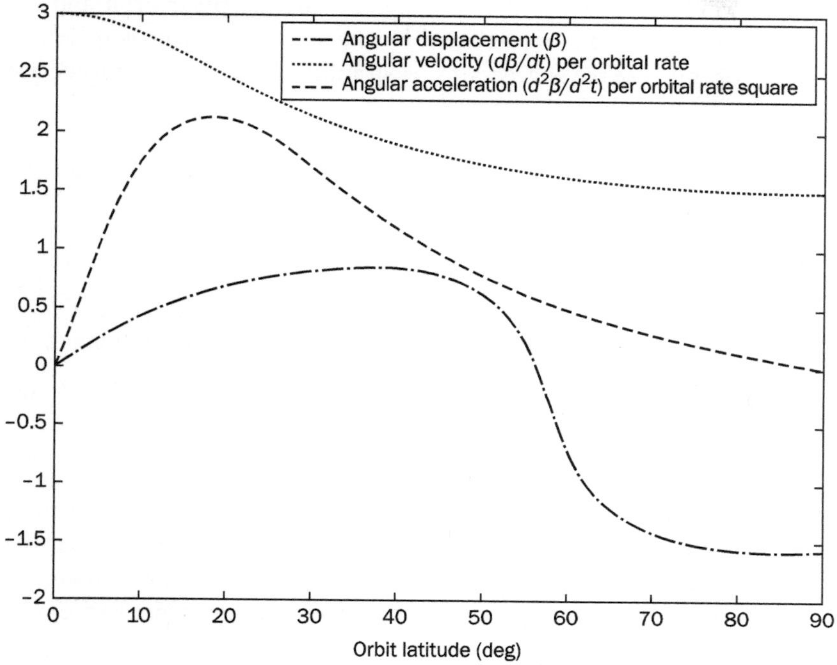

acceleration ($\ddot{\beta}_{max}$) is $3n^2$ (rad/sec^2) at the magnetic latitude of 18 degrees. The angular acceleration is 0 (rad/sec^2) at the poles.

With a single magnetic torque, the satellite can damp out its natural oscillation. By adding more magnetic rods, the satellite can damp out the oscillation in a short period of time. The only restrictions for magnetic damping are the location of the torque rods in the satellite and how close the torque rods are to each other. These actuators can also be used to maintain the orientation of the satellite but are constrained by the power available in the satellite [10].

6.4 Suggested problems

Problem 6.1. For the right circular cylinder explained in Problems 3.6 and 3.7, determine the moment for a total absorbing body, N^+, and total reflective body, N^-. Answer: Both moments are zero.

Problem 6.2. Develop an expression for the angle, β, between the magnetic (dipole) axis of a spacecraft and the Earth's dipole axis for the case where the spacecraft is moving in a circular polar orbit and the Earth's magnetic field is represented by the spin-axis oriented magnetic dipole. For this case the spacecraft's geographic and magnetic latitude are the same. Answer: Equation (6.68a).

Problem 6.3. Show that at a latitude of $\pi/2$ radians (North Pole) the angle, β, in Problem 6.2 is 180°.

Problem 6.4. Develop expressions for $\frac{d\beta}{dt}$ and $\frac{d^2\beta}{d^2t}$.

Problem 6.5. Show, using the techniques of calculus, that the maximum value of $\frac{d^2\beta}{d^2t}$ occurs at a magnetic latitude between 17° and 18° (see Figure 6.16).

Problem 6.6. Repeat problem 6.2 for the case where the spacecraft is moving in a circular equatorial orbit. What can you conclude about the magnetic attitude stabilization of a spacecraft in a circular equatorial orbit as compared with a spacecraft in a circular polar orbit? Explain.

7

Continuous and digital control systems

Abstract: Control systems are used to maintain the stability of the translational and rotational motion of a body. In certain cases, an analysis can be performed to determine the conditions under which the stability of a body can be maintained. The control systems can be analyzed in the discrete and continuous domain. The continuous state describes the equations of motion and the controller in the time domain. Once the signal in the continuous state is sampled, the system is discretized and can be represented in the digital domain. In both domains, pole placement techniques and optimal control schemes can be used to determine the control laws. The objective of this chapter is to show the reader how to develop any type of control system in the continuous and discrete domain.

Key words: continuous control, transfer functions, stability conditions, Routh–Hurwitz criteria, second order systems, pole placement techniques, optimal control, discrete control, Z-transform, state vectors, Ackermann's formula, linear quadratic regulator, hierarchical control scheme.

7.1 Introduction to methods of designing continuous and discrete control systems

One of the most interesting topics is the stabilization of mechanical systems, space vehicles, and any other body that translates and/or rotates about a Space. The system used to stabilize a body in motion is called the control process. A control process or system behaves very similarly to the motion of a ball through a plane with many curves; in addition, the motion of the ball depends on what you want to do with the ball within the plane. If the ball is on a flat surface, the ball can stay in the same location or move to a different place by applying an outside force. On the other hand, the ball can be in a valley; in this case, the ball cannot move up or down because the control system has reached a minimum state where the motion of the ball is stable. Also, the ball can reach the maximum peak of a hill. If a small amount of force is applied to the ball, the ball rolls down hill. Because of this, a stable state can be found so that the applied force to the ball can stabilize the motion on the top of the hill; and, this is the work of a control system. There are different analyses that can be performed to design a control system. A control process can be composed of continuous, discrete, and adaptive schemes. These control schemes can have different block diagrams but use very similar techniques to determine their gain constants (coefficients) to stabilize the body. These techniques can be divided in two large categories which define continuous and discrete control systems. The continuous systems analyze the differential equations by obtaining the transfer functions and state vectors in the continuous time domain. When the system is discretized, an analog to digital converter transforms the signal into the discrete domain by sampling the signal every certain interval of time. In a similar manner, the discrete control systems can have transfer functions associated with the differential equations. For the continuous and discrete control systems, the desired control functions can be determined with pole placement techniques, optimal control theory, and enforcing stability conditions.

These methods can be compiled in a set of procedures to analyze the stability of the motion of a body. These rules can be written and can be described as follows:

a. Integration of the nonlinear system of differential equations.
b. Obtain the closed-form solution to determine the steady-state, transient, stability boundaries, among other things; closed form solution is usually not available, and a numerical integration routine is required to obtain the solution of the system during the application of the control system.
c. Linearization of the equations of motion about the stability point which is to be investigated.

d. This defines the equations in a linear sense to determine the characteristics of the system that provides the desired conditions for stability.

e. Determine if the roots of the system characteristic equation have positive real parts.

f. Apply the Routh–Hurwitz criteria for stability in case the determination of the roots becomes complicated; on the other hand, write the equation in terms of the state vectors and solve the eigenvalue problem.

g. Determine the control input function by either using the pole placement technique or optimal control theory.

These techniques can be applied to the continuous and discrete domain. The difference is in the type of equations that are used.

7.2 Ares V equations of motion for first stage flight

Before developing the control formulation, a set of equations of motion are required. In the previous chapters, the Ares V equations of motion are written for the space vehicle. The equations of motion for the Ares V rocket are described for the first stage (FS) flight. The following equations of motion describe the FS flight for the Ares V rocket:

$$J\dot{\vec{\varpi}} + \vec{\varpi} \times J\vec{\varpi} = \vec{N}_A + \vec{N}_{SRB} + \vec{N}_{RS\text{-}68b} + \vec{N}_{RoCS} \tag{7.1a}$$

where,

$$\vec{\varpi} = \begin{bmatrix} \dot{\varphi} - \dot{\psi}\sin\theta \\ \dot{\theta}\cos\varphi + \dot{\psi}\sin\varphi\cos\theta \\ -\dot{\theta}\sin\varphi + \dot{\psi}\cos\varphi\cos\theta \end{bmatrix} \tag{7.1b}$$

$$N_A = L_{CP}\begin{bmatrix} 0 & 0 \\ -\cos\psi\sin\theta\cos\varphi - \sin\psi\sin\varphi & -\cos\theta\cos\varphi \\ \cos\psi\sin\theta\sin\varphi - \sin\psi\cos\varphi & \cos\theta\sin\varphi \end{bmatrix}\begin{bmatrix} A_F \\ N_F \end{bmatrix} \tag{7.1c}$$

$$N_{SRB} = -2T_{SRB}L_{GP}\begin{bmatrix} 0 \\ \cos\psi\sin\theta\cos\varphi + \sin\psi\sin\varphi \\ \cos\psi\sin\theta\sin\varphi - \sin\psi\cos\varphi \end{bmatrix} \tag{7.1d}$$

$$N_{RS-68b} = -6T_{RS-68b}L_{GP}\begin{bmatrix} 0 \\ \sin\beta \\ -\cos\beta\sin\delta \end{bmatrix} \tag{7.1e}$$

$$\vec{N}_{RoCS} = n_T T_{RoCS} R_{AresV}\hat{\imath} \qquad T_{RoCS} = -gI_{SP}^{RoCS}\dot{m}_{RoCS} \tag{7.1f}$$

$$T_{RS\text{-}68b} = -gI_{SP}^{RS\text{-}68b}\dot{m}_{RS\text{-}68b} \qquad T_{SRB} = -gI^{SRB}{}_{SP}\dot{m}_{SRB} \tag{7.1g}$$

$$N_F = QC_NS \qquad A_F = QC_AS \tag{7.1h}$$

7.3 Continuous control formulation

The first step to understanding the control systems is to develop the continuous formulation. A continuous formulation refers to the definition of the equations of motion in terms of its time derivatives. To solve this problem, it is required to integrate the differential equations to obtain the solution for the motion of the space vehicle; in some cases, the solutions of the differential equations are not available; for this reason, this section concentrates on the different steps that are required to determine the appropriate stability conditions and control formulation.

First, the linearized formulation is explained. Throughout this text, some equations have been linearized to determine certain characteristics of the space vehicle. Also, the Laplace formulation and state vector formulation are defined to understand the different methods of obtaining the solution of the equations of motion. Second, the stability conditions are defined such that the necessary conditions for maintaining a stable motion of the space vehicle can be obtained. With these conditions, the control formulation is developed based on different types and arrangements of actuators. These techniques are known as active control schemes which depend on the positions and velocities of the vehicle.

Finally, the control schemes are developed to maintain the translational and rotational motion of the space vehicle. In order to understand this formulation, the equations of motion for the Ares V rocket are used. These equations of motion provide different characteristics about the ascent trajectory of the vehicle.

7.3.1 Linearized formulation

After obtaining the equations of motion for a vehicle, the system of equations is linearized such that certain characteristics can be obtained. For

Figure 7.1 **Sine function**

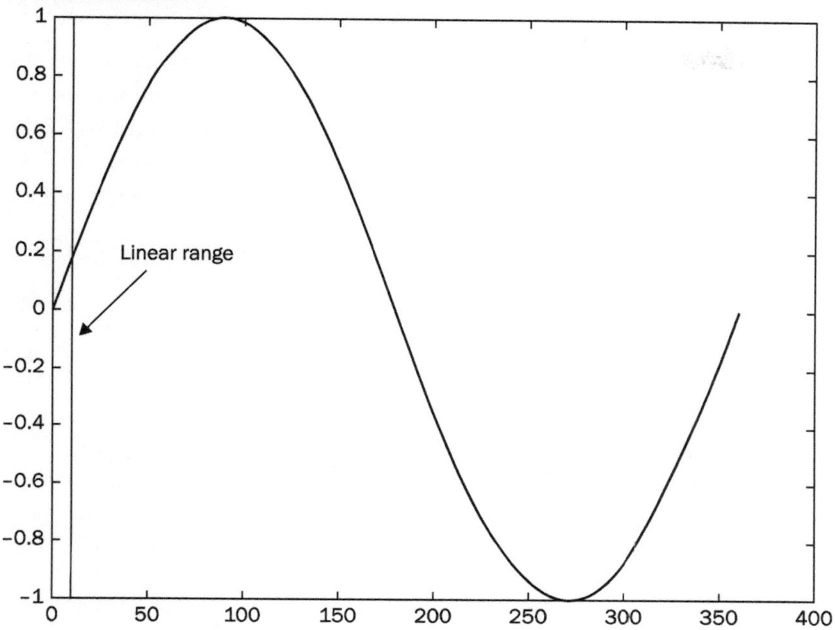

a translational motion, the linearization of the equations can be determined by eliminating the higher order terms in the equations of motion. For the attitude motion, the linearization is performed, considering the sine and cosine functions. Linearization refers to the characteristics of a function in which for a small range, the function behaves as a linear function. One good example is the sine function as shown in Figure 7.1. For angles between 0 and 10 degrees, the sine function acts as a linear function.

For angles larger than 10 degrees, the slope of the line keeps varying and is required to perform other frame rotations in order to understand the behavior of those points.

The Taylor's series expansion [**13**] for the sine and cosine function about zero can be written as,

$$\sin x = x - \frac{x^3}{3!} + \frac{x^5}{5!} - \cdots \tag{7.2a}$$

$$\cos x = 1 - \frac{x^2}{2!} + \frac{x^4}{4!} - \cdots \tag{7.2b}$$

Small angle assumption refers to angles close to zero. From equations (7.2), the sine and cosine functions can be reduced to,

$$\sin x \approx x \tag{7.3a}$$

$$\cos x \approx 1 \tag{7.3b}$$

If two sine functions are multiplied by each other, the result of the multiplication can also be approximated to zero. These definitions can be used for the FS equations of motion presented in equations (7.1). Substituting the definitions for a linearized function in equations (7.3), the linear equations of motion for the Ares V function can be written as,

$$\ddot{\varphi} = b_1 T_{RoCS} \tag{7.4a}$$

$$\ddot{\theta} = -a_1\theta + b_2\beta - d_1 \tag{7.4b}$$

$$\ddot{\psi} = -a_2\psi + a_3\varphi - b_3\delta \tag{7.4c}$$

where,

$$a_1 = \frac{L_{CP}A_F - 2gl_{SP}^{SRB}\dot{m}_{SRB}}{J_2} \qquad a_2 = \frac{L_{CP}A_F - 2gl_{SP}^{SRB}\dot{m}_{SRB}}{J_3} \qquad a_3 = \frac{L_{CP}N_F}{J_3}$$

$$b_1 = \frac{n_T R_{AresV}}{J_1} \qquad b_2 = \frac{6gl_{SP}^{RS\_68b}\dot{m}_{RS\_68b}L_{GP}}{J_2} \qquad b_3 = \frac{6gl_{SP}^{RS\_68b}\dot{m}_{RS\_68b}L_{GP}}{J_3}$$

$$d_1 = \frac{L_{CP}N_F}{J_2}$$

Equations (7.4) describe the motion of the vehicle when the angle variations are close to zero. The controller is added to these equations of motion to determine the best control gains and parameters to maintain a stable motion. After the linear equations are analyzed, the nonlinear equations are used again with the controller. For different initial conditions, the controller acts on the vehicle to maintain its desired orientation and velocity; in other words, the controller reduces the error between the desired and actual states back to zero. The definitions of the different controllers are explained in the following subtopics, but the transfer functions and stability conditions are first explained to understand how the equations can be analyzed.

7.3.2 Actuator and control function definitions

After developing the linearized functions, the actuator or control functions are defined. The actuators are the systems that interact with the environment and the vehicle to cause variations in the motion and rotation of a space vehicle. On the other hand, the control functions are the equations that specify how the actuators behave depending on the sensed location and/or velocity of the space vehicle. The formulation for the different control functions is explained in the following topics, but the actuator functions are explained here. The first step to understanding an actuator is to ask the following question: What do I physically use in my space vehicle to cause a translational or rotational motion? This question is answered in Chapter 6 but is expanded here to identify other characteristics about the actuators.

The selection of the actuators depends on the type of space vehicle used and the environment in which the vehicle is moving. There are two principal types of space vehicles already flying within and outside of the Earth's atmosphere: rockets and satellites. One good example is the Ares V rocket equations shown in equations (7.4). To cause a change in the translation of the vehicle, the thruster motors such as the RS-68b and SRB can be used. For the rockets, there usually is no interest in varying the thrust force because the rocket uses a maximum thrust force to move against the gravitational acceleration. It is required to change the rotation of the vehicle such that the rocket can achieve the insertion point in a short period of time without expending a lot of fuel. In order to rotate the vehicle, the motors are gimbaled about a point to force a rotation. For the Ares V, the RS-68b motors can provide torques about two axes to allow pitch and yaw corrections. As shown in equations (7.4b) and (7.4c), the gimbal angles β and δ control the rotation of the vehicle about two of the axes. On the other hand, motion about the third axis of rotation in equation (7.4a) is controlled by a thruster. Within the thruster, there is a motor that can control the amount of mass flow rate that changes the applied force causing the vehicle to rotate about the roll axis.

These details about the variations of the mass flow rate and gimbal angles should be classified depending on the space vehicle. In Chapter 6, the different actuators are explained for a satellite. Table 7.1 presents a summary of different actuator functions. The maximum values shown in the third column of Table 7.1 are included because actuators are constrained based on the design and on the maximum allowed forces or boundaries on the motion amplitudes.

The actuator functions are expanded in terms of unit values, $u(t)$; in this way, the actuator is not constrained to a specific system but can be designed based on the conditions of the problem. Continuing with the Ares V equations of motion, the actuator is given by the β gimbal angle in equation (7.4b). The β gimbal angle can be represented as follows,

$$\beta(t) = \frac{\beta_{max}}{\beta_{max}}\beta(t) = \beta_{max}u_A(t) \tag{7.5a}$$

Table 7.1 Actuator functions summary

Actuator	Function	Units	Approximate maximum value	Maximum actuator function	Reason
Thruster – gimbal angle	$\beta(t)$	Degrees	Approximately 8 degrees	β_{max}	High specific impulses provide enough thrust to translate the vehicle and require small variations in the gimbaled angles to cause a rotation.
Thruster – force	$T(t)$	Newtons	Depends on the type of propellant and specific impulse [**10**]	T_{max} or $\dot{m}_{max}$	The force can be varied depending on the valve hole opening.
Magneto-torquer	$D(t)$	$A - m^2$	Maximum dipole moment 100 $A - m^2$	D_{max} or I_{max}	The current controls the intensity of the induced field but should be maintained to a minimum. For relatively large magnetic dipoles, the mass of the magnetic torquer increases.
Reaction wheels	$N_{RW}(t)$	$N - m$	Depends on the voltage specified by the manufacturer	N_{max} or E_{max}	The applied voltage specifies the amount of additional angular momentum to cause a rotation to the space vehicle.

where,

$$u_A(t) = \frac{\beta(t)}{\beta_{max}} \tag{7.5b}$$

$u_A(t)$ is the actuator function which is constrained to values between –1 and 1. When $u_A(t) = 0$, the actuator is in its original position causing no motion of the vehicle. When $u_A(t) = 1$ or –1, the actuator is located in a position in which a force is applied along or against the movement of the vehicle, respectively.

Equation (7.5b) specifies the maximum (or saturation) values that the actuator can have but does not define the characteristics of the actuator and the connection to the control functions. When a control signal is sent to the actuator, the actuator does not move instantly because the actuator has a series of mechanisms to turn on. This causes time delays in the computational process that can be expressed in the equations. The actuator response to a control signal can be expressed in terms of a second order function but mainly depends on the hardware. As shown in Reference [60], the mechanism that results in a motor gimbal angle of the rocket can be modeled as a second order function as,

$$\ddot{u}_A(t) + 2\xi\omega_n\dot{u}_A(t) + \omega_n^2 u_A(t) = \omega_n^2 u(t) \tag{7.6}$$

where ξ is the damping function, ω_n is the natural frequency, and $u(t)$ is the control function. In practice, equation (7.6) can be used to describe the actuator dynamics.

Normally, the control books [76] [78] only consider an ideal actuator for the first analysis of the control system. An ideal actuator means that $u_A(t) = u(t)$. This is not a bad assumption to understand the behavior of the vehicle. After the controller is identified, the actuator formulation (shown in equation (7.6)) should be used to include other characteristics of the system. The following topics are explained assuming an ideal actuator, but the actual behavior of the actuator should also be considered at some point in the analysis.

7.3.3 Transfer function formulation

The transfer functions are used to simplify the solution of the equations of motion and understand the stability conditions. The first step to obtain the transfer function is to solve the Laplace transform [76]. The Laplace transform of a function is written as,

$$\mathcal{L}\{f(t)\} = F(s) = \int_0^\infty f(t)e^{-st}\,dt \tag{7.5}$$

There are common Laplace transforms for different functions as shown in Appendix D. To demonstrate one solution, the Laplace transform of equation (7.4a) is obtained. The Laplace transform of the left hand side (LHS) of equation (7.4a) equals to,

$$\mathcal{L}\{J_1\ddot{\varphi}\} = J_1\mathcal{L}\{\ddot{\varphi}\} = J_1 \int_0^\infty \ddot{\varphi}(t)e^{-st}dt$$

Integrating by parts,

$$\mathcal{L}\{J_1\ddot{\varphi}\} = J_1 \left[-\ddot{\varphi}(0) + \int_0^\infty s\ddot{\varphi}(t)e^{-st}dt\right]$$

Integrating by parts once again,

$$\mathcal{L}\{J_1\ \ddot{\varphi}\} = J_1 \left[-\ddot{\varphi}(0) - s\varphi(0) + s^2 \int_0^\infty \varphi(t)e^{-st}dt\right]$$

Thus, the Laplace equation of the LHS of equation (7.4a) is written as,

$$\mathcal{L}\{J_1\ddot{\varphi}\} = J_1 s^2\Phi(s) - J_1 s\varphi(0) - J_1\ddot{\varphi}(0) \tag{7.6a}$$

where,

$$\Phi(s) = \int_0^\infty \varphi(t)e^{-st}dt \tag{7.6b}$$

Equation (7.6b) is the Laplace transform of the function which is the definition shown in equation (7.5). Equation (7.6a) defines the motion of the vehicle in the s domain (or the imaginary plane). Equation (7.4a) defines the gimbal angle (β) that controls the direction of the thrust force to cause a pitch rotation to the vehicle. The Laplace transformation is performed to the right hand side (RHS) of equation (7.4a) which equals to,

$$J_1 s^2\Phi(s) - J_1 s\varphi(0) - J_1\dot{\varphi}(0) = b_1\tilde{T}_{RoCS}(s) \tag{7.7}$$

where,

$$\tilde{T}_{RoCS}(s) = \int_0^\infty T_{RoCS}e^{-st}dt$$

Assuming the initial conditions equal to zero,

$$J_1 s^2\Phi(s) = b_1\tilde{T}_{RoCS}(s)$$

A transfer function is described as the output function divided by the input function. In the previous equation, $\Phi(s)$ is the output function while $\tilde{T}_{RoCS}(s)$ is the input function. Then, the transfer function is written as,

$$G_p(s) = \frac{\Phi(s)}{\tilde{\upsilon}_1(s)} = T_{max}\frac{b_1/J_1}{s^2} \tag{7.8}$$

where

$$\tilde{v}_1(s) = \frac{\tilde{T}_{RoCS}(s)}{T_{max}}$$

Equation (7.8) is known as the plant or process transfer function. Normally, the plant function is described with the following variable, $G_p(s)$. The transfer function for the actuator system is defined as,

$$G_A(s) = \frac{u_{RoCS}(s)}{\tilde{v}_1(s)} = \frac{\omega_n^2}{s^2 + 2\xi\omega_n s + \omega_n^2} \tag{7.9}$$

This actuator transfer function describes the physical behavior of the actuator system when the control signal is transmitted by the computer. The control signal is described by the controller transfer function as $G_C(s) = \frac{u_{RoCS}(s)}{x_E(s)}$. $x_E(s)$ is the difference transfer function between the command function ($x_C(s)$) and the actual function ($x_A(s)$). The command function defines the desired parameters, and the actual function is the parameters measured from the different sensors located along the body of the system.

The process, actuator, control, command, and actual transfer functions presented here can be shown in a diagram. The diagram is called a closed-loop system and is shown in Figure 7.2a. In Figure 7.2a, the system can be commanded to a desired location and/or orientation. Also, there is another transfer function that is added to the system called the sensor

Figure 7.2 **(a) Closed loop control and (b) open loop control**

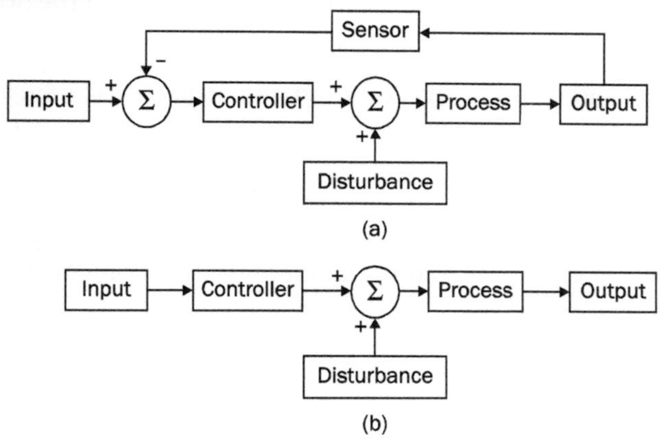

transfer function, $G_S(s)$. This transfer function describes the sensor formulation associated with the determination of the orientation and/or location of the vehicle in space. When the sensor information is no longer used, Figure 7.2b shows the diagram called the open-loop control system. The open-loop control is used to force the vehicle to move in a desired translational and rotational motion.

As shown in this section, the closed and open loop systems are important in the analysis of a control system. These systems help the designer to determine the desired transfer functions which are then used to calculate the stability condition for the system.

7.3.4 State vector formulation

Another format to write the differential equations is known as the state vectors. A state refers to a condition in the motion of a body; in other words, the state refers to the position, angular position, velocity, and/or angular velocity of a body. On the other hand, there are other cases in which the state could be referred to the actuator function. Mathematically, the state equations are described as the reduction of higher order differential equations to a set of first order differential equations [**12**]. The following steps describe how to write a state vector formulation:

1. Identify the order of the differential equations.
2. Write the state variables up to the highest order of the differential equation.
3. Rewrite the second order differential equations as first order differential equations.
4. Complete the set of first order differential equations if the order of the equations of motion is higher than 2.

To understand the state vector formulation, equations (7.4a) and (7.4c) are used as examples. Following the steps, equations (7.4a) and (7.4c) are second order differential equations. For the second step, the state variables can be identified with any variable, but commonly x is used. To identify every state, a subscript is added to the state variable because of the differential equations and the number of equations of motion. The state variables are written as follows,

First State: *Angular Position*

$$x_1(t) = \varphi(t) \qquad x_2(t) = \psi(t) \tag{7.10a}$$

Second State: *Angular Velocity*

$$\dot{x}_1(t) = x_3(t) = \dot{\varphi}(t) \qquad \dot{x}_2(t) = x_4(t) = \dot{\psi}(t) \tag{7.10b}$$

Third (Final) State: *Angular Acceleration*

$$\dot{x}_3(t) = \ddot{\varphi}(t) \qquad \dot{x}_4(t) = \ddot{\psi}(t) \tag{7.10c}$$

The first state identifies the lowest order of the differential equations described by the angular positions. The time derivative of equation (7.10a) describes the angular velocity of the body and is the second state for the two differential equations. These new states are identified with a new state variable because the highest order of the differential equations is two. Once more, the time derivative of equation (7.10b) is obtained and represents the angular acceleration of the body. This is the highest order of the differential equations; it is not necessary to add another state variable. In fluid [79] and thermodynamic [80] problems, the order of the differential equations can be up to four. In satellite motion, the highest order is two.

The next step is to write the differential equations as a set of first order differential equations. Using equations (7.10), the state equations can be defined as follows,

$$\dot{x}_3 = b_1 T_{max} \tilde{\upsilon}_1 \tag{7.11a}$$

$$\dot{x}_4 = -a_2 x_2 + a_3 x_1 - b_3\, \delta_{max} \tilde{\upsilon}_3 \tag{7.11b}$$

where,

$$\tilde{\upsilon}_1 = \frac{T_{RoCS}}{T_{max}} \qquad \tilde{\upsilon}_3 = \frac{\delta}{\delta_{max}} \tag{7.11c}$$

In equations (7.11), the right hand side should not have any time derivative because of the expression used in equations (7.10). A set of state variables are added to explain the angular velocity of the vehicle; then, the complete set of state variables are written as,

$$\dot{x}_1(t) = x_3(t) \qquad \dot{x}_2(t) = x_4(t) \tag{7.12a}$$

$$\dot{x}_3(t) = b_1 T_{max} \tilde{\upsilon}_1 \tag{7.12b}$$

$$\dot{x}_4(t) = -a_2 x_2(t) + a_3 x_1(t) - b_3 \delta_{max} \tilde{\upsilon}_3 \tag{7.12c}$$

In practice, the state equations are written in matrix format to visualize other characteristics of the equations. In addition, those equations can be used to determine the stability conditions of the equations of motion. In matrix format [12], the state equations are written as,

$$\begin{bmatrix} \dot{x}_1(t) \\ \dot{x}_2(t) \\ \dot{x}_3(t) \\ \dot{x}_4(t) \end{bmatrix} = \begin{bmatrix} 0 & 0 & 1 & 0 \\ 0 & 0 & 0 & 1 \\ 0 & 0 & 0 & 0 \\ a_3 & -a_2 & 0 & 0 \end{bmatrix} \begin{bmatrix} x_1(t) \\ x_2(t) \\ x_3(t) \\ x_4(t) \end{bmatrix} + \begin{bmatrix} 0 & 0 \\ 0 & 0 \\ b_1 T_{max} & 0 \\ 0 & -b_3 \delta_{max} \end{bmatrix} \begin{bmatrix} \tilde{v}_1 \\ \tilde{v}_3 \end{bmatrix} \tag{7.13}$$

Equation (7.13) is also represented in the following format,

$$\dot{\vec{x}}(t) = \boldsymbol{A}\vec{x}(t) + \boldsymbol{B}\vec{u}(t) \tag{7.14a}$$

where,

$$\boldsymbol{A} = \begin{bmatrix} 0 & 0 & 1 & 0 \\ 0 & 0 & 0 & 1 \\ 0 & 0 & 0 & 0 \\ a_3 & -a_2 & 0 & 0 \end{bmatrix} \tag{7.14b}$$

$$\boldsymbol{B} = \begin{bmatrix} 0 & 0 \\ 0 & 0 \\ b_1 T_{max} & 0 \\ 0 & -b_3 \delta_{max} \end{bmatrix} \tag{7.14c}$$

$$\vec{x}(t) = \begin{bmatrix} x_1(t) \\ x_2(t) \\ x_3(t) \\ x_4(t) \end{bmatrix} \tag{7.14d}$$

$$\vec{u}(t) = \begin{bmatrix} \tilde{v}_1 \\ \tilde{v}_3 \end{bmatrix} \tag{7.14e}$$

Equation (7.14a) is known as the state vector equation. $\vec{x}(t)$ is the state vector, and $\vec{u}(t)$ is the control input vector. As shown previously, the control input vector can be substituted by the solution of the actuator equation. Through this analysis of the equations, the control input vector is assumed to have an ideal actuator. $\boldsymbol{A}$ is known as the state matrix, and $\boldsymbol{B}$ is the control matrix. In general, the state and control matrix have $n \times n$ and $n \times m$ dimensions, respectively. n and m refers to the number of state variables and control inputs used to represent the equations of motion, respectively. The state and control input vectors, respectively, has dimensions $n \times 1$ and $m \times 1$.

As shown in equations (7.14), the state vector formulation can be easily used to obtain the solution of the equations of motion by numerical integration [14]. If the transfer function is used, the inverse Laplace transform [76] is used to obtain the function in terms of time. Also, the transfer function becomes complicated for cases in which the equations of motion are coupled. The equations are coupled when two or more equations have variables in common with each other. By observing equations (7.14), the roll equation is decoupled from the yaw equation, but the yaw equation is coupled with the roll equation. If a solution is desired by taking the Laplace transform, a system of equations must be solved to obtain the equations of motion. If the state vector format is used, a simple integration method can be used to determine the solution of the equations of motion without obtaining the actual function in time.

On the other hand, the state vector equation can also be transformed into a transfer function. The Laplace transform of equation (7.14) can be written as,

$$s\vec{X}(s) - \vec{X}(0) = \boldsymbol{A}\vec{X}(s) + \boldsymbol{B}\vec{U}(s) \tag{7.15}$$

Rearranging equation (7.15) and assuming the initial conditions equal to zero,

$$\vec{X}(s) = (s\boldsymbol{I} - \boldsymbol{A})^{-1}\boldsymbol{B}\vec{U}(s) \tag{7.16}$$

Equation (7.16) can be used to represent the transfer function for the equations of motion and is useful in obtaining the stability conditions.

The formulations for the transfer function and the state vector equation are useful representations for the understanding of the equations of motion and the controller. The designer can use any of the two formulations to analyze the motion of the vehicle, but both formulations can be used together to determine the best control system to maintain the motion and/or orientation of the vehicle.

7.3.5 Stability conditions

The next step before developing a control system is to determine the conditions under which the motion of a body is stable. To determine the stability conditions, the Routh–Hurwitz and eigenvalue problem can be used. The Routh–Hurwitz method [76] is used to understand the stability of the equations of motion by using the denominator of the transfer function. In a similar manner, the eigenvalue problem [81] obtains the solution of the equations of motion by solving the characteristic equation which is obtained from the state matrix. Both methods provide the same solution to determine the stability conditions using different formulations.

As part of the stability analysis, the second order responses are explained to understand different stability conditions.

7.3.5.1 Second order response for mechanical systems

The stability conditions of a vehicle can be explained through second order responses. A second order response refers to the solutions obtained from a second order differential equation. The second order responses are also used to determine how the controller enhances the control function of the vehicle to maintain a desired translational and rotational motion. Also, many of the solutions obtained from the Routh–Hurwitz criteria and the eigenvalue problem are understood through the second order responses.

The second order responses are commonly used to express the motion of a vehicle. The solution of the differential equations [82] is described in terms of two functions – natural and forced response. The natural response is the response of the system when there is no force applied to the system. Generally, this function is also called the homogeneous solution of the differential equation [12]. The forced response is another function that describes a response to a force applied to the vehicle and is also known as the particular solution of the differential equation [12]. The state equations can also be used to explain these two different functions. If the control input vector is zero, the solution of the differential equations is the system's natural response. When the control input function is defined by a function, the control input function is the forced response.

To analyze the transient response of the system, a unit step function is applied as the forcing function. A unit step function is shown in Figure 7.3 and is used to determine the time response of the system; in this way, different response characteristics can be used to understand the response of the homogeneous part of the differential equations. The second order differential equations are commonly used to describe the physical motion of a body. A second order differential equation provides four different responses based on the roots of the transfer function. An example of a second order function can be described by the mass-spring-damper system shown in Figure 7.4. The transfer function for the mechanical system in Figure 7.4 can be described as follows,

$$G(s) = \frac{1/M_b}{s^2 + \frac{f_v}{M_b}s + K/M} \qquad (7.17)$$

Figure 7.3 Unit step function

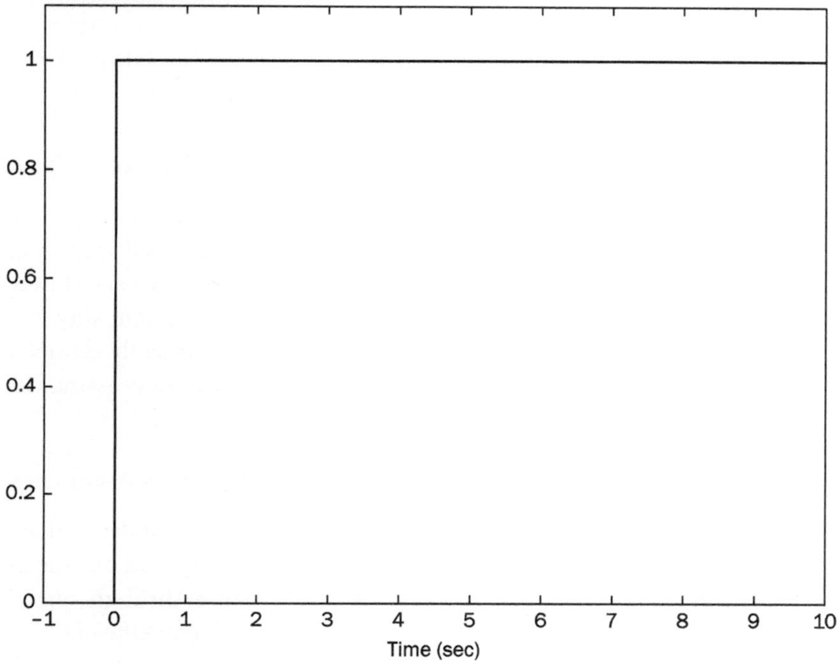

Figure 7.4 Mass-spring-damper system

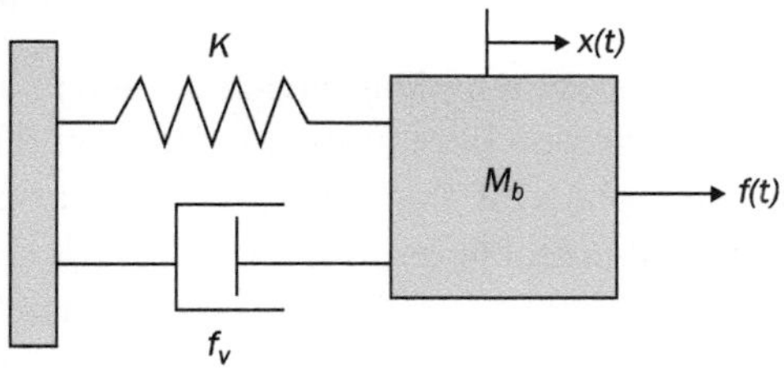

where M_b is the mass of the block, f_v is the damper constant, and K is the spring constant. To characterize the second order responses, the roots of the denominator of equation (7.17) are determined. Applying the quadratic formula [13], the roots of the denominator in equation (7.17) are equal to,

$$s_{1,2} = \frac{1}{2}\left[-\frac{f_v}{M_b} \pm \sqrt{\left(\frac{f_v}{M_b}\right)^2 - \frac{4K}{M_b}}\right] \tag{7.18}$$

The roots shown in equation (7.18) are also called poles. The second order response depends mainly on the solution of the radical. The solution of the radical in the quadratic formula provides different responses as shown in Figure 7.5 for the case where $M_b = 4$, and $K = 1$, while f_v varies, to show the second order responses and is described as follows:

a. $\left(\frac{f_v}{M_b}\right)^2 - \frac{4K}{M_b} \geq 0$: Equation (7.18) provides two real poles without imaginary parts. This solution provides two characteristic responses of the system. If $f_v > 4$, the system shows an overdamped response. The overdamped response shows that the response rises very slowly toward one without crossing the limit. If $f_v = 4$, the system shows a similar response but is called critically damped. A critically damped function is the maximum response of the system without crossing the limit.

b. $\left(\frac{f_v}{M_b}\right)^2 - \frac{4K}{M_b} < 0$: The solution of equation (7.18) is two complex poles and is known as an underdamped solution. The underdamped response rises and settles to the final value in a short period of time as shown in Figure 7.5. The crossing of the response through one is known as the overshoot and depends on the damping coefficient, f_v.

c. $f_v = 0$: The solution of equation (7.18) is two complex poles without real parts and is known as the critically damped response. The critically damped response can be described by a sine function and resembles the solution of the simple harmonic oscillator [**82**].

By determining the roots of the polynomial as shown in equation (7.18), Figure 7.6 shows that the poles are always located on the left hand side (LHS) of the complex plane. If the poles are located on the imaginary axis, the system provides a stable response which is the critically damped response. If there are any poles located on the right hand side (RHS) of the complex plane, the system is unstable because the function defining the solution of the second order response is expressed by an exponential function that increases with time (i.e. $f(t) = Ae^{bt}$ where A and b are constants, and $b > 0$).

If the second order system is stable, there are measurements that can be used to characterize the time response of the system and are described as follows: a. rise time, b. settling time, c. natural frequency, d. damping ratio, e. peak time, and f. percent of overshoot. These measurements can only be performed for overdamped, critically damped, and underdamped responses.

Figure 7.5 **Second order response**

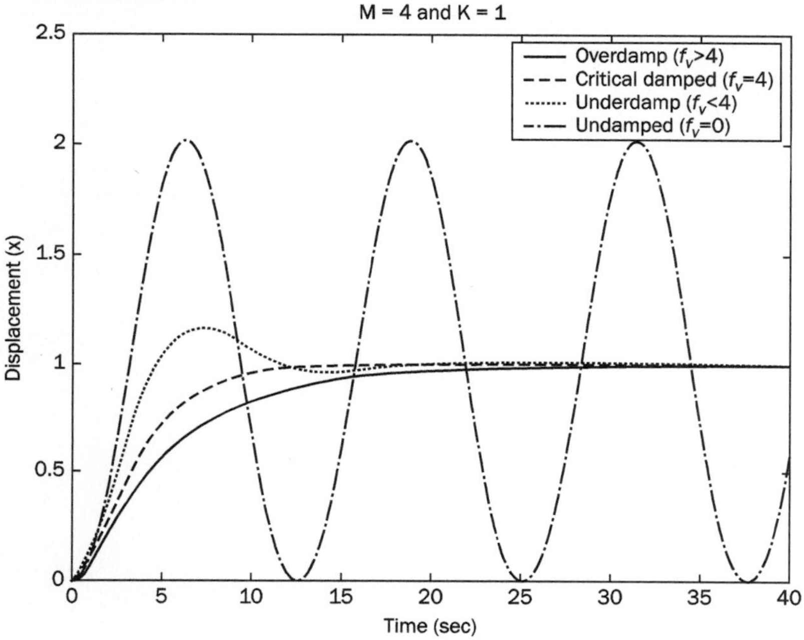

Figure 7.6 **Complex plane showing poles location**

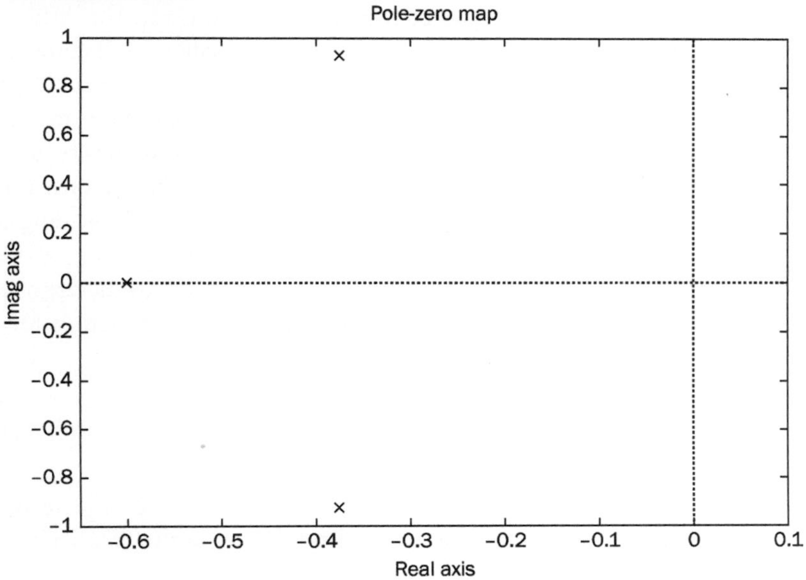

Figure 7.7 **Second order response characteristics**

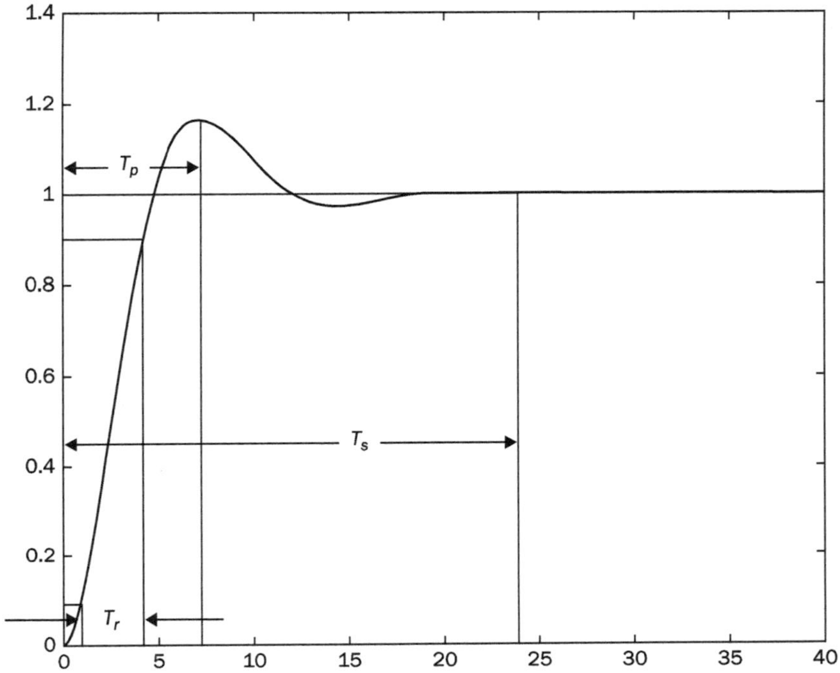

Figure 7.7 shows an underdamped response to demonstrate the different measurements that can be obtained for second order systems. In Figure 7.7, the rise time (T_r) is defined as the time for the waveform to go from 10% to 90% of its final value. The rise time is used to determine how fast the system responds to a force input. In a similar manner, the system is required to reach a stable motion when the input force is applied. The response time for the waveform to reach and stay within 2% (or can be 5% depending on the design) of its final value is called the settling time (T_s).

The natural frequency (ω_n) and the damping ratio (ξ) are determined from the transfer function. In general form, the transfer function for a second order system can be written as,

$$G(s) = \frac{b}{s^2 + as + b} = \frac{\omega_n^2}{s^2 + 2\xi\omega_n s + \omega_n^2} \tag{7.19}$$

The RHS of equation (7.19) is used in practice to describe the second order response for a system. The natural frequency is the frequency of oscillation of a system without damping and is equal to,

$$\omega_n = \sqrt{b} \tag{7.20}$$

The damping ratio is the ratio of the exponential decay frequency to the natural frequency and is defined as,

$$\xi = \frac{a}{2\omega_n} \tag{7.21}$$

In Figure 7.7, the peak time (T_p) is the time to reach the maximum peak in the waveform. The peak time only happens for the underdamped responses because for critically damped and overdamped responses, the system slowly reaches the final value without exceeding the final value. The percent of overshoot is the amount that the waveform overshoots the steady-state final value. This value is determined from the amplitude of the waveform at the peak time. From reference [76], the equations defining the settling time, time peak, and settling time can be written respectively as,

$$T_s = \frac{4}{\xi\omega_n} \tag{7.22a}$$

$$T_p = \frac{\pi}{\omega_n\sqrt{1-\xi^2}} \tag{7.22b}$$

$$\%OS = \exp\left(-\xi\pi \Big/ \sqrt{1-\xi^2}\right) \times 100 \tag{7.22c}$$

$\%OS$ is the percent of overshoot and is the percentage of the difference between the maximum amplitude at peak time and one. From equation (7.22c), the percent of overshoot is related to the damping ratio. Using equations (7.20) through (7.22), the system response can be either characterized or designed. In practice, this information is used to design the system response of the control input function.

7.3.5.2 Routh–Hurwitz criterion

The second order responses provide an idea of how a system can behave depending on the input force. If the polynomial in the denominator of the transfer function is n^{th} order, the determination of the time response for the system is complicated. There is an alternate method of determining the stability of a linear system with constant coefficients. This method is known as the Routh–Hurwitz criterion [83]. The Routh–Hurwitz criterion provides the necessary and sufficient conditions to determine if the system has a steady-state motion. The Routh–Hurwitz method does not require the solution of the transfer function to determine the stability of motion of the system.

Let the characteristic equation for the denominator of the transfer function be written as,

$$F(s) = a_0 s^N + a_1 s^{N-1} + \ldots + a_N \tag{7.23}$$

where N is the degree of the polynomial, a_i $(i = 0,1, \ldots, N)$ are the coefficients of the polynomial. Using the following determinant format,

$$\Delta_1 = a_1$$

$$\Delta_2 = \begin{vmatrix} a_1 & a_0 \\ a_3 & a_2 \end{vmatrix}$$

$$\Delta_3 = \begin{vmatrix} a_1 & a_0 & 0 \\ a_3 & a_2 & a_1 \\ a_5 & a_4 & a_3 \end{vmatrix}$$

$$\Delta_4 = \begin{vmatrix} a_1 & a_0 & 0 & 0 \\ a_3 & a_2 & a_1 & a_0 \\ a_5 & a_4 & a_3 & a_2 \\ a_7 & a_6 & a_5 & a_4 \end{vmatrix}$$

$$\ldots$$

$$\Delta_N = \begin{vmatrix} a_1 & a_0 & \cdots & 0 \\ a_3 & a_2 & \cdots & 0 \\ \vdots & \vdots & \cdots & \vdots \\ a_{2N-1} & a_{2N-2} & \cdots & a_N \end{vmatrix}$$

Then,

1. All the roots of $F(s)$ have negative real parts if and only if all the Δ_i $(i = 1,2,3, \ldots, N)$ are larger than zero. All the coefficients written as a_{-N} are replaced by zeros. This is a necessary and sufficient condition for stability of the system.
2. All the coefficients for $F(s)$ are not zero and have the same sign; a stability condition can be obtained. This is a necessary condition but is not sufficient to determine the stability of the system. For this reason, this application of the Routh–Hurwitz criterion provides the necessary conditions to determine the stability of the vehicle.

Reference [76] provides another formulation using the Routh table to determine the stability of the transfer function.

Consider the following transfer function,

$$F(s) = s^4 + s^2\,(\beta^2 - 2\alpha^2) + \alpha^4$$

Let $a_0 = 1$, $a_1 = 0$, $a_2 = \beta^2 - 2\alpha^2$, $a_3 = 0$, and $a_2 = \alpha^4$; then, the determinants are equal to: $\Delta_1 = \Delta_2 = \Delta_3 = \Delta_4 = 0$. In the Routh–Hurwitz sense, the system is such that not all the roots have negative real parts which lead to an unstable motion. Consider the mass-spring-damper system in equation (7.17),

$$F(s) = s^2 + {}^{f_v}\!/_{M}\,s + K/M$$

The coefficients are non-zero and are defined as: $a_0 = 1$, $a_1 = f_v/M$, $a_2 = K/M$. The determinants in the Routh–Hurwitz criterion are equal to,

$$\Delta_1 = f_v/M > 0 \rightarrow f_v > 0$$

$$\Delta_2 = \begin{vmatrix} a_1 & a_0 \\ a_3 & a_2 \end{vmatrix} = \begin{vmatrix} {}^{f_v}\!/_{M} & 1 \\ 0 & {}^{K}\!/_{M} \end{vmatrix} = {}^{Kf_v}\!/_{M^2} > 0 \rightarrow Kf_v > 0$$

If $f_v > 0$ and $K > 0$, the system is stable, and the roots of the characteristic equation have negative real parts. This is shown in the previous section when the second order response is explained. The difference between the Routh–Hurwitz criteria and the second order response solution is the calculation of the variables to stabilize the system; but both solutions are used to understand the stability of motion of a body.

7.3.5.3 Eigenvalue problem to determine the stability conditions

Another method to determine the stability of the system is to obtain the eigenvalue problem of the state equations. The state equations are written in matrix form as,

$$\dot{\vec{x}}(t) = \boldsymbol{A}\vec{x}(t) + \boldsymbol{B}\vec{u}\,(t)$$

The transfer function of the state equations is defined as,

$$\vec{X}(s) = (s\boldsymbol{I} - \boldsymbol{A})^{-1}\,\boldsymbol{B}\vec{U}(s)$$

The inverse of a matrix is written as,

$$\boldsymbol{A}^{-1} = \frac{adj(\boldsymbol{A})}{|\boldsymbol{A}|}$$

where *adj*( ) is known as the adjunct of a matrix [**42**], and | | is the determinant of a matrix. The denominator of the inverse of a matrix defines certain characteristics about the solution. If $|A| = 0$, the matrix is singular; otherwise, the matrix is non-singular. The determinant of the inverse function can be written as,

$$|sI - A| = 0 \tag{7.24}$$

Calculating the determinant of equation (7.24), a characteristic equation for the matrix A can be determined. If the Routh–Hurwitz criterion is used, the stability conditions can be determined. If the roots can be determined for the polynomial, the location of the roots defines the stability of the system. On the other hand, the roots can be determined by developing an algorithm for the eigenvalue problem as shown in References [**14**] and [**84**]. The algorithm outputs the roots for the state matrix which can be used to assess the stability conditions for the problem. The algorithm can be easily developed in a computer to determine the eigenvalues for the state matrix. Using the algorithm in Appendix E, the location of the poles in the imaginary plane can be determined to know the stability of motion of the body.

7.3.6 Continuous control formulation

The stability conditions provide the necessary information to understand the mechanical system. In a satellite, it is possible to determine certain conditions to stabilize the attitude and/or translational motion. The stability conditions can be a function of the mass of the satellite, moments of inertia, point mass locations, among other things. By carefully selecting the conditions, it is possible to correct the attitude motion of the satellite; but, the motion of the body cannot be completely stabilized by using the stability criteria. In order to stabilize the body, the actuators are used as a control system.

A control system is basically a mathematical formulation that can be used to correct the translational and/or attitude motion of a body. These mathematical equations are based on gains applied to different measurements such as the position and velocity for the angular and translational motion of a body. One of these control laws based on gains on certain states is called the proportional-integral-derivative (PID) control law. As the name suggests, the control law is used to determine a set of gains that are multiplied to the proportional, integral, and derivative state of a variable. There are different controllers that can be developed depending on the desired transient response and steady-state error [**76**].

There are other control systems that use conditions in the control law to perform a correction and are known as optimal control laws [85]. The optimal control laws are based on minimum (or maximum) criteria of a cost function which is affected by a series of constraints. By combining all these equations, the necessary conditions to obtain an optimal control problem are obtained. Many of the control schemes are determined from optimal control theory. This theory is explained to understand how a problem can be formulated to determine an admissible control function. One interesting concept about this theory is the solution of two point boundary value problems (TPBVP) [86]. TPBVP have two constraints based on the initial and final state variables and time [85]. The difference between the PID control law and optimal control is the end condition. Both control laws have an initial condition that is satisfied. The end conditions for the PID control law are satisfied when the final time is infinity; for the TPBVP, the end conditions for the state variables are defined at the final time. The solutions of TPBVP are demonstrated with two known solutions for a satellite system [87].

This section provides the foundation for the development of control systems for the translational and rotational motion of a body and is an important theory for the development of control systems for space vehicles. In summary, the theory is explained such that it can be easily applied to any known mechanical problem beginning from the linear equations.

7.3.6.1 Control gain laws and pole placement techniques

After the stability of the vehicle is assessed, a control law can be selected based on the design requirements. The control gain laws are based on the gains applied to some state variables. This control law can be written either in terms of the transfer function or the state equation. For these control laws, Figure 7.8 shows the block diagram for the corresponding control system. The objective of the control law is to aid (in a way) the process to achieve the desired response. If a system is commanded to achieve a desired direction or orientation, the controller provides the necessary capability to achieve the desired command by showing an underdamped and/or overdamped response. This desired response can be obtained by placing the correct poles and/or zeros in the complex plane.

There are different schemes to place the poles and/or zeros in the complex plane. These techniques are known as pole placement techniques [76]. This scheme determines the gains to achieve a second order response. The pole placement techniques [88] can also be applied to the state vector system.

Figure 7.8 **Control system for controller design**

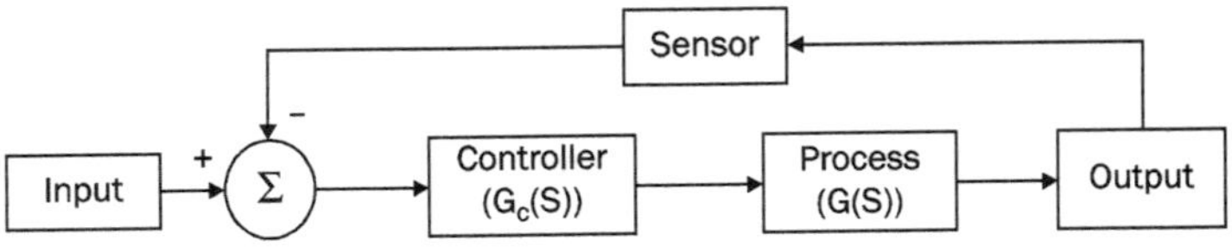

7.3.6.1.1 Pole placement via transfer functions

To begin the development of a control function, the control gain laws based on transfer functions are first explained. The control system that is used for this explanation is the closed-loop system shown in Figure 7.8; and the sensor and actuator models are assumed to be ideal. The process is explained by equation (7.4b) as,

$$\ddot{\theta} = -a_1\,\theta + b_2\,\beta - d_1 \tag{7.25}$$

Assume $a_1 = 2$, $b_2 = 1$, and $d_1 = 0$. The transfer function explaining the process can be written as,

$$G_p(s) = \frac{\Theta(s)}{\mathrm{B}(s)} = \frac{1}{s^2+2} \tag{7.26}$$

By obtaining the roots of the polynomial in the denominator, the poles are located at $\pm\sqrt{2}i$ making the system stable. Remember that $\theta(t)$ and $\beta(t)$, respectively, is the output and input transfer functions in the system. If a step function is input to the system, the solution is shown in Figure 7.9. The solution is stable but is an undamped response.

One of the most complicated steps is to determine which control law is used to obtain the desired command. Table 7.2 shows a comprehensive table containing all the transfer functions for the control laws that can be used to compensate the system [76]. In practice, the commonly used controllers are the proportional-integral (PI), proportional-derivative (PD), and proportional-integral-derivative (PID). The lead, lag, and lead-lag controllers can also be used to compensate the system. The transfer functions are the same, but the locations of the zero and the pole on the LHS of the complex plane are different. For the lag controller, the zeros are close and to the left of the pole. For the lead controller, the pole is more negative and to the left of the controller. In Table 7.2, the controller either improved the steady-state error, the transient response, or both. The steady-state error describes the behavior of the system as time tends to infinity. The main objective of the steady-state error is to determine if the function can be maintained within 2% of the final value.

Figure 7.9 Unit step response without compensation

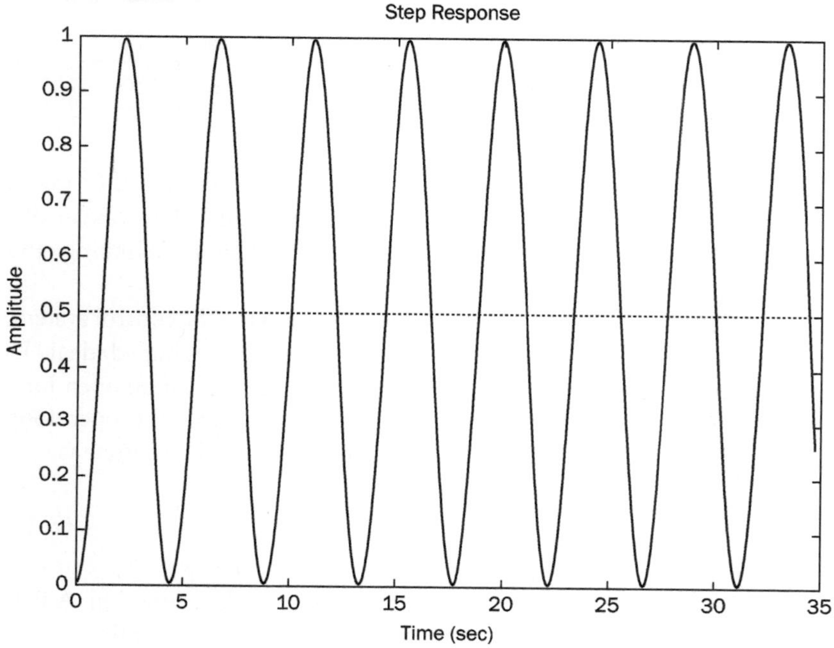

Table 7.2 Control laws for compensation of the process

Control law	Control characteristics	Transfer function $G_c(s)$
Proportional-Integral (PI)	Improve steady-state error	$\frac{K(s+z_1)}{s}$
Lag	Improve steady-state error	$\frac{K(s+z_1)}{(s+p_1)}$
Proportional-Derivative (PD)	Improve transient response	$K(s+z_1)$
Lead	Improve transient response	$\frac{K(s+z_1)}{(s+p_1)}$
Proportional-Integral-Derivative (PID)	Improve transient response and steady-state error	$\frac{K(s+z_1)(s+z_2)}{s}$
Lag-lead	Improve transient response and steady-state error	$\frac{K(s+z_1)(s+z_2)}{(s+p_1)(s+p_2)}$

Figure 7.10 Closed-loop control for the pitch equation

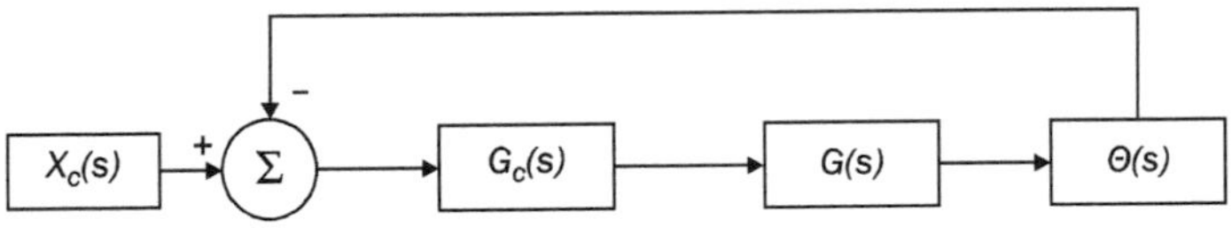

On the other hand, the transient response describes the initial characteristics of the function as described in Section 7.3.5.1. Basically, the controller improves the rise time, percent of overshoot, damping, and natural frequency.

Figure 7.10 shows the closed-loop block diagram for the control system. In Figure 7.10, the sensor and the actuator systems are assumed ideal. To actually analyze the block diagram, the transfer function for an open-loop system is obtained to analyze the poles of the system. The open-loop transfer function [76] for the system in Figure 7.10 can be written as,

$$G_{OL}(s) = \frac{\Theta(s)}{X_c(s)} = \frac{G_c(s)G_p(s)}{1+G_c(s)G_p(s)} \tag{7.27}$$

where $X_c(s)$ is the Laplace transform of the commanding function. A PID controller is used and can be written in the time domain as follows,

$$u(t) = K_P\theta(t) + K_D\dot{\theta}(t) + K_I \int_0^\tau \theta(\tau)\, d\tau \tag{7.28a}$$

Or taking the Laplace transform,

$$G_c(s) = \frac{U(s)}{\Theta(s)} = \frac{K_D s^2 + K_P s + K_I}{s} \tag{7.28b}$$

Equation (7.28) uses a set of gains to describe the pole locations. K_P is the proportional gain, K_D is the derivative gain, and K_I is the integral gain. A PID controller improves the transient response as well as the steady-state error. Because of these characteristics, the PID controller takes care of outside perturbations affecting the motion of the system as shown in equation (7.25).

The PD controller is selected to perform the control analysis. The PD control transfer function is written as,

$$G_c(s) = \frac{U(s)}{\Theta_c(s)} = K_D s + K_P \tag{7.29}$$

Substituting equations (7.26) and (7.29) into equation (7.27), the open loop transfer function can be written as,

$$G_{OL}(s) = \frac{\Theta(s)}{\Theta_c(s)} = \frac{K_D s + K_P}{s^2 + K_D s + (K_P + 2)} \quad (7.30)$$

If the Routh–Hurwitz criterion is used, the control gains provide the following stability condition,

$$K_D > 0 \quad K_P > -2 \quad (7.31)$$

The denominator of equation (7.30) is a second order system. Because of this, the denominator of equation (7.30) can also be described as,

$$F(s) = s^2 + 2\xi\omega_n s + \omega_n^2 \quad (7.32)$$

Comparing equation (7.32) with equation (7.30), the control gains are equal to,

$$K_D = 2\xi\omega_n \quad K_P = -2 + \omega_n^2 \quad (7.33)$$

Equation (7.33) satisfies the condition in equation (7.31). The difference is that equation (7.33) depends on the location of the poles in the LHS of the complex plane; in addition, the transient response of the second-order response can be used to obtain the control gains. To demonstrate the solution of both approaches, the gains selected from equation (7.31) are the following: $K_D = 1$ and $K_p = 2.5$. To calculate the control gains in equation (7.33), $\xi = 1.0$ and $\omega_n = 2$; then, Figure 7.11 shows the solutions. In Figure 7.11, the gains with the Routh–Hurwitz criterion provided a larger final amplitude to the system than with the gains calculated with the second order response. This happens because the gains with the second order response are determined due to the characteristics of the system. By doing this, the poles are placed in a desired location that provides an improvement to the transient response. If the closed-loop response for the PID controller is calculated, an additional pole is multiplied in equation (7.32). The additional pole to be multiplied in equation (7.32) can be written as: $s + z_p$ where z_p is the pole. If the pole is located far to the left in the complex plane, the system improves its steady-state solution.

7.3.6.1.2 Pole placement technique via state vector

The state vector formulation provides a simpler solution when the equations of motion are coupled. For state vectors, the control input function is different as compared to the transfer function. The state vector formulation is written as,

$$\dot{\vec{x}}(t) = \mathbf{A}\vec{x}(t) + \mathbf{B}\vec{u}(t)$$

Figure 7.11 **Control response for the pitch equation**

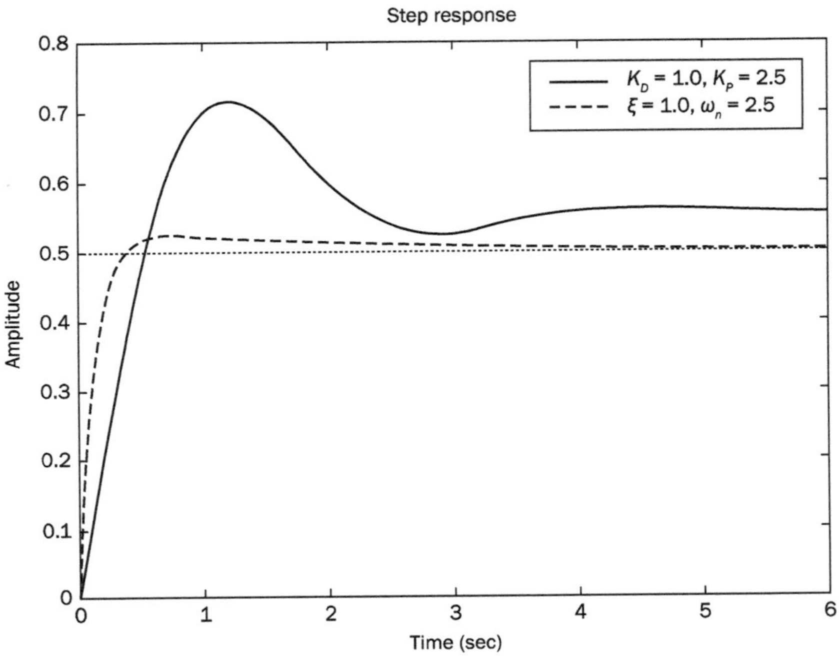

The control input equation that is used to provide a closed-loop control is defined as follows,

$$\vec{u} = -K\vec{x} \tag{7.34}$$

where K is the control gain matrix and has dimensions $m \times n$. Substituting equation (7.34) into the state equation, the closed-loop form of the sate equation is equal to,

$$\dot{\vec{x}}(t) = (A - BK)\vec{x}(t) \tag{7.35}$$

Solving the eigenvalue problem for the matrix between parentheses, the closed loop poles can be determined. A stable solution is obtained by calculating the entries for the control gain matrix. The block diagram for the state vector closed-loop equation is shown in Figure 7.12. Figure 7.12 shows a simplified version in comparison with the transfer function in Figure 7.8. The main advantage of using the state vector formulation is the integration procedure to determine the states.

Continuing with the same pitch equation, equation (7.25) can be written in state vector format as,

Figure 7.12 **Closed-loop block diagram for the state vector equation**

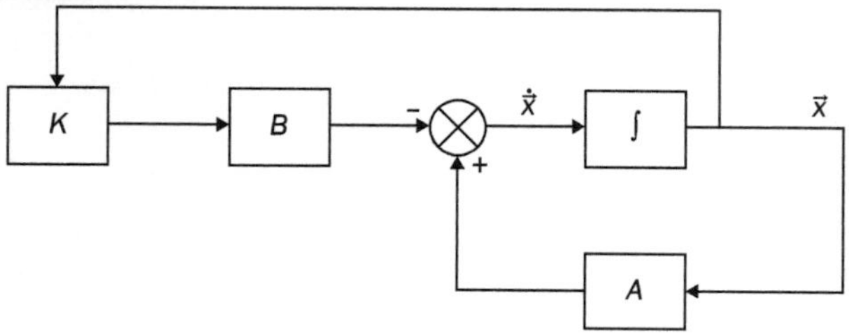

$$\dot{\vec{x}}(t) = \boldsymbol{A}\vec{x}(t) + \boldsymbol{B}\beta(t) + \vec{\psi} \tag{7.36a}$$

where,

$$\boldsymbol{A} = \begin{bmatrix} 0 & 1 \\ -a_1 & 0 \end{bmatrix} \quad \boldsymbol{B} = \begin{bmatrix} 0 \\ b_2 \end{bmatrix} \tag{7.36b}$$

$$\vec{x}(t) = \begin{bmatrix} \theta(t) \\ \dot{\theta}(t) \end{bmatrix} \quad \vec{\psi} = \begin{bmatrix} 0 \\ d_1 \end{bmatrix} \tag{7.36c}$$

Using $a_1 = 2$, $b_2 = 1$, and $d_1 = 0$, the closed-loop matrix can be written as,

$$\boldsymbol{A}_{CL} = \boldsymbol{A} - \boldsymbol{BK} = \begin{bmatrix} 0 & 1 \\ -2 & 0 \end{bmatrix} - \begin{bmatrix} 0 \\ 1 \end{bmatrix} \begin{bmatrix} K_{11} & K_{12} \end{bmatrix} = \begin{bmatrix} 0 & 1 \\ -2 - K_{11} & -K_{12} \end{bmatrix} \tag{7.37}$$

Solving the eigenvalue problem for the closed loop matrix, the characteristic equation can be written as,

$$s^2 + K_{12}s + (K_{11} + 2) = 0 \tag{7.38}$$

Equation (7.38) is the same as the denominator of the open loop transfer function in equation (7.30). The results presented in Figure 7.11 are the same if the same weights are used. As an exercise, the solution of equation (7.14) using the Routh–Hurwitz criteria and/or the state vector format is left to the reader.

There is another pole placement technique to provide a stable control system. This technique is called the Ackermann's formula. The Ackermann's formula [76] uses the controllability matrix to determine the gain matrix. The controllability matrix is used to determine if the system can be controllable such that a stable motion can be obtained. The controllability matrix is defined as,

$$\boldsymbol{C} = [\boldsymbol{B} \quad \boldsymbol{AB} \quad \boldsymbol{A}^2\boldsymbol{B} \quad \ldots \quad \boldsymbol{A}^{n-2}\boldsymbol{B} \quad \boldsymbol{A}^{n-1}\boldsymbol{B}] \tag{7.39}$$

If the controllability matrix is linearly independent, the system is controllable. Using this fact, the Ackermann's formula can be used to determine the gain matrix, $\boldsymbol{K}$. The first step to obtain the gain matrix is to define a desired characteristic polynomial from the desired pole locations as,

$$F(s) = s^N + a_{N-1}s^{N-1} + \ldots + a_0$$

Second, evaluate the state matrix into the previous equation,

$$\boldsymbol{F}(\boldsymbol{A}) = \boldsymbol{A}^N + a_{N-1}\boldsymbol{A}^{N-1} + \ldots + a_0\boldsymbol{I}$$

Finally, calculate the gain matrix with the following equation [76],

$$\boldsymbol{K} = [\,0 \quad 0 \quad 0 \quad \ldots \quad 0 \quad 1]\boldsymbol{C}^{-1}\boldsymbol{F}(\boldsymbol{A}) \tag{7.40}$$

Equation (7.40) is only used for multi-input-single-output systems (MISO) so that the matrix inverse of equation (7.39) is obtained.

In equation (7.38), the solution of the state vector is a PD controller. To add the integral term, an additional state variable is added containing the integral of the state variables. This new state variable can be written as,

$$y(t) = \int x_1(\tau)\, d\tau \rightarrow \dot{y}(t) = x_1(t)$$

Then,

$$\dot{\vec{z}}(t) = \boldsymbol{A}\vec{z}(t) + \boldsymbol{B}\beta(t) + \vec{\psi} \tag{7.40a}$$

where,

$$\boldsymbol{A} = \begin{bmatrix} 0 & 1 & 0 \\ 0 & 0 & 1 \\ 0 & -a_1 & 0 \end{bmatrix} \qquad \boldsymbol{B} = \begin{bmatrix} 0 \\ 0 \\ b_2 \end{bmatrix} \tag{7.40b}$$

$$\vec{z}(t) = \begin{bmatrix} y(t) \\ \theta(t) \\ \dot{\theta}(t) \end{bmatrix} \qquad \vec{\psi} = \begin{bmatrix} 0 \\ 0 \\ d_1 \end{bmatrix} \tag{7.40c}$$

Assuming $a_1, = 2$, $b_2 = 1$, and $d_1 = 0$, the closed loop matrix is equal to,

$$\boldsymbol{A}_{CL} = \boldsymbol{A} - \boldsymbol{BK} = \begin{bmatrix} 0 & 1 & 0 \\ 0 & 0 & 1 \\ -K_{11} & -2-K_{12} & -K_{13} \end{bmatrix} \tag{7.41}$$

Solving the eigenvalue problem, the characteristic equation for equation (7.41) is written as,

$$F(s) = s^3 + K_{13}s^2 + (2 + K_{12})s + K_{11} = 0 \quad (7.42)$$

Using the Routh–Hurwitz criteria, the sufficient conditions to determine the gain matrix for a stable motion are expressed as,

$$K_{13} > 0 \qquad K_{12} > \frac{K_{11}}{K_{13}} - 2 \quad (7.43)$$

Selecting $K_{11} = 1.5$, $K_{12} = 4.0$, and $K_{13} = 1.5$ and using the following initial conditions,

$$\vec{z}_0(t) = \begin{bmatrix} 0 \\ 1 \\ 0 \end{bmatrix} \quad (7.44)$$

Figure 7.13 shows the output for the three state variables. In Figure 7.13, the system comes into a steady-state solution in approximately 6 seconds. Once the system reaches 10 seconds, the system has completely settled to a zero state. The integral term does not settle at 10 seconds because the integral term can be considered as the accumulation of the angular

Figure 7.13 State vector solution for the pitch equation

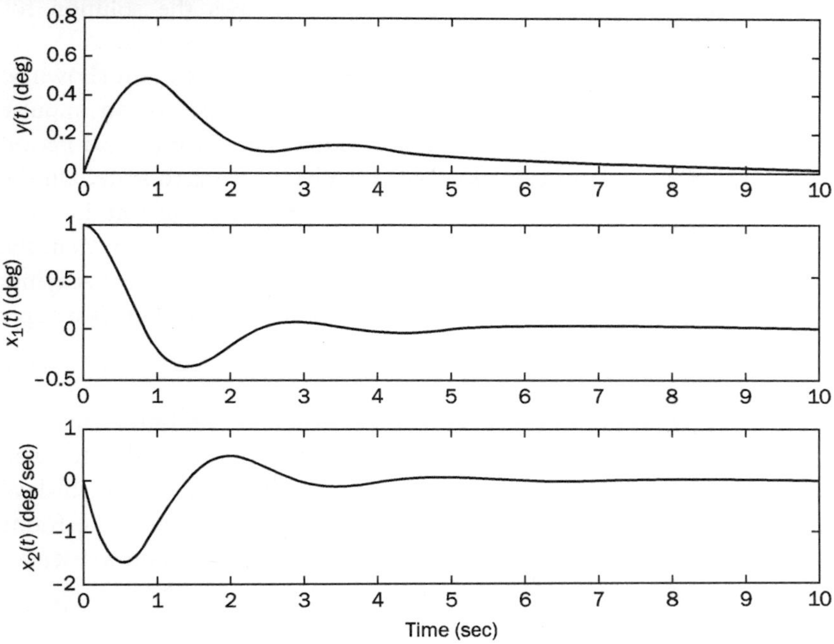

Figure 7.14 **Control input function for the pitch equation**

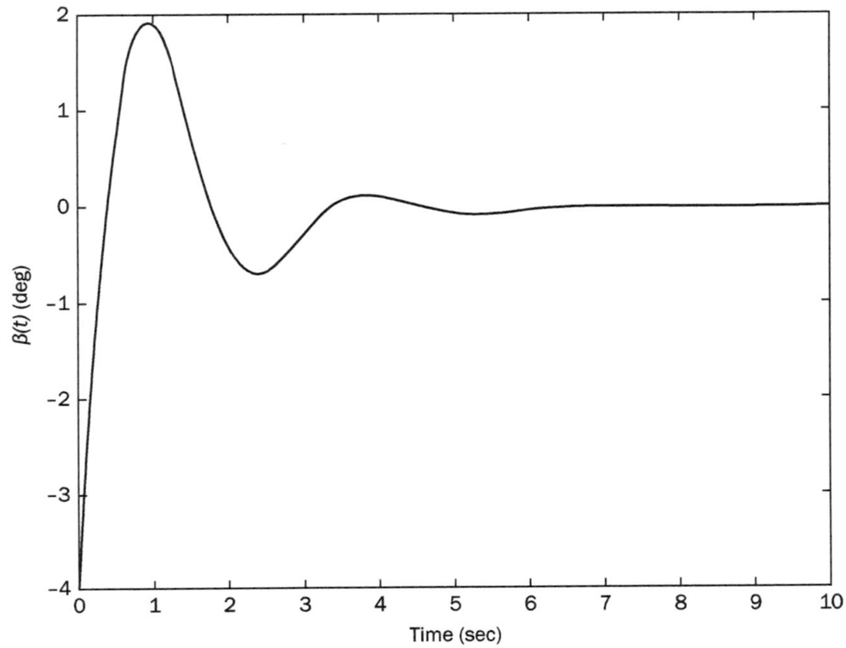

position function through time. Figure 7.14 shows the solution for equation (7.34) which also settles in approximately 6 seconds.

Up to this point, there is a difference between the solution shown in Figure 7.11 and 7.13. Figure 7.13 shows the response for an impulse [89]. An impulse is a large force applied to the system for a short period of time; then the system is released to come back to a zero state. On the other hand, Figure 7.11 shows the response for a step function. In order to demonstrate the solution for a desired state in state vector format, the state vector equations have to be solved using optimal control schemes [87] which lead to two point boundary value problems. This is a topic that is introduced in the following subsection.

7.3.6.1.3 Quaternion feedback control law with integral terms

The PID control law explained for angular positions and rates also applies for quaternions. Before explaining the quaternion control law, the small angle assumptions are explained for quaternions. The frame rotations of a body are explained in terms of a unit quaternion written as,

$$q = \cos\frac{\theta}{2} + \hat{q}\sin\frac{\theta}{2} \tag{7.45}$$

where $\hat{q}$ is a unit vector associated with the orientation of the rotation. Using the small angle approximations, the unit quaternion for frame rotations is approximated to,

$$q \approx 1 + \hat{q}\frac{\theta}{2} \tag{7.46}$$

where $q_4 = 1$, and $\vec{q} = \hat{q}\frac{\theta}{2}$. This approximation is used to determine the gains of a control system [47] [90].

The quaternion feedback laws are similar to the PID control laws explained in previous sections. The PD quaternion control law can be expressed as follows [47],

$$\vec{u}(t) = K_P\vec{q} + K_D\vec{\omega} \tag{7.47}$$

Adding the integral term, the PID control law can be written as follows,

$$\vec{u}(t) = K_P\vec{q} + K_D\vec{\varpi} + K_I\int_0^t \vec{q}(\tau)\, d\tau \tag{7.48}$$

Equation (7.47) is used to correct the transient response of the system while equation (7.48) can be used to correct the steady-state response. In many attitude problems, the system can be commanded to reach a desired attitude pointing which is determined between the difference of the commanding and actual quaternion. This error quaternion is written as follows [90],

$$\begin{bmatrix} q_{1e} \\ q_{2e} \\ q_{3e} \\ q_{4e} \end{bmatrix} = \begin{bmatrix} q_{4c} & q_{3c} & -q_{2c} & -q_{1c} \\ -q_{3c} & q_{4c} & q_{1c} & -q_{2c} \\ q_{2c} & -q_{1c} & q_{4c} & -q_{3c} \\ q_{1c} & q_{2c} & q_{3c} & q_{4c} \end{bmatrix} \begin{bmatrix} q_1 \\ q_2 \\ q_3 \\ q_4 \end{bmatrix} \tag{7.49}$$

where q_e and q_c are the error quaternion and the commanding quaternion, respectively. If $q_{4c} = 1$ and $\vec{q}_c = 0$, the error quaternion is the same as the actual quaternion, and the PID quaternion law is used. When the commanding quaternion is defined by another attitude motion, the PID quaternion law can be written as follows,

$$\vec{u}(t) = K_P\vec{q}_e + K_D\vec{\varpi}_e + K_I\int_0^\tau \vec{q}_e(\tau)\, d\tau \tag{7.50}$$

where K_P, K_D and K_I, respectively, are the principal diagonal matrices containing the proportional, derivative, and integral gains for the roll, pitch, and yaw angles. Once the small angle assumptions are obtained, the Routh–Hurwitz criteria or the transfer function approach for pole placement techniques can be used to determine the control gains for every rotation. As shown in the previous equations, the quaternions can be used to simplify the calculation process for the attitude control of the system. On the other hand, the quaternion is also used to solve two point boundary value problems as shown in Reference [91].

7.3.6.2 Optimal control

As mentioned earlier, a control system based on the state equations cannot be used to input a desired state for the system. In order to obtain a desired state, a two point boundary value problem (TPBVP) should be solved. A TPBVP has a set of initial and final conditions that should be satisfied. There are different techniques that can be applied such as the Euler–Lagrange equations [85], the Pontryagin minimum principles [92], and the Bolza problem [93] to solve the TPBVP. Any of these techniques can provide the control input function that can satisfy the initial and final states.

This section shows three different techniques that can be used to solve the problem. In addition, two commonly used examples in optimal controls are presented to demonstrate some of these mathematical techniques. The final topic that is discussed in this section is the linear quadratic regulator (LQR) technique which is commonly used to obtain an active control scheme based on the state equations.

7.3.6.2.1 Optimal control problem, cost functional, and constraints

Before explaining the problems and techniques, it is necessary to explain the composition of an optimal control problem. An optimal control problem contains two principal characteristics: 1) the cost function, and 2) the constraints. These two characteristics are explained here to provide the reader with an understanding of the different problems that can be established.

Assume that the equations of motion can be written in terms of the state vectors as,

$$\dot{\vec{x}}(t) = \vec{f}(\vec{x}(t), \vec{u}(t), t) \tag{7.51}$$

$\vec{x}(t)$ and $\vec{u}(t)$ is the state and control vectors containing i and k variables, respectively. Also, assume that f_i is twice differentiable composed of $i = 1,2,3, \ldots, n$ first order differential equations. The control function must satisfy the physical constraints on the system such that $u_k(t)$ $k = 1,2,3, \ldots, r$ can be of the form $\vec{u}(t) \in U$; and U is a closed set in an

k dimensional space which is the size of the number of control input functions. When $r = 1$, the control function is considered to be subjected to saturation. When the control input function is saturated, it requires that $|\vec{u}(t)| \leq 1$. For this case, U has closed interval $[-1,1]$ on the real axis. In addition to the condition in $\vec{u}(t)$ it is required that each $u_k(t)$ is a piecewise continuous function of time. Any control function that satisfies these two conditions is considered as an admissible control function.

The optimal control problem is established as follows: *Find $\vec{u}(t)$ such that the system is taken in accordance with equation (7.40) from some initial state $\vec{x}(t_0) = \vec{x}_0$ to some final state $\vec{x}(t_1) = \vec{x}_1$ such that the system performance is optimum in some specified sense.* This is the basic definition for an optimal control problem, but the optimum performance of a system is measured through a cost function described as,

$$\mathcal{J} = \int_{1_{t_0}}^{t} \mathcal{F}(\vec{x}, \vec{u}, t)\, dt \tag{7.52}$$

The system performance is optimum when equation (7.52) becomes a minimum with respect to all the admissible controls, $\vec{u}(t)$; and the particular control $\vec{u}^*(t)$ that realizes this minimum control function (if it exists) will be called the optimal control of the problem.

The total time transition, $T = t_1 - t_0$ from initial to final state can either be an unknown quantity or specified constraint in the problem. Because of these initial and end conditions, the problem is known as a two point boundary value problems (TPBVP). These boundary value problems are not only constrained by end conditions; they can be constrained depending on the desired paths and functions. Reference [**94**] provides a comprehensive look at the problem and provides a table with the different optimal problems that can be generated. There are three main problems in optimal control applications that are shown graphically in Figure 7.15 and are explained as follows:

Fixed End Conditions – This is the principal condition for TPBVP and is shown in Figure 7.15a. For fixed end conditions, the problem has to satisfy the initial time and final time desired states of the system; in other words, $\vec{x}(t_0) = \vec{x}_0$ and $\vec{x}(t_1) = \vec{x}_1$. The problem can take any path to reach the final state knowing the initial condition.

One Fixed Condition – This condition considers that if the initial condition is known and the final time is known, the final state can be free. The term 'free' refers to that either the time or the states can have any value and is described in Figure 7.15b. Mathematically, this problem can be represented as follows: $\vec{x}(t_0) = \vec{x}_0$ and $\vec{x}(t_1) = \text{free}$ or $\vec{x}(t_0) = \text{free}$ and $\vec{x}(t_1) = \vec{x}_1$.

Free Conditions – These conditions consider that the initial state and final state can be free as well as the time. These conditions can be written

Figure 7.15 Possible end conditions for optimal control problem

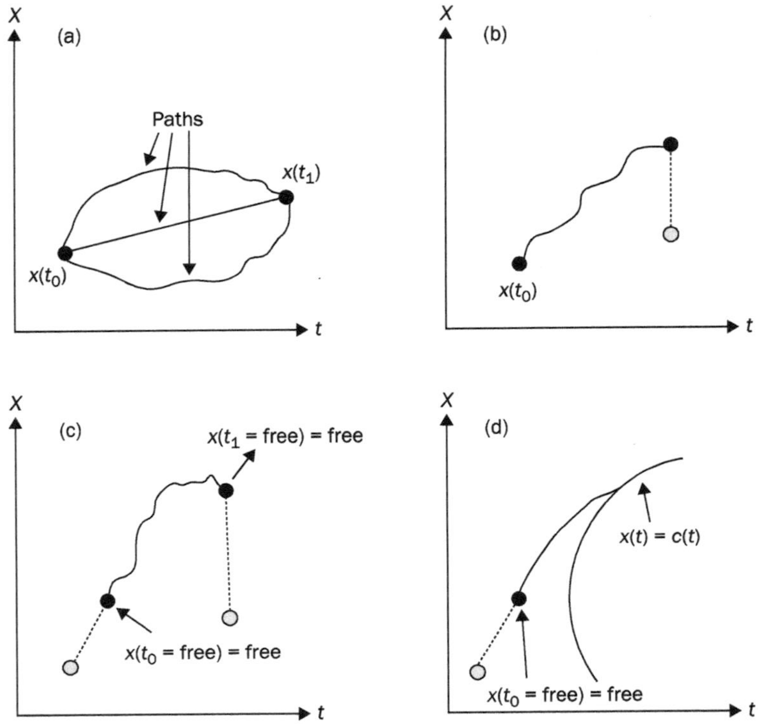

as: $\vec{x}(t_0 = \text{free}) = \text{free}$ and $\vec{x}(t_1 = \text{free}) = \text{free}$. There are cases in which either the initial and final condition can lie on a curve. If one of the end conditions lies on a curve, it can be represented as $\vec{x}(t) = c(t)$. Both conditions are shown in Figure 7.15c and d.

There are constraint equations that should be satisfied depending on the optimal control problem. The main objective of the optimal control techniques is to minimize (or maximize) the cost function constrained to the state equation and the initial and/or the end conditions of the problem.

7.3.6.2.2 Euler–Lagrange equation

The Euler–Lagrange equation is used to minimize the cost function depending on the conditions of the problem. The solution obtained from these equations are called extremals [85] because it calculates the minimum path that the system has to take to obtain the desired or final state.

The Euler–Lagrange equation is developed through the calculus of variations [**11**]. Assuming the cost function written in equation (7.52), the Euler–Lagrange equations can be written as,

$$\frac{\partial F}{\partial x} - \frac{d}{dt}\left(\frac{\partial F}{\partial \dot{x}}\right) = 0 \tag{7.53}$$

In addition to the equations obtained from equation (7.53), it is necessary to use another set of equations to solve the problem. These equations are obtained from the transversality conditions. A transversality condition is determined from the end conditions of the problem and can be used to actually obtain the solution of the problem. The transversality conditions used for every condition in the problem are written as follows:

a. Fixed End Conditions – Solve equation (7.53) for every $t \in [t_0, t_1]$ and the end conditions $\vec{x}(t_0) = \vec{x}_0$ and $\vec{x}(t_1) = \vec{x}_1$.

b. One Fixed Condition – Solve equation (7.53) for every $t \in [t_0, t_1]$ Depending on the fixed condition, the transversality condition can be written as,

$$\left.\frac{\partial F}{\partial \dot{x}}\dot{x}\right|_{t=t_0} = 0 \text{ or } \left.\frac{\partial F}{\partial \dot{x}}\dot{x}\right|_{t=t_1} = 0 \tag{7.54}$$

c. Free Conditions – Solve equation (7.53) for every $t \in [t_0, t_1]$. Depending on the fixed condition, the transversality condition can be written as,

$$\left.\frac{\partial F}{\partial \dot{x}}\dot{x}\right|_{t=t_0} = 0 \text{ or } \left.\frac{\partial F}{\partial \dot{x}}\dot{x}\right|_{t=t_1} = 0 \tag{7.55a}$$

And,

$$\left.F - \frac{\partial F}{\partial \dot{x}}\dot{x}\right|_{t=t_0} = 0 \text{ or } \left.F - \frac{\partial F}{\partial \dot{x}}\dot{x}\right|_{t=t_1} = 0 \tag{7.55b}$$

If the end conditions are constrained to a surface, equation (7.55b) is written as,

$$\left.F + \frac{\partial F}{\partial \dot{x}}(c(t) - \dot{x})\right|_{t=t_0} = 0 \text{ or } \left.F + \frac{\partial F}{\partial \dot{x}}(c(t) - \dot{x})\right|_{t=t_1} = 0 \tag{7.55c}$$

where $c(t)$ is the surface on which one of the end conditions is constrained.

The Euler–Lagrange equation and the transversality conditions can be used to minimize the cost function; in this way a minimum path can be obtained for the system.

7.3.6.2.3 Pontryagin maximum (or minimum) principle

Corresponding to the control problem explained in Section 7.3.6.2.1, the Hamilton–Jacobi equation [11] is introduced to provide a solution for a conservative system by obtaining an admissible set of control input functions. Knowing the cost function in equation (7.52) and the state vector system in equation (7.51), the Hamilton–Jacobi equation can be written as follows,

$$\mathcal{H}(\vec{x}, \vec{u}, \vec{\lambda}, t) = \vec{\lambda}^T \vec{f} - \mathcal{F}(\vec{x}, \vec{u}, t) \tag{7.56}$$

where $\vec{\lambda}$ is the co-state variable. In Reference [11], $\vec{\lambda}$ is also known as a Lagrange undetermined multiplier. This multiplier is used to relate the non-holonomic systems. With these multipliers, it is possible to maximize (or minimize) the cost function based on the equations of motion. If the Euler–Lagrange equations are used to obtain the maximum problem with equation (7.56), the necessary conditions can be written as,

$$\dot{\vec{\lambda}} = -\frac{\partial \mathcal{H}}{\partial \vec{x}} \tag{7.57a}$$

$$\dot{\vec{x}} = \frac{\partial \mathcal{H}}{\partial \vec{\lambda}} \tag{7.57b}$$

$$\frac{\partial \mathcal{H}}{\partial \vec{u}} = 0 \tag{7.57c}$$

The boundary condition depends on the problem to be solved by the designer. Even if $\vec{u}(t)$ is given as a function of time, equation (7.57a) is not uniquely defined for $\vec{\lambda}(t)$ because the boundary conditions of $\vec{\lambda}(t)$ have not been specified.

Now, let $\vec{u}(t)$ be some admissible control, and $\vec{x}(t)$ be the solution of equation (7.51) emanating from the initial condition; then, Pontryagin stated [92] the maximum principle as: *If $\vec{u}^*(t)$ is the optimal control, then these exists a vector $\vec{\lambda}^*(t)$ satisfying equation (7.57a) such that at every time, $t_0 \leq t \leq t_1$*

$$\mathcal{H}(\vec{x}^*; \vec{u}^*; \vec{\lambda}^*; t) \geq \mathcal{H}(\vec{x}, \vec{u}, \vec{\lambda}, t) \tag{7.58}$$

with respect to all admissible control; in other words,

$$\mathcal{H}(\vec{x}^*; \vec{u}^*; \vec{\lambda}^*; t) = \min_{\vec{u} \in U} \mathcal{H}(\vec{x}, \vec{u}, \vec{\lambda}, t)$$

In general, the maximum principle gives only a necessary condition for optimality of the admissible control function, $\vec{u}^*(t)$. In most practical cases, it gives sufficient information to uniquely define the optimal control problem.

There are problems in which the cost function is to be minimized. This problem is treated in the same manner with exception of the Hamiltonian. Because of this condition, the Hamiltonian for minimum principles are written as,

$$\mathcal{H}(\vec{x}, \vec{u}, \vec{\lambda}, t) = \mathcal{F}(\vec{x}, \vec{u}, t) - \vec{\lambda}^T \vec{f} \tag{7.59a}$$

And the Pontryagin problem for minimum conditions are written as,

$$\mathcal{H}(\vec{x}^*; \vec{u}^*; \vec{\lambda}^*; t) = \min_{\vec{u} \in U} \mathcal{H}(\vec{x}, \vec{u}, \vec{\lambda}, t)$$

7.3.6.2.4 Bolza problem

The Pontryagin maximum (or minimum) principle can be used to determine the solution for TPBVP. This problem refers to the solution of the fixed end constraints. When there are other constraints involved in the optimal problem such as weights on the end conditions, surfaces, and desired final times, the Bolza problem takes into account all those effects on the Hamiltonian.

To incorporate these conditions, the cost function can be reformulated as follows,

$$\mathcal{J} = \mathcal{M}[\vec{x}(t_1), t_1] + \int_{t_0}^{t_1} \mathcal{F}(\vec{x}, \vec{u}, t)\, dt \tag{7.60}$$

where $\mathcal{M}[\vec{x}(t_1), t_1]$ is a condition associated with a desired state that should be achieved by the system and is measured through the cost function to maintain a maximum (or minimum) solution. Knowing the state vector system written in the form of equation (7.60) and the number of constraints in the problem, the Hamilton–Jacobi equation is written as,

$$\mathcal{H}(\vec{x}, \vec{u}, \vec{\lambda}, t) = \mathcal{F}(\vec{x}, \vec{u}, t) + \vec{\lambda}^T \vec{f} \tag{7.61}$$

where $u(t) \in U$ and $t \in [t_0, t_1]$. The solution of equation (7.61) through the Euler–Lagrange equations provides the same solution as in equations (7.57). Using the same conditions in the Pontryagin minimum principles, a set of admissible functions are determined to minimize the Hamilton–Jacobi equation. Up to this point, this problem is very similar to the one stated in equation (7.58). The main difference is the initial or end constraints to the problem that can be written in the following form,

$$\vec{\mathcal{N}}[\vec{x}(t_1), t_1] = 0 \tag{7.62}$$

Equation (7.62) describes the conditions that are established depending on the desired surface to be reached by the system at a desired time. $\vec{\mathcal{N}}$ has a dimension of q. Including the differential equations, the conditions on $\vec{\mathcal{N}}$, the end constraints on the cost function, and the co-state variables, the number of unknowns are $2n + q + 1$. In order to solve the problem, the following boundary conditions can be used,

$$\vec{x}(t_0) = \vec{x}_0 \tag{7.63a}$$

$$\vec{\mathcal{N}}\,[\vec{x}(t_1), t_1] = 0 \tag{7.63b}$$

$$\frac{\partial \mathcal{M}}{\partial t_1} + \frac{\partial \mathcal{N}^T}{\partial t_1}\vec{\vartheta} + H = 0 \; at \; t = t_0 \; or \; t = t_1 \tag{7.63c}$$

$$\frac{\partial \mathcal{M}}{\partial \vec{x}} + \frac{\partial \mathcal{N}^T}{\partial \vec{x}}\vec{\vartheta} - \vec{\lambda} = 0 at \; t = t_0 \; or \; t = t_1 \tag{7.63d}$$

Equations (7.63) provide the necessary conditions to determine the admissible control functions to minimize the cost function. $\vec{\vartheta}$ is a column vector of unknown variables that are related to the end conditions of the system. t_1 is also assumed to be unkown. In addition, these equations provide the number of known functions to solve the optimal control problem.

7.3.6.2.5 Minimum time problem

It is necessary to solve some examples to actually understand the different optimal control theories. A minimum time problem happens when the cost function is written as,

$$\mathcal{J} = \int_0^{t_1} 1 \; dt = t_1 \tag{7.64}$$

Writing the roll equation in state vector format, equation (7.4a) can be defined as,

$$\dot{\vec{x}}(t) = \boldsymbol{A}\vec{x}(t) + \boldsymbol{B}u(t) \tag{7.65a}$$

where,

$$\boldsymbol{A} = \begin{bmatrix} 0 & 1 \\ 0 & 0 \end{bmatrix} \qquad \boldsymbol{B} = \begin{bmatrix} 0 \\ T_{max}/J_1 \end{bmatrix} \tag{7.65b}$$

$$\vec{x}(t) = \begin{bmatrix} \varphi(t) \\ \dot{\varphi}(t) \end{bmatrix} \qquad u(t) = \frac{T_{RoCS}}{T_{max}} \qquad \vec{x}(t_0) = \begin{bmatrix} 1 \\ 1 \end{bmatrix} \qquad \vec{x}(t_1) = \begin{bmatrix} 0 \\ 0 \end{bmatrix} \tag{7.65c}$$

where T_{max} is the maximum thrust developed by the motors, J_1 is the moment of inertia about the roll axis, and $|u(t)| \leq 1$. T_{max}/J_1 is assumed to be equal to 2. The cost function in equation (7.64) resembles the Bolza problem because of the constraint in the cost function. In addition, t_1 is assumed to be unkown. On the contrary, the Pontryagin maximum (or minimum) principle is used if the final time is known [95].

The Hamilton–Jacobi equation for equation (7.64) and (7.65) is written as,

$$\mathcal{H}(\vec{x}, \vec{u}, \vec{\lambda}, t) = \vec{\lambda}^T (\boldsymbol{A}\vec{x}(t) + \boldsymbol{B}u(t)) \tag{7.66}$$

Applying the minimum conditions, the co-state variables (equation (7.57a)) are equal to,

$$\dot{\lambda}_1 = 0 \qquad \dot{\lambda}_2 = -\lambda_1 \tag{7.67a}$$

The state equations are defined as,

$$\dot{x}_1 = x_2 \qquad \dot{x}_2 = 2u(t) \tag{7.67b}$$

The minimum of the control input function (equation (7.57c)) is expressed as,

$$\frac{\partial \mathcal{H}}{\partial u} = \boldsymbol{B}^T\vec{\lambda} = 2\,\lambda_2 \tag{7.68}$$

Note that $\mathcal{H}$ is a maximum if the term $2\lambda_2 > 0$ for $t \in [0, t_1]$; hence, the admissible control function that provides the minimum for the cost function is written as,

$$u^*(t) = sgn\,(2\lambda_2) = \begin{Bmatrix} 1 & 2\lambda_2 > 0 \\ -1 & 2\lambda_2 < 0 \end{Bmatrix} \tag{7.69}$$

The initial conditions for the state equations are known, but the conditions for the co-state variables are unknown. Because of this restriction, one more condition is required and is obtained from equation (7.63c) at the initial time as follows,

$$1 + \lambda_1 x_2 + 2\lambda_2 u = 0 \tag{7.70}$$

Equation (7.70) can be evaluated either at t_0 or t_1. Equation (7.63d) cannot be used because it provides a nonsense solution.

The co-state equations can be integrated because they are not dependent on the control input function; thus,

$$\lambda_1 = K_1 \qquad \lambda_2 = -K_1 t + K_2 \tag{7.71}$$

The state equations are integrated depending on the control input function. Evaluating equation (7.70) at t_0 with equations (7.71) and using $u(t) = -1$, the following relation can be obtained,

$$K_1 = 2K_2 - 1 \tag{7.72}$$

Substituting the relation in equations (7.71), the co-state equations become,

$$\lambda_1 = 2K_2 - 1 \qquad \lambda_2 = K_2(1 - 2t) + t \tag{7.73}$$

At some point, λ_2 in equation (7.73) is equal to zero. This crossing through zero creates a switching in the control input function from –1 to 1. The time at which the value of the control input function changes is called the switching time. From equation (7.73), the switching time is defined as,

$$t_s = \frac{K_2}{2K_2 - 1} = \frac{K_1 + 1}{2K_1} \tag{7.74}$$

Integrating equations (7.67b) for $u(t) = -1$ and using the initial conditions, the state equations are equal to,

$$x_1(t) = -t^2 + t + 1 \tag{7.75a}$$

$$x_2(t) = -2t + 1 \tag{7.75b}$$

At the switching time, the end state conditions become,

$$x_1(t_s) = -t_s^2 + t_s + 1 \tag{7.76a}$$

$$x_2(t_s) = -2t_s + 1 \tag{7.76b}$$

Equations (7.76) are the initial conditions for the system when the control input function is switched to one. Integrating equations (7.67b) and using as initial condition equations (7.76), the state equations for the new control input equals to,

$$x_1(t) = t^2 + t - 2t_s^2 - 2t_s + 1 \tag{7.77a}$$

$$x_2(t) = 2t - 4t_s + 1 \tag{7.77b}$$

There is one unknown and is the switching time. Evaluating equation (7.77b) at $t = t_1$ and using the end condition, equation (7.77b) defines t_1 as,

$$t_1 = 2t_s - 1/2 \tag{7.78}$$

Substituting equation (7.78) into equation (7.77a) evaluated at $t = t_1$, the polynomial describing the switching time is equal to,

$$2t_s^2 - 2t_s + 3/4 = 0 \tag{7.79}$$

Solving the quadratic polynomial for the switching time, $t_s = (2 \pm \sqrt{10})/4 \approx$ 1.29 sec; and substituting the solution for t_s into equation (7.78), $t_1 = 2.08$ sec. The admissible control function that minimizes the control input function can be written as,

$$u^*(t) = \begin{Bmatrix} -1 & 0 \le t_s < 1.29 \\ 1 & 1.29 \le t_s < 2.08 \\ 0 & t_s \ge 2.08 \end{Bmatrix} \tag{7.80}$$

Using equation (7.80) as the control input function, the integration of equation (7.67b) is shown in Figure 7.16. The vehicle moves through two arcs that are caused by the change of the control input function. For $0 \le t_s < 1.29$, the system decreases the angular position and velocity of the vehicle through time. Once the angular position and rate crosses to the negative side, the system switches the control input function at 1.29 seconds. By switching the input function, the angular position and velocity of the satellite returns to zero as required by the end condition.

Figure 7.16 Phase plane relation for minimum time control

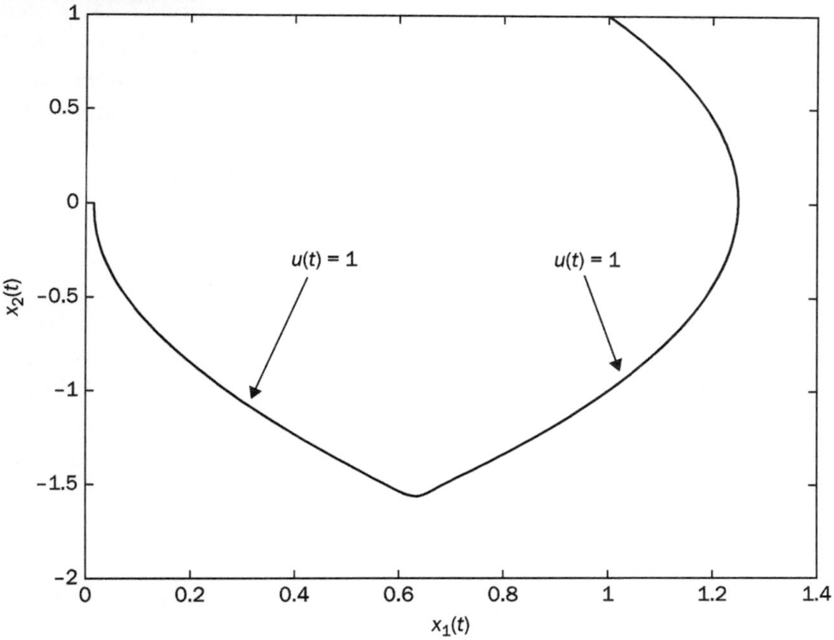

Figure 7.17 Optimal control relation with respect to co-state variable

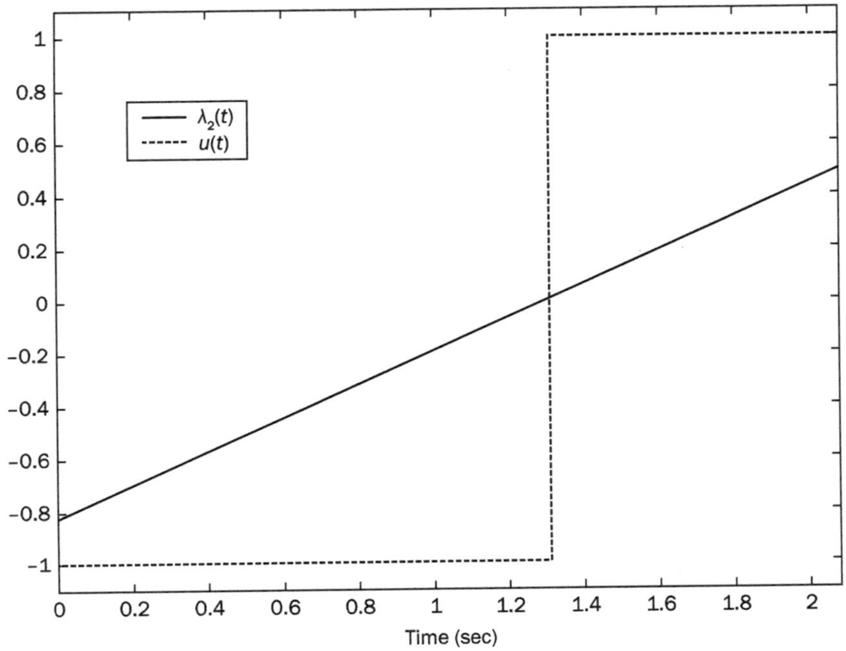

In order to understand the control input function, equation (7.74) is used to calculated the constants for the co-state variables; thus, $K_1 = -0.632$ and $K_2 = -0.82$. Figure 7.17 clearly shows the relation between λ_2 in equation (7.71) and $u(t)$ in equation (7.80). This solution demonstrates the capability of using optimal control systems to satisfy the initial and end conditions in a problem. The optimal controller shown here is known as a bang-bang controller.

7.3.6.2.6 Minimum gas consumption problem

The minimum gas consumption problem can be stated as follows: *Given some initial state $x(t_0)$, apply some control torques to the vehicle via thrusters so that at some time $t > 0$, the final state is $x(t_1) = 0$ and the gas expenditure is zero*. To understand the gas consumption in a thruster, the thrust equation explained in Section 6.2.3 is used. Equation (7.1g) defines the thrust equation in terms of the mass flow rate as,

$$T_{RoCS} = -gI_{SP}^{RoCS}\dot{m}_{RoCS}$$

By controlling the mass flow rate, the amount of force applied can be controlled; then, the control equation can be written as,

$$T_{RoCS}(t) = -gI_{SP}^{RoCS}\,\dot{m}_0 u(t) \tag{7.81a}$$

where,

$$u(t) = \frac{\dot{m}_{RoCS}(t)}{\dot{m}_0} \tag{7.81b}$$

$\dot{m}_0$ is the maximum flow rate that can be achieved through the exit nozzle of a thruster [10]. The cost functional that minimizes the amount of thrust consumed in a thruster can be written as,

$$\mathcal{J} = \int_0^{t_1} u(t)\,dt \tag{7.82}$$

t_1 is assumed a fixed value. Using equation (7.65) and the cost function in equation (7.82), the Hamiltonian that can be used to maximize the thrust consumption can be written as,

$$\mathcal{H} = \lambda^T(\mathbf{A}\mathbf{x} + \mathbf{B}u) - u(t) \tag{7.83}$$

where,

$$\boldsymbol{A} = \begin{bmatrix} 0 & 1 \\ 0 & 0 \end{bmatrix} \quad \boldsymbol{B} = \begin{bmatrix} 0 \\ -gI_{SP}^{RoCS}\,\dot{m}_0 / J_1 \end{bmatrix}$$

Using the maximum conditions shown in equation (7.57a), the co-state variable differential equations are equal to,

$$\dot{\lambda}_1 = 0 \quad \dot{\lambda}_2 = -\lambda_1 \tag{7.84a}$$

And,

$$\lambda_1 = K_1 \quad \lambda_2 = -K_1 t + K_2 \tag{7.84b}$$

where K_1 and K_2 are unknown constants. The maximum consumption of thrust is obtained through equation (7.57c) as,

$$\frac{\partial \mathcal{H}}{\partial u} = -2\lambda_2 - 1 \tag{7.85}$$

For $0 \le t \le t_1$, the maximum principle to the Hamiltonian provides the following admissible control function,

$$u^*(t) = \begin{Bmatrix} -1 & 2\lambda_2 \ge 1 \\ 0 & |-2\lambda_2| < 1 \\ 1 & 2\lambda_2 \le -1 \end{Bmatrix} \tag{7.86}$$

Figure 7.18 **Minimum gas consumption (bang-off-bang) controller**

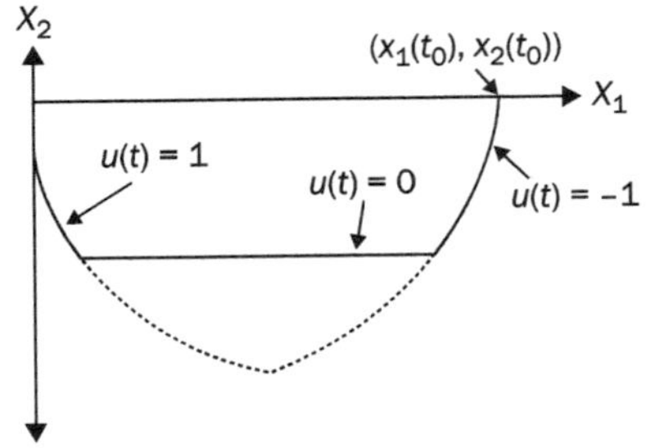

This optimal control must be an on-off-on type which is commonly known as a bang-off-bang optimal controller. Figure 7.18 shows the behavior of the admissible control function. From λ_2 in equation (7.83), the sign of λ_2 can reverse at most one time. Corresponding to any initial state which can be restored to the origin in an interval of time equal or less than t_1, there is one and only one control having the required admissible control function to transfer the initial state to the origin. Figure 7.18 shows the curve produced by the bang-off-bang optimal controller. Beginning at an initial condition, the admissible controller takes on successive values of either –1, 0, or 1. The loci of points at which the admissible controller changes values are composed of parabolas as shown in Figure 7.15. At some time, the admissible control becomes zero, and the system continues its path without showing any changes. Once the satellite reaches the other switching time, the admissible control takes on another successive value opposite to the first value selected which directs the system to the desired state. In the minimum time controller, the system uses longer arcs to reach the desired state. In the minimum consumption, the thruster requires three arcs and less time to reach the desire state.

One of the problems in solving the optimal controllers is the solution of the co-state variables. Because of the unknown co-state variables, the optimal controller has been solved with shooting methods [86]. A shooting method integrates the equations of motion and the co-state variables by guessing the initial state of the co-state variables such that the control input function can be determined. If the final time t_1 becomes unknown, the Bolza problem can be used to solve the optimal controller. On the other hand, there are unconstrained methods [96] that can be used to solve TPBVP. This method requires some iterations to obtain a solution to the problem. Reference [94] shows a computational method to determine the solution of TPBVP using the conditions of the Bolza problem.

7.3.6.2.7 Linear quadratic regulator problem

Pontryagin minimum (or maximum) principles have been used by different authors [91] [97] [98] to obtain an admissible control which leads to an optimal control that maintains a desired state for a single or group of satellites. Through this section, the linear quadratic regulator (LQR) is developed to satisfy a minimum-time problem by minimizing the control input. As the name of the problem states, the linear quadratic regulator uses the square of the states and the control input to define the cost function. The cost function is written as follows,

$$J = \frac{1}{2}(\vec{x}(t_1) - \vec{x}_D)^T \boldsymbol{F}(\vec{x}(t_1) - \vec{x}_D) + \frac{1}{2}\int_{t_0}^{t_1} ((\vec{x} - \vec{x}_D)^T \boldsymbol{Q}(\vec{x} - \vec{x}_D) + \vec{u}^T \boldsymbol{R}\vec{u})\, dt \quad (7.87)$$

$\boldsymbol{Q}$ and $\boldsymbol{R}$, respectively, are the weighting matrices for the state and the control input. $\boldsymbol{Q}$ has dimensions $n \times n$ while $\boldsymbol{R}$ has dimensions $m \times m$. $\boldsymbol{F}$ is the weighting matrices for the final state and has the same dimensions as the state weighting matrix. $\vec{x}_D$ is the desired state and has dimensions $n \times 1$. The weighting matrices are used to increase the effects of the controller. If the state vectors are weighted more, the control input is high to achieve the desired state in a short period of time. If the control inputs are weighted more, the control input uses a low consumption of thrust while the state takes a longer period of time to achieve the desired state. By having a suitable selection of the weighting matrices, the system can achieve a desired state. In practice, the weighting matrices are guessed, but the following chapter shows a technique that can be used to determine the weighting matrices based on the desired transient response.

The LQR optimal control is implemented with the use of the cost function defined by equation (7.87). The solution of the LQR problem leads to an optimal feedback system with the property that the components of the state vector $\vec{x}(t)$ are kept near the desired state vector $\vec{x}_D$ without excessive expenditure of control energy [87]. The existence of the optimal control is obtained from the solution of the Hamilton–Jacobi equation which is defined everywhere to obtain a minimum-time problem. Consider the time-varying system that is defined by equation (7.25) in the following form,

$$\dot{\vec{x}} = \boldsymbol{A}\vec{x} + \boldsymbol{B}\vec{u} + \vec{\psi} \quad (7.88)$$

and the cost functional defined by equation (7.87). $\vec{\psi}$ is a $n \times 1$ column vector containing the perturbation terms of the equations of motion.

Assume that for any initial state, there exists an optimal control function that can obtain a desirable minimum consumption problem. The Hamiltonian for the system (7.88) and the cost function (7.87) can be defined as:

$$\mathcal{H} = \frac{1}{2}((\vec{x} - \vec{x}_D)^T \boldsymbol{Q}(\vec{x} - \vec{x}_D) + \vec{u}^T \boldsymbol{R}\vec{u}) + \vec{\lambda}^T(\boldsymbol{A}\vec{x} + \boldsymbol{B}\vec{u} + \vec{\psi}) \tag{7.89}$$

The minimum principles are used to obtain the necessary conditions for the optimal control [87]. The co-state vector $\vec{\lambda}(t)$ is the solution of the vector differential equations:

$$\dot{\vec{\lambda}} = -\frac{\partial \mathcal{H}}{\partial \vec{x}} = -\boldsymbol{Q}\vec{x} + \boldsymbol{Q}\vec{x}_D - \boldsymbol{A}^T\vec{\lambda} \tag{7.90}$$

The optimal trajectory is obtained as,

$$\frac{\partial \mathcal{H}}{\partial \vec{u}} = \vec{u}\boldsymbol{R} + \boldsymbol{B}^T\vec{\lambda} = 0 \rightarrow \vec{u}(t) = -\boldsymbol{R}^{-1}\boldsymbol{B}^T\vec{\lambda} \tag{7.91}$$

Using equation (7.88) and substituting equation (7.91), the time-varying state equations can be rewritten as:

$$\dot{\vec{x}} = \boldsymbol{A}\vec{x} - \boldsymbol{S}\vec{\lambda} + \vec{\psi} \tag{7.92a}$$

where,

$$\boldsymbol{S} = \boldsymbol{B}\boldsymbol{R}^{-1}\boldsymbol{B}^T \tag{7.92b}$$

$\boldsymbol{S}$ is a square matrix which has a dimension of $n \times n$. Using equations (7.90) and (7.92), a combination of the state vectors and canonical equations are represented as:

$$\begin{bmatrix} \dot{\vec{x}} \\ \dot{\vec{\lambda}} \end{bmatrix} = \begin{bmatrix} \boldsymbol{A} & -\boldsymbol{S} \\ -\boldsymbol{Q} & -\boldsymbol{A}^T \end{bmatrix} \begin{bmatrix} \vec{x} \\ \vec{\lambda} \end{bmatrix} + \begin{bmatrix} \vec{\psi} \\ \boldsymbol{Q}\vec{x}_D \end{bmatrix} \tag{7.93}$$

In equation (7.93), the first row is a system of $2n$ linear differential equations while the second row is a column matrix of $2n$ components. A unique solution can be obtained from this system of linear differential equations if the boundary conditions are known. The initial boundary conditions $\vec{x}(t_0)$ are given at t_0, but, at t_1, the remaining boundary conditions are furnished by the transversality conditions for the co-state variables at the end time, $\vec{\lambda}(t_1)$. Reference [87] shows the process of how to obtain the transversality conditions that explain the solution of the co-state variables and is shown in Appendix F. The equation describing the co-state variable determined from the transversality conditions is equal to,

$$\vec{\lambda}(t) = \boldsymbol{P}(t)\vec{x}(t) + \vec{G}(t) \tag{7.94}$$

If equation (7.94) is differentiated with respect to time, it can be found that:

$$\dot{\vec{\lambda}}(t) = \dot{\boldsymbol{P}}(t)\vec{x}(t) + \boldsymbol{P}(t)\dot{\vec{x}}(t) + \dot{\vec{G}}(t) \tag{7.95}$$

Equations (7.90) and (7.95) are the same, and can be written as,

$$\dot{\boldsymbol{P}}(t)\vec{x}(t) + \boldsymbol{P}(t)\dot{\vec{x}}(t) + \dot{\vec{G}}(t) = -\boldsymbol{Q}\vec{x} + \boldsymbol{Q}\vec{x}_D - \boldsymbol{A}^T\vec{\lambda} \tag{7.96}$$

Substituting equation (7.92a) and (7.90) into equation (7.96), the following equation is obtained as:

$$\begin{aligned}&\dot{\boldsymbol{P}}(t)\vec{x}(t) + \boldsymbol{P}(t)\,(\boldsymbol{A}\vec{x} - \boldsymbol{SP}(t)\vec{x}(t) - \boldsymbol{S}\vec{G}(t) + \vec{\psi}) + \dot{\vec{G}}(t)\\ &= -\boldsymbol{Q}\vec{x} + \boldsymbol{Q}\vec{x}_D - \boldsymbol{A}^T\boldsymbol{P}(t)\vec{x}(t) - \boldsymbol{A}^T\vec{G}(t)\end{aligned} \tag{7.97}$$

Equation (7.97) can be grouped into terms in which two different equations are obtained,

$$\dot{\boldsymbol{P}}(t) = -\boldsymbol{P}(t)\boldsymbol{A} - \boldsymbol{A}^T\boldsymbol{P}(t) + \boldsymbol{P}(t)\boldsymbol{SP}(t) - \boldsymbol{Q} \tag{7.98a}$$

$$\dot{\vec{G}}(t) = -(\boldsymbol{A}^T - \boldsymbol{P}(t)\boldsymbol{S})\vec{G}(t) + \boldsymbol{Q}\vec{x}_D - \boldsymbol{P}(t)\vec{\psi} \tag{7.98b}$$

Equation (7.98a) is the Riccati equation (RE), equation (7.98b) is known as the adjoint Riccati equation (ARE). The RE and ARE must be solved backwards in time as explained in reference **[66]**. This system can be solved using Runge–Kutta methods **[14]**, but this method always runs forward in time; for this reason, Euler's method can be used to obtain an approximate solution because it can be applied backwards in time **[81]**. Euler's method must be applied until a stable solution is obtained. From the transversality conditions, the final conditions for equation (7.98) are defined as,

$$\boldsymbol{P}(t_1) = \boldsymbol{F} \quad \vec{G}(t_1) = \boldsymbol{F}\vec{x}_D \tag{7.99}$$

The process to solve the LQR control problem can be specified as follows,

1. Define the initial condition for the state vector.
2. Integrate the RE backward in time using the end condition $\boldsymbol{P}(t_1) = \boldsymbol{F}$.
3. Integrate the ARE backward in time using the end condition $\vec{G}(t_1) = \boldsymbol{F}\vec{x}_D$ and the solution of the RE.

4. Integrate equation (7.92) forward in time using the initial condition for the state vector and substituting the solution of the RE into equation (7.94).
5. Calculate the control input vector in equation (7.91) with equation (7.94).
6. Plot the results for the state vector and control input vector.

It is possible to use a second order equation to show the LQR problem. Instead, a first order differential equation is used to show the effects of the RE and ARE when the weighting matrices are varied. The following first order differential equation is used,

$$\dot{x}(t) = x(t) + u(t) + 2 \tag{7.100}$$

For the LQR problem, assume that $F = 1$, $Q = 1$, $t_1 = 1$, $x(0) = 2$, and $x_D = 0$. Using R as a parameter, Figure 7.19 shows the solution of the RE and ARE for various control weighting matrices. As R decreases, the RE solution decreases toward zero; in a similar manner, the ARE also shows

Figure 7.19 Riccati equation solutions for varying control weighting matrix

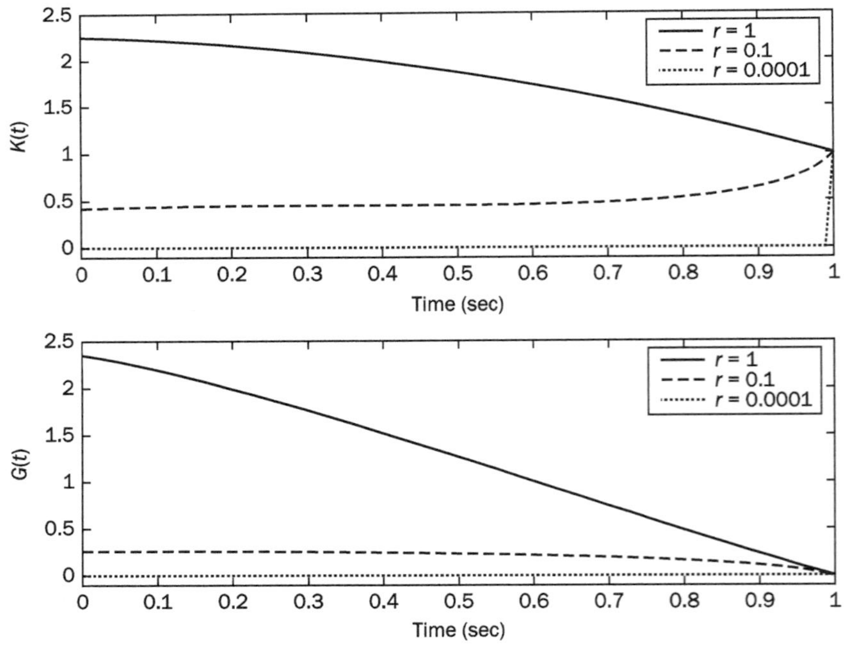

a similar pattern. Because of the end condition, the RE reaches one but has a stable value once it is approaching the initial time. The ARE shows a similar pattern, but the end condition is the desired state. These characteristics for the RE and ARE can be used later to determine a steady-state response solution. Also, the RE is always a positive definite matrix such that the solution of the control input function can always compensate the state equations. This condition is obtained from the quadratic multiplication in the Riccati equation. Figure 7.20 shows the solution for the state equations and the control input function. As R decreases, the required control effort also decreases to satisfy the desired state, but the state equation shows different responses to satisfy the desired state.

If the same simulations are executed for $R = 0.001$, for $t_1 = 0.5$ and for a varying Q, the solutions for the RE and ARE are shown in Figure 7.21. In Figure 7.22, the solutions for the state and control input equations are shown. By incrementing Q, the transient response for the state equations is improved while the control input solution is maintained closed to each other. In other words, the rise time is reduced with an increment of Q without expending control effort.

Figure 7.20 State equation solution for varying control weighting matrix

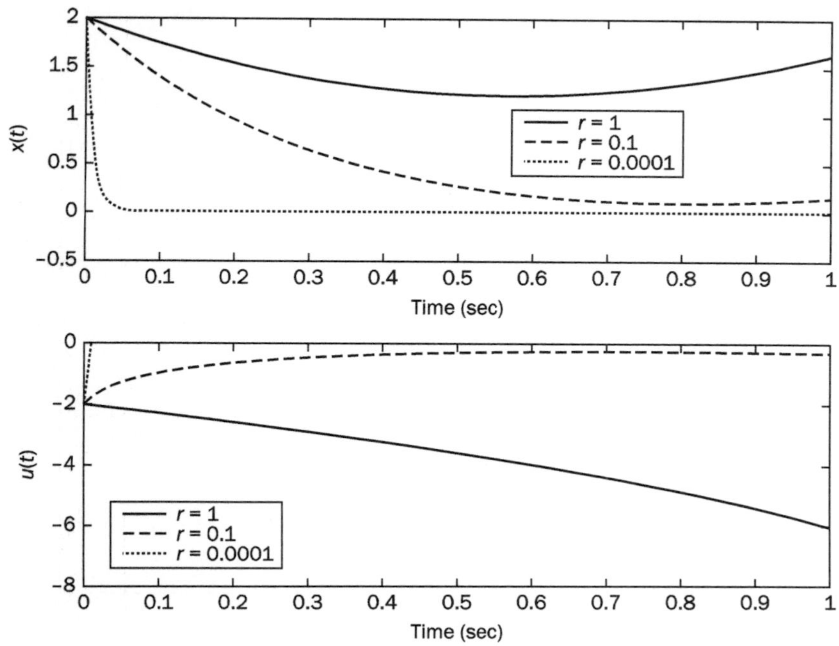

Figure 7.21 Riccati equation solutions for varying state weighting matrix

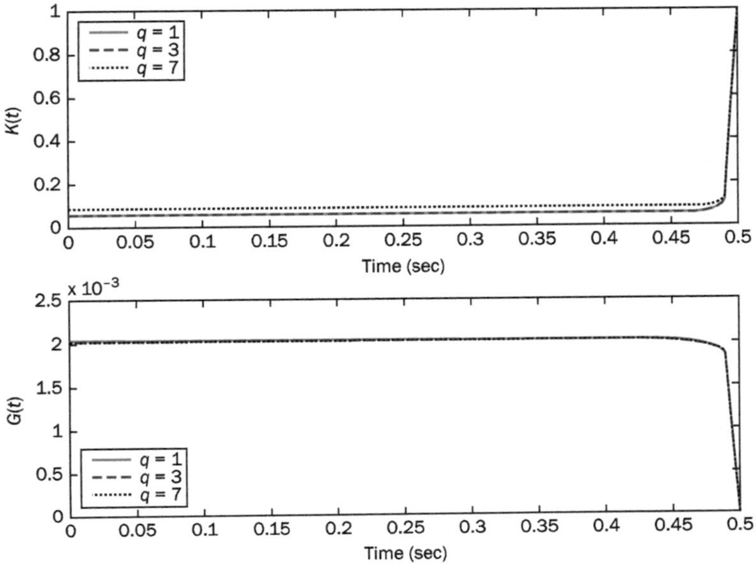

Figure 7.22 State equation solution for varying state weighting matrix

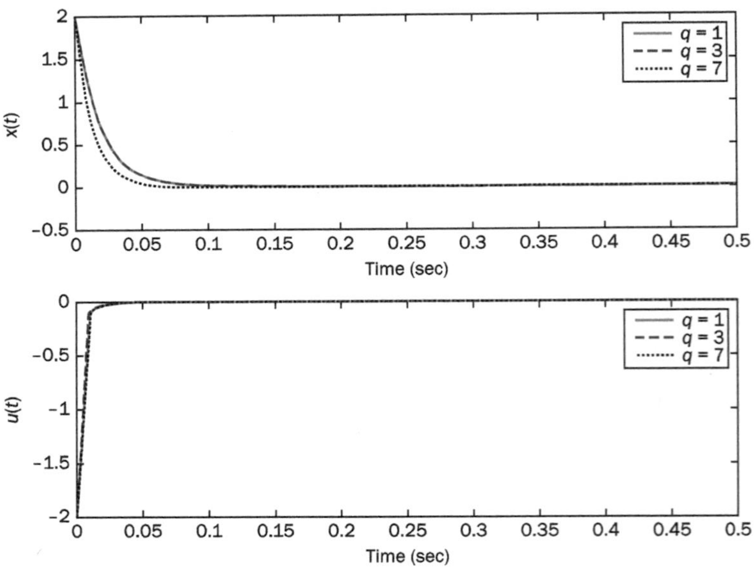

By observing these solutions, the weighting matrix for the end conditions has not been analyzed to determine if it causes a variation to the solution. The same problem is used for the following conditions: $R = 0.001$, $t_1 = 0.5$ and $Q = 3$. F is considered as a parameter. Figure 7.23 shows the solution for the RE and ARE. For $F = 0$, the RE and ARE shows a slow variation to a constant value as compared to $F = 1$ or $F = 3$ when the graph is analyzed backwards in time. As time tends to zero, the RE and ARE have the same solution regardless of the final weighting matrix observed at $t = 0.5$. The solution of the state equations in Figure 7.24 has a very similar trend as in Figure 7.22. By incrementing the weighting of the end state, the desired state is satisfied. This desired state is obtained by expending more control effort at the initial time.

Thus, this is the trade-off of the LQR control problem which is the selection of the weighting matrices. In practice, the best choice of the weighting matrices is to maintain the ***R*** matrix as an indentity matrix while the values of the ***Q*** matrix are selected. Because of the chosen weighting matrices, the control input maintains a minimum consumption of control effort while the states are corrected in minimum time. If the value of the control weighting matrix is chosen to be less than one and

Figure 7.23 **Riccati equation solutions for varying final state weighting matrix**

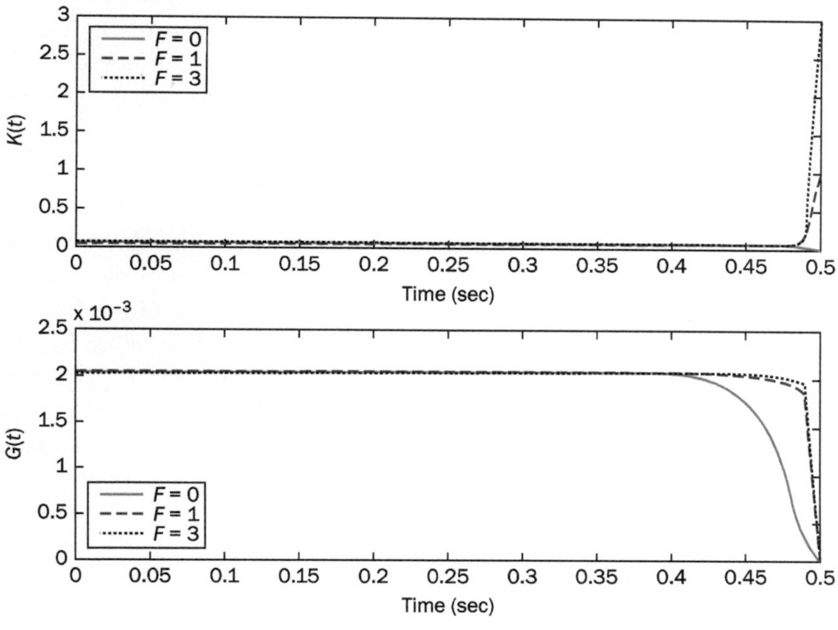

Figure 7.24 State equation solution for varying final state weighting matrix

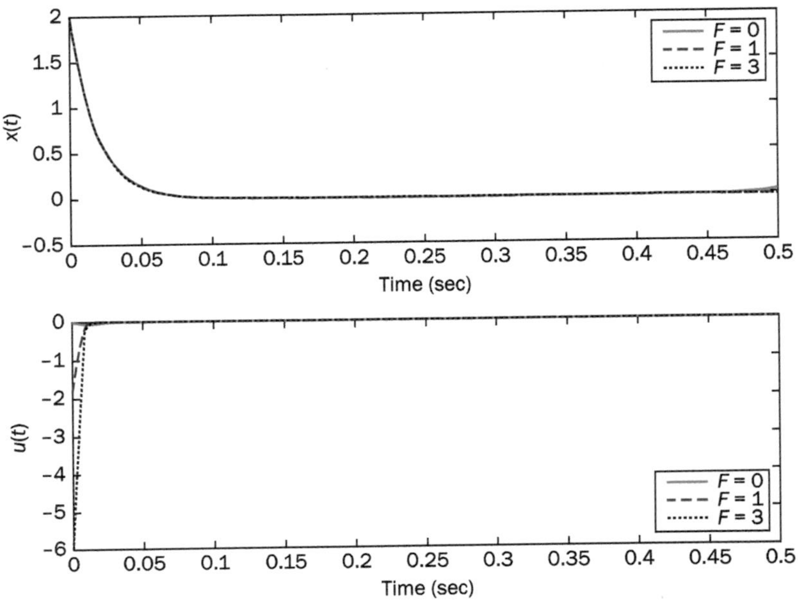

the state weighting matrix is incremented while the final state weighting matrix is maintained to an identity matrix, the solution of the states satisfies the desired end condition and corrects the perturbation vector; in this way, a minimum control effort can be obtained to satisfy the two point boundary value problem.

7.3.6.2.8 Steady-state linear quadratic regulator

What is the difference between the steady-state LQR and the previous LQR? The main difference of the problem is the solution of the Riccati equations. Figure 7.19 shows an important characteristic of the Riccati equations. Between the initial time and before the Riccati solution begins to move toward zero, a constant value is observed for the Riccati equations. This constant value of the Riccati equations results in the state equation reaching a zero state. Using this fact, the end time can be taken to infinity such that the Riccati solution can have a constant value. Because of this reason, the cost function can be written as,

$$\mathcal{J} = \frac{1}{2}\int_{t_0}^{\infty} ((\vec{x} - \vec{x}_D)^T \boldsymbol{Q}(\vec{x} - \vec{x}_D) + \vec{u}^T \boldsymbol{R}\vec{u})\, dt \qquad (7.101)$$

The end conditions are not necessary in the problem because the final time is inifinity. The minimum conditions can be applied to the cost functional in equation (7.101). The solution is the same solution as shown in the previous section for the state equation, control input, and co-state variable. The difference is the method to obtain the stable solution for the RE. The system can be defined continuously with respect to the time domain so that a solution is stable. The following approximation can be applied for the system:

$$\lim_{\Delta t\to\infty} \dot{P}(t) = \lim_{\Delta t\to\infty} \dot{\vec{G}}(t) = \lim_{\Delta t\to\infty} \frac{P(t+\Delta t) - P(t)}{\Delta t}$$
$$= \lim_{\Delta t\to\infty} \frac{\vec{G}(t+\Delta t) - \vec{G}(t)}{\Delta t} = 0 \qquad (7.102)$$

Equations (7.87) become as,

$$\boldsymbol{P}_\infty(t)\boldsymbol{A} + \boldsymbol{A}^T \boldsymbol{P}_\infty(t) - \boldsymbol{P}_\infty(t)\boldsymbol{S}\boldsymbol{P}_\infty(t) + \boldsymbol{Q} = 0 \qquad (7.103a)$$

$$\vec{G}_\infty(t) = (\boldsymbol{A}^T - \boldsymbol{P}(t)\boldsymbol{S})^{-1} (\boldsymbol{Q}\vec{x}_D - \boldsymbol{P}(t)\vec{\psi}) \qquad (7.103b)$$

where $\boldsymbol{P}_\infty(t)$ and $\vec{G}_\infty(t)$ are the steady-state RE and ARE, respectively. The RE can be solved as a system of linear equations; and the ARE uses the solution of the RE to determine the solution. Using this formulation, the control input function can be written in steady-state format as,

$$\vec{u}(t) = -\boldsymbol{R}^{-1} \boldsymbol{B}^T \vec{\lambda} = \boldsymbol{K}_\infty \vec{x}(t) + \vec{K}^P_\infty \qquad (7.104a)$$

where,

$$\boldsymbol{K}_\infty = -\boldsymbol{R}^{-1} \boldsymbol{B}^T \boldsymbol{P}_\infty(t) \qquad (7.104b)$$

$$\vec{K}^P_\infty = -\boldsymbol{R}^{-1} \boldsymbol{B}^T \vec{G}_\infty(t) \qquad (7.104c)$$

$\boldsymbol{K}_\infty$ is the steady-state control gain, and $\vec{K}^P_\infty$ is the perturbation gain which is given by the desired state and the perturbation vector. Equations (7.104) are substituted into the state vector equation which can be solved using a numerical integration scheme such as the Runge–Kutta method forward in time [**14**]. Substituting equation (7.104) into (7.88), the state vector is defined as:

$$\dot{\vec{x}} = (\boldsymbol{A} + \boldsymbol{B}\boldsymbol{K}_\infty)\, \vec{x} + \vec{K}^P_\infty + \vec{\psi} \qquad (7.105)$$

The following procedure is adapted to obtain a solution for the RE, the state vector, and the control vector:

1. Use the form of equations (7.103) to obtain a solution for the Riccati equations.

2. Substitute the solution of equations (7.103) into equations (7.104).
3. Integrate forward in time equation (7.105) using the initial conditions $x(t_0) = x_0$.
4. Determine the control input function in equation (7.104a).

The steady-state LQR problem is very simple to solve and always provides stability if certain conditions are met. The conditions to use this form of the LQR controller are the following [87]:

1. The open loop state equations are required to have at least one pole in the left hand side of the imaginary plane.
2. The system must have zero dynamics (i.e. there must exist a solution such that the closed loop system is stable – eigenvalues have negative real parts).

By satisfying these two conditions, the LQR produces a stable solution. Reference [93] provides other characteristics to measure the stability of the LQR control problem.

Using equation (7.100), assuming $Q = 3$, $R = 0.001$, and $x(0) = 2$, and using the procedure to solve the steady-state LQR problem, Figure 7.25 shows the solution for the state and control vector. Comparing

Figure 7.25 Steady-state solution for the LQR problem

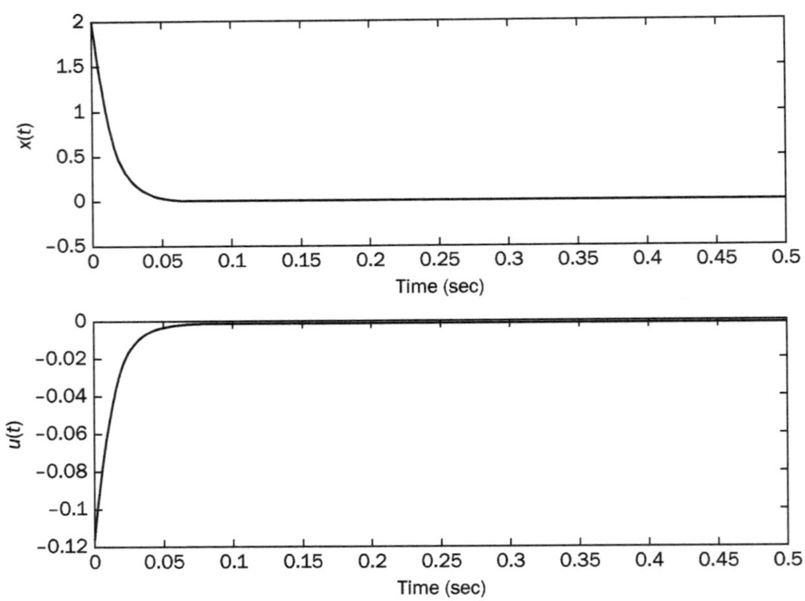

Figure 7.22 to Figure 7.25, the control input vector solution is lower in amplitude in Figure 7.25 than in Figure 7.22. This happens because the solution for the steady-state LQR controller is constant through the complete time interval. The backward integration process of the RE and ARE makes the solution change with respect to time and use more control effort than the steady-state LQR control problem. In practice, the steady-state LQR controller is commonly used because it does not require an integration procedure to solve the problem backwards in time.

7.4 Discrete control formulation

The continuous control formulation provides a basis for the understanding of the discrete control formulation. The difference between formulations is the development of the transfer functions and state equations. In the continuous domain, the equations of motion are expressed as a set of differential equations that are required to be integrated with a numerical method [**14**]. The discrete continuous formulation transforms the equations of motion to be analyzed at a specified sampling time. Because of this difference, the discrete domain simplifies the solution process for the controller. On the other hand, the stability criteria methods, controller designs, and optimal controllers are the same but are defined in the discrete domain.

This section focuses in the development of the discrete formulation that is important in the implementation process on a computer. First, the Z-transform is studied because it is the equivalent of the Laplace transform. The final Z-transfer function is written in terms of the state vectors to obtain the solution. On the other hand, the state vector formulation is also redefined in the discrete domain. This information provides the reader with a practical formulation that can be easily implemented on the computer. Finally, the optimal control formulation in the discrete domain is introduced. The interesting fact about this formulation is the definition of the cost function and the solution of the co-state variables. In the literature, this formulation is not commonly shown because the optimal controls are generally solved in the time domain instead of the discrete domain.

7.4.1 Z-transform formulation

The Laplace transform is determined from the differential equations. In a similar manner, the Z-transform is obtained from the Laplace transform since the complex plane is substituted in the discrete domain. To work in the discrete domain, it is necessary to change the analog signal by sampling the signal every certain period of time. This is commonly known as the sampling time which is a constant value in some cases. The sampling time is determined from the number of samples taken in a period of time. This description provides the basis for the development of analog to digital converters (A/D) [99]. In a similar manner, the A/D converter can work in vice versa; in other words, the signal can be reconstructed from the discrete domain to the analog domain. This device is known as the digital to analog (D/A) converter. If a process is run between the A/D and D/A converters, Figure 7.26 shows the block diagram for a discrete control system, and the dashed box shows the work done by the computer.

One of the advantages of using the discrete formulation is the number of samples used to analyze the process. If the number of samples is reduced during the process, the computational time can be reduced. On the other hand, the disadvantage is that the signal conversion has a time delay that must be analyzed in the formulation. In order to analyze the signal conversion into the discrete domain, the Z-transform is used. The Z-transform is formulated as,

$$G_p(z) = \frac{z-1}{z} Z\left\{\frac{G_p(s)}{s}\right\} \tag{7.106}$$

The Z-transform uses the Laplace transform of the differential equation to determine the function in the discrete domain. In addition, the time

Figure 7.26 Digital closed-loop control system

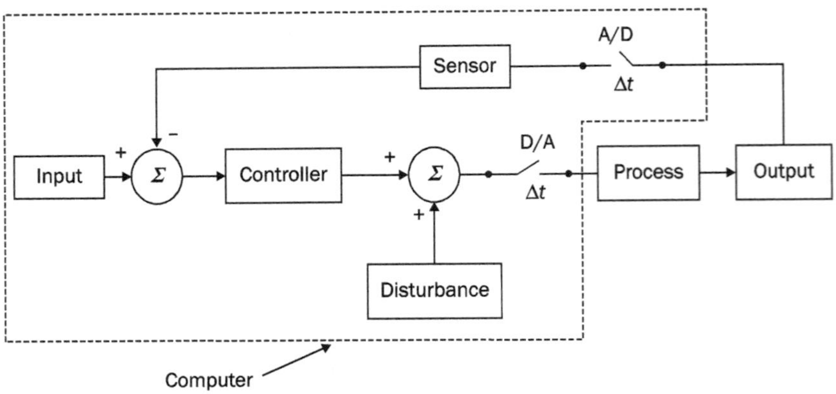

delay is due to the A/D and D/A converters described by the discrete function as $(z-1)/z$. In the discrete formulation, the variable z substitutes the s in the Laplace transform. This is a modified version for the Z-transform because the integral is more complex [78]. Appendix D shows some common Z-transforms for certain functions.

To demonstrate this formulation, equation (7.8) can be used to determine the Z-transform for the roll equation of the Ares V rocket. The Z-transform of equation (7.8) can be written as,

$$Z\left\{\frac{G_p(s)}{s}\right\} = Z\left\{\frac{c}{s^3}\right\} = \frac{\Delta t^2 z(z+1)}{2(z-1)^2} \qquad (7.107)$$

where,

$$G_p(s) = \frac{c}{s^2} \qquad c = \frac{T_{max} b_1}{J_1}$$

Δt is known as the sampling time. Substituting equation (7.107) into equation (7.106), the Z-transform of the plant process in the discrete domain can be written as,

$$G_p(z) = \frac{\Delta t^2 (z+1)}{2(z-1)} \qquad (7.108)$$

Knowing this transformation, the control problem can be easily analyzed with the techniques explained in the previous sections.

7.4.2 State vector formulation

As shown in Section 7.3.4, the state vector formulation in the continuous domain can be used to simplify the integration and control formulation process. In the discrete domain, there are three ways to determine the state vector formulation: 1) Z-transfer function [78], 2) discrete state equations [78], 3) Euler approximation [100]. Any of these three state vector formulations can provide the solution of the equations of motion.

7.4.2.1 Z-transfer function

The Z-transfer function can be used to obtain the solution of the differential equation without solving the inverse Laplace transform. Through this formulation, the discrete system becomes a simple iterative procedure to determine the solution for the problem. In order to understand this formulation, the following Z- transfer function is used as an example,

$$G(z)=\frac{Y(Z)}{U(z)}=\frac{z^2+2z+1}{z^3+2z^2+z+1} \tag{7.109}$$

The first step in the formulation is to use an auxiliary variable to separate equation (7.109). Including this auxiliary variable, equation (7.109) can be written as,

$$G(z)=\frac{Y(Z)}{U(z)}=\frac{z^2+2z+1}{z^3+2z^2+z+1}\frac{E(z)}{E(z)} \tag{7.110}$$

where $E(z)$ is the auxiliary variable. Equation (7.110) can be separated as,

$$Y(Z) = (z^2 + 2z + 1)E(z) \tag{7.111a}$$

$$U(z) = (z^3 + 2z^2 + z + 1)E(z) \tag{7.111b}$$

The real translation property [78] is used to perform a correspondence between the sample taken at a sampling time (see Appendix D). By using this property, the following correspondence can be done for the auxiliary variable,

$$E(z) \to e(k) \quad zE(z) \to e(k + 1) \quad z^2E(z) \to e(k + 2) \ldots$$

Using this correspondence, equations (7.111) are written respectively as,

$$e(k + 3) + 2e(k + 2) + e(k + 1) + e(k) = u(k) \tag{7.112a}$$

$$y(k) = e(k + 2) + 2e(k + 1) + e(k) \tag{7.112b}$$

The state variable rules shown in Section 7.3.4 can be applied in a similar manner for the discrete system. Before assimilating the rules, the state variables can be written from this correspondence as,

$$x_1(k) = e(k) \quad x_2(k) = x_1(k + 1) = e(k + 1) \quad x_3(k) = x_2(k + 1) = e(k + 2) \ldots$$

From Section 7.3.4, the equations in state vector format are written for the highest derivative. In the discrete domain, the state vector is written up to the highest sample time. For equation (7.112a), the highest sample time is $e(k + 3)$. Also, the state vector for the discrete time is expressed in the present time, k, while the resultant state vector is expressed in the future time, $k + 1$. Thus, equation (7.112) are written as,

$$x_3(k + 1) = -2x_3(k) - x_2(k) - x_1(k) + u(k) \tag{7.113a}$$

$$y(k) = x_3(k) + 2x_2(k) + x_1(k) \tag{7.113b}$$

Equation (7.113a) is completed with the following equations,

$$x_1(k+1) = x_2(k) \quad x_2(k+1) = x_3(k) \tag{7.113c}$$

Then,

$$\vec{x}(k+1) = \hat{\boldsymbol{A}}\vec{x}(k) + \hat{\boldsymbol{B}}\vec{u}(k) \tag{7.114a}$$

$$\vec{y}(k+1) = \hat{\boldsymbol{C}}\vec{x}(k) \tag{7.114b}$$

where,

$$\vec{x}(k) = [x_1(k) \quad x_2(k) \quad x_3(k)]^T \quad \vec{u}(k) = u(k)$$

$$\hat{\boldsymbol{A}} = \begin{bmatrix} 0 & 1 & 0 \\ 0 & 0 & 1 \\ -1 & -1 & -2 \end{bmatrix} \quad \hat{\boldsymbol{B}} = \begin{bmatrix} 0 \\ 0 \\ 1 \end{bmatrix} \quad \hat{\boldsymbol{C}} = [1 \quad 2 \quad 1]$$

Equation (7.114a) is the discrete format of the state vector equation, and equation (7.114b) is known as the output equation. $\hat{\boldsymbol{A}}$, $\hat{\boldsymbol{B}}$, and $\hat{\boldsymbol{C}}$ are the discrete state matrix, control matrix, and output matrix respectively. $\vec{x}(k+1)$ is known as the future state vector, and $\vec{x}(k)$ is the present state vector. In the computer, the system can be easily solved by solving equations (7.114) in an iterative manner. The variable k is known as the sample. The matrices of Equation (7.114a) have the same size as the matrices in equation (7.14a).

7.4.2.2 Discrete state equations

The state vector equation (shown in equation (7.14a)) has a closed form solution as shown in References [76] and [78]. The state vector equation in continuous format can be written as,

$$\dot{\vec{x}}(t) = \boldsymbol{A}\vec{x}(t) + \boldsymbol{B}\vec{u}(t) \tag{7.115}$$

As shown in Reference [78], the state vector in continuous format can be written in discrete format with the following transformation,

$$\vec{x}(k+1) = \hat{\boldsymbol{A}}\vec{x}(k) + \hat{\boldsymbol{B}}\vec{u}(k) \tag{7.116a}$$

where,

$$\hat{\boldsymbol{A}} = \boldsymbol{I} + \boldsymbol{A}\Delta t + \boldsymbol{A}^2 \frac{\Delta t^2}{2!} + \boldsymbol{A}^3 \frac{\Delta t^3}{3!} + \ldots \tag{7.116b}$$

$$\hat{B} = \left[I\Delta t + A\,\frac{\Delta t^2}{2!} + A^2\,\frac{\Delta t^3}{3!} + \ldots \right] B \tag{7.116c}$$

This formulation adds an extra step to the calculation. For $\Delta t > 1$, more terms are required to obtain an accurate solution for the discrete equations. For $\Delta t < 1$, many of the terms become higher order terms and can be truncated.

The same state vector equation in the continuous domain can be linearized in the discrete format. Euler's theorem is used to represent the partial derivative with respect to time. Euler's theorem can be written as [81],

$$\dot{x} = \frac{x(k+1) - x(k)}{\Delta t} \tag{7.117}$$

To represent Equation (7.117), equation (7.116) is truncated for second order terms to determine the discrete form of the state vector equations as,

$$\vec{x}(k+1) = \hat{A}\vec{x}(k) + \hat{B}\vec{u}(k) \tag{7.118a}$$

where,

$$\hat{A} = I + A\Delta t \tag{7.118b}$$

$$\hat{B} = \Delta t B \tag{7.118c}$$

Equations (7.118) simplify the process shown in equations (7.116). If Δt is small, the partial derivative can be determined to its approximate correct value. In the discrete control scheme, the value of Δt makes a difference in the solution of the equations of motion.

7.4.3 Stability conditions in the discrete domain

The interesting fact about the Z-transfer function is the use of the stability conditions in the discrete domain which are the same techniques shown in Section 7.3. The second order responses, the Routh–Hurwitz criteria, and the eigenvalue problem can be applied in the discrete domain as in the continuous domain.

The Routh–Hurwitz criteria explained in Section 7.3.5.2 can be applied to the Z-transfer function, but the characteristic polynomial is written as a function of z. By applying the Routh–Hurwitz criteria a stable system can be obtained. Similarly, the solution of the eigenvalue problem can be performed to determine the location of the poles in the complex plane for the discrete system. This eigenvalue problem can be performed to determine the pole location in the complex plane. For discrete systems, the eigenvalue problem is written as,

$$|z\mathbf{I} - \hat{\mathbf{A}}| = 0$$

For large matrices, the algorithm explained in Appendix E can also be applied to this problem.

7.4.4 Control formulation

In discrete controls, the system is discretized such that it can be easily implemented in the computer onboard the space vehicle. In a similar manner, many of the techniques applied to the continuous functions can also be applied to the discrete controllers. To understand the discrete controllers, the closed-loop diagram for digital control systems should be explained. Figure 7.26 shows the closed-loop diagram for discrete control systems. In Figure 7.26, the computer is controlling the vehicle and reading the sensors in the discrete time. In practice, the sensors are read at a sampling frequency of about 10 to 30 Hz. The controller can output information at the same rate as the process. Because of this selection, an A/D converter is set before the sensor to begin reading the sensors in the discrete domain, and a D/A converter is located after the controller to output the control input function. Because of the location of these converters, the computer can speed up the process.

The sensors, controller, and input function are written with the Laplace transform; but with the Z-transform, the system can be expressed in the discrete domain. To actually do a complete analysis of the process, the Z-transform is also applied to the transfer function of the process. The open-loop Z-transfer function of equation (7.27) becomes,

$$G_{OL}(z) = \frac{\Theta(z)}{X_c(z)} = \frac{G_c(z)G_p(z)}{1+G_c(z)G_p(z)} \tag{7.119}$$

The big difference between equation (7.27) and (7.119) is the transfer function of the controller. This section explains how the transfer function for the controller is described. In addition, this section demonstrates how the sampling time makes a difference to the control input function and the output of the process.

Another formulation presented in this section is the discrete optimal controls. The discrete function obtained in this section looks completely different from the differential equations obtained in the previous section. The cost function and Hamiltonian equation are described in a different way for discrete systems. These optimal formulations are only developed for the linear quadratic regulator, but this can be demonstrated for the other minimum time and gas consumption problem. These problems are left as an exercise to the reader but can be solved to obtain the same solution in the discrete domain.

7.4.4.1 Pole placement technique for discrete controls

In the continuous domain, there are different transfer functions that can be used to describe the controller. As shown in Table 7.2, the control transfer functions can be used to stabilize and control the system to a desired state. These functions are described in the s-domain but can be transformed into the Z-domain with equation (7.106). When these control equations are used, the Z-transfer function shown in equation (7.119) is solved. By using the Routh–Hurwitz criteria or the second order responses to place poles, the stable control gains can be determined to stabilize the process.

On the other hand, the discrete state vector system shown in equation (7.118) can also be used to determine a suitable control function. The control input function is described very similarly as in equation (7.34) but is written for the discrete state vector as,

$$\vec{u}(k) = -K\vec{x}(k) \tag{7.120}$$

Equation (7.120) shows the control gain matrix but is described in the present time. Substituting equation (7.120) into equation (7.118), the discrete closed loop equation is defined similarly as the continuous closed loop equation. As a matter of fact, the eigenvalue problem can be solved to determine the stable control gain matrix. This is achieved by solving either the Routh–Hurwitz criteria or the desired transfer functions. The determination of the gain matrix can also be done through the Ackermann's formula. The same function can also be used to determine the control gain matrix; but the discretized state and control matrices are used.

One question remains: how the sampling time affects the control gains and the control input function? To understand this, let's use the same conditions in equation (7.36) and (7.37). To obtain the control matrix, the Ackermann's formula shown in (7.40) is used. To work this problem, the state and control matrix are transformed to the discrete domain with equations (7.118). For this problem, the desired function used for this problem is the following:

$$F(z) = z^2 + 2\rho\omega_n z + \omega_n^2 \tag{7.121}$$

Assuming $\rho = 1$ and $\omega_n = 2$, Figure 7.27 shows the solution for different sampling times. When $\Delta t = 0.2$, the pitch angle shows few points on the graph such that the graph looks like it is only composed of lines. When the sampling time is increased, the solution for the pitch equation has more points and is approaching the continuous state. On the other hand, the control input function solution increases as the sampling time increases because of the increment in the control gain matrix. Due to this condition, it is necessary to choose a sampling time that does not require a relatively large control effort to satisfy the problem. A sampling time of 0.1 seconds (or sampling frequency of 10 Hz) provides enough information to control and stabilize the process.

Figure 7.27 Discrete control solution for different sampling times

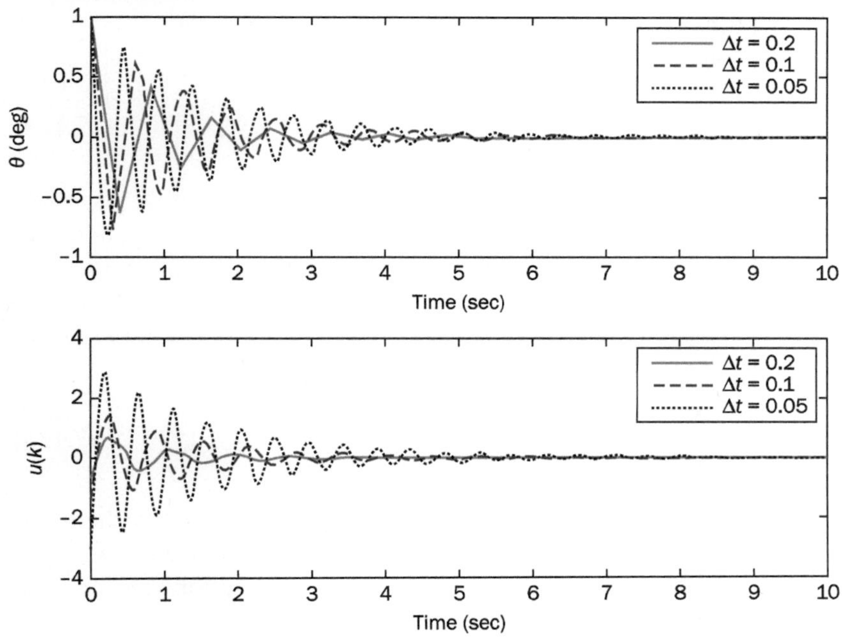

7.4.4.2 Discrete optimal controls

The transversality conditions, control indexes, and function evaluations have been shown in the continuous domain to demonstrate the use of optimal controls. Euler–Lagrange equations, Pontryagin minimum (or maximum) principles, and the Bolza problem provide the necessary minimum (or maximum) conditions to achieve a desired state for the process. The same principles can be used for discrete control systems.

As explained in Section 7.3.6.2.1, the plant process is written in terms of the state variables, but the discrete system is given as,

$$\vec{x}(k+1) = \vec{f}(\vec{x}(k), \vec{u}(k), k) \tag{7.122}$$

where the initial condition is given as follows $\vec{x}(t_0) = \vec{x}_0$. In order to minimize equation (7.122), a performance index is necessary. The performance index for the continuous system is described in equation (7.52) as follows,

$$\mathcal{J} = \int_{t_0}^{t_1} \mathcal{F}(\vec{x}, \vec{u}, t)\, dt$$

Mathematically, an integral can be approximated by,

$$\int_{t_0}^{t_1} \mathcal{F}\, dt = \Delta t \sum_{k=0}^{N_{f-1}} \mathcal{F}(\vec{x}(k), \vec{u}(k), k) = \mathcal{J}(k) \tag{7.123}$$

k ranges from 0 to $N_f - 1$ where N_f is the last sample obtained in the solution of the control process defined at the final time and is expressed as,

$$N_f - 1 = \frac{t_1 - t_0}{\Delta t} \tag{7.124}$$

where t_1 and t_0 is the final and initial time, respectively. As in the continuous problem, the optimal control is determined by minimizing $\mathcal{J}(k)$ constraint to $\vec{x}(k+1)$ such that a minimum control input function, $\vec{v}(k)$, can be obtained to minimize the cost index performance. To achieve this problem, a discrete Hamiltonian can be written to obtain the desired problem,

$$\mathcal{H}(k) = \sum_{k=0}^{N_{f-1}} \Delta t \mathcal{F}(\vec{x}(k), \vec{u}(k), k) + \vec{\lambda}^{T}(k+1) \\ [\vec{x}(k+1) - f(\vec{x}(k), \vec{u}(k), k)] \tag{7.125}$$

where $\vec{\lambda}\,(k+1)$ is the discrete co-state variable expressed in the future time. Let $\vec{u}^*(k)$ be the admissible control that can minimize $\mathcal{J}^*(k)$ to the smallest achievable value. Also, the following assumptions can be made about the other variables in equation (7.125),

$$\vec{x}(k) = \vec{x}^*(k) + \epsilon\vec{\chi}(k) \tag{7.126a}$$

$$\vec{u}(k) = \vec{u}^*(k) + \delta\vec{u}(k) \tag{7.126b}$$

$$\vec{\lambda}(k) = \vec{\lambda}^*(k) + \gamma\vec{\xi}(k) \tag{7.126c}$$

where ξ, η, and γ are small error functions in the variables. Equations (7.126) can be substituted into equation (7.125) to rewrite the discrete Hamiltonian as,

$$\begin{aligned}\mathcal{H}(k) = {} & \textstyle\sum_{k=0}^{N_f-1} \Delta t\mathcal{F}(\vec{x}^*(k) + \epsilon\vec{\chi}(k), \vec{u}^*(k) + \delta\vec{u}(k), k) \\ & + (\vec{\lambda}^*(k) + \gamma\vec{\lambda}(k))^T [\vec{x}^*(k+1) + \epsilon\vec{\chi}(k+1) - f(\vec{x}^*(k) \\ & + \epsilon\vec{\chi}(k), \vec{u}^*(k) + \delta\vec{u}(k), k)] \\ \mathcal{H}(k) = {} & \textstyle\sum_{k=0}^{N_f-1} \widehat{\mathcal{H}}(\vec{x}^*(k), \vec{u}^*(k), \vec{\lambda}^*(k), k)\end{aligned} \tag{7.128}$$

Using a Taylor series expansion, equation (7.128) can be expanded and reduced to,

$$\begin{aligned}\mathcal{H}(k) - \mathcal{H}^*(k) = {} & \epsilon\sum_{k=0}^{N_{f-1}}\left[\vec{\chi}^T(k)\frac{\partial\mathcal{H}^*(k)}{\partial\vec{x}^*(k)} + \vec{\chi}^T(k+1)\frac{\partial\mathcal{H}^*(k)}{\partial\vec{x}^*(k+1)}\right] \\ & + \delta\sum_{k=0}^{N_{f-1}}\vec{u}^T(k)\frac{\partial\mathcal{H}^*(k)}{\partial\vec{u}^*(k)} + \gamma\sum_{k=0}^{N_{f-1}}\vec{\xi}^T(k+1)\frac{\partial\mathcal{H}^*(k)}{\partial\vec{\lambda}^*(k+1)}\end{aligned} \tag{7.129}$$

To obtain the necessary minimum condition, $\Delta\mathcal{H}(k) = 0$. Because of this condition, ϵ, δ, and γ cannot be equal to zero because these are real constants; then, the necessary conditions to minimize the problem can be described as,

$$\vec{\chi}^T(k)\frac{\partial\mathcal{H}^*(k)}{\partial\vec{x}^*(k)} + \vec{\chi}^T(k+1)\frac{\partial\mathcal{H}^*(k)}{\partial\vec{x}^*(k+1)} = 0 \tag{7.130a}$$

$$\vec{u}^T(k)\frac{\partial\mathcal{H}^*(k)}{\partial\vec{u}^*(k)} = 0 \tag{7.130b}$$

$$\vec{\xi}^T(k+1)\frac{\partial\mathcal{H}^*(k)}{\partial\vec{\lambda}^*(k+1)} = 0 \tag{7.130c}$$

Equations (7.130) can be also analyzed in which $\vec{\chi}(k)$, $\vec{u}(k)$, and $\vec{\xi}(k+1)$ cannot be equal to zero. Also, equation (7.130a) can be expanded to determine the transversality conditions for the problem. Because of this analysis, the conditions to solve the discrete optimal control problem can be written as,

$$\frac{\partial\mathcal{H}(k)}{\partial\vec{x}(k+1)} + \frac{\partial\mathcal{H}(k+1)}{\partial\vec{x}(k+1)} = 0 \tag{7.130a}$$

$$\frac{\partial \mathcal{H}(k)}{\partial \vec{u}(k)} = 0 \tag{7.130b}$$

$$\frac{\partial \mathcal{H}(k)}{\partial \vec{\lambda}(k+1)} = \vec{x}(k+1) - f(\vec{x}(k), \vec{u}(k), k) = 0 \tag{7.130c}$$

Equation (7.130a) is the discrete Euler–Lagrange equation. Equation (7.130b) provides the minimum conditions to obtain an admissible control function, and equation (7.130c) is the constraint equation imposed on the problem. The ( )* description is dropped out at this point. Furthermore, equation (7.130a) provides the transversality condition for the problem as,

$$\vec{\chi}^T(k) \frac{\partial \mathcal{H}(k)}{\partial \vec{x}(k)}\bigg|_{k=N_f} - \vec{\chi}^T(k) \frac{\partial \mathcal{H}(k)}{\partial \vec{x}(k)}\bigg|_{k=0} = 0 \tag{7.131}$$

In addition, equation (7.130a) can be rewritten to show a similar solution as in equation (7.57a). Equation (7.125) can be substituted into equation (7.130a) to describe the co-state variables as,

$$\vec{\lambda}(k) = \frac{\partial \mathcal{H}(k)}{\partial \vec{x}(k)} \tag{7.132}$$

Equation (7.132) is solved together with equation (7.130b) and (7.130c) to resemble the Pontryagin maximum principle (equation (7.56)) in the discrete domain. These equations can be used to solve the maximum time and maximum consumption problem which are left to the reader as an exercise. It is an interesting procedure to solve the discrete control problem to provide a similar solution as in the continuous domain.

7.4.4.2.1 Discrete linear quadratic regulator

One problem developed here is the discrete linear quadratic regulator (DLQR). The DLQR is not commonly used in practice but can be used to reduce many of the computational schemes for control systems. As shown in Section 7.3.6.2.7, the LQR has to be solved with a Runge–Kutta [14] routine. On the other hand, the DLQR uses an iterative procedure based on the present and future value of the state, co-state variables, and the control input function to solve the control problem.

The cost function of the DLQR control is written as,

$$J(k)=\frac{1}{2}(\vec{x}(N_f)-\vec{x}_D)^T \boldsymbol{F}(\vec{x}(N_f)-\vec{x}_D)$$
$$+\frac{\Delta t}{2}\sum_{k=0}^{N_{f-1}}(\vec{x}(k)-\vec{x}_D)^T \boldsymbol{Q}(\vec{x}(k)-\vec{x}_D)+\vec{u}^T(k)\boldsymbol{R}\vec{u}(k) \tag{7.133}$$

The existence of the optimal control is obtained from the solution of the discrete Hamilton–Jacobi equation which is defined everywhere to obtain the minimum conditions for the DLQR control problem.

The system of discrete linear equations can be defined from equations (7.116) but is expanded to include additional disturbing terms as,

$$\vec{x}(k+1)=\hat{\boldsymbol{A}}\vec{x}(k)+\hat{\boldsymbol{B}}\vec{u}(k)+\Delta t\vec{\psi}[\vec{x}^L(k)] \tag{7.134a}$$

$$\vec{x}^L(k) \in \vec{x}(k) \tag{7.134b}$$

where $\hat{\boldsymbol{A}}$ is the state matrix, $\hat{\boldsymbol{B}}$ is the control matrix, and $\vec{\psi}[\vec{x}^L(k)]$ is a $n \times 1$ matrix expressing the nonlinear terms. In equation (7.134b), $\vec{x}(k)$ provides a basis for the vector $\vec{x}^L(k)$ which is used to reflect the nonlinear terms in the perturbation vector, $\vec{\psi}[\vec{x}(k)]$. $\vec{x}^L(k)$ plays the same role as $\vec{x}(k)$; but, for the hierarchical control scheme, these two vectors are used differently. In the first run, $\vec{x}^L(k)$ is set to zero to solve the system of equations in the linear range; in this way, the solution of the control problem takes advantage of the linear form of the state equations. In the following executions, $\vec{x}^L(k)$ is updated with the known states (or $\vec{x}(k)$) to correct for the nonlinear terms in the state equation. For this reason, the difference between $\vec{x}(k)$ and $\vec{x}^L(k)$ is used in the discrete Hamilton–Jacobi equation to determine if the error between them is small in the following runs. If the errors between $\vec{x}(k)$ and $\vec{x}^L(k)$ are less than a tolerance value, the hierarchical control scheme has achieved a suboptimal solution. This error is added into the discrete Hamilton–Jacobi equation by multiplying $\vec{\gamma}(k+1)$ to this difference in which $\vec{\gamma}(k+1)$ is a Lagrange multiplier. $\vec{\gamma}(k+1)$ is known as the error vector which is used to correct the nonlinearities in the state equations in a piecewise manner. Thus, the discrete Hamiltonian for a hierarchical control system can be written as:

$$\mathcal{H}(k)=\frac{\Delta t}{2}(x(k)-x_D)^T \boldsymbol{Q}(x(k)-x_D)+\frac{\Delta t}{2}u^T(k)\boldsymbol{R}u(k)$$
$$+\vec{\lambda}^T(k+1)(\hat{\boldsymbol{A}}\vec{x}(k)+\hat{\boldsymbol{B}}\vec{u}(k)+\Delta t\vec{\psi}[\vec{x}^L(k)]) \tag{7.135}$$
$$+\vec{\gamma}^T(k+1)(\vec{x}(k)-\vec{x}^L(k))$$

$\vec{y}(k+1)$ is the error vector. The hierarchical control scheme uses the perturbation technique [86] to solve the two point boundary value problems (TPBVP). This hierarchical control scheme is used by Hassan et al. [101] [102] for the control of a turbo machine. In addition, the perturbation technique is also used by Minter et al. [96] to solve the TPBVP for an unconstrained problem defined continuously in time. In this present formulation, the perturbation technique is used to solve a constrained TPBVP.

The solution of the DLQR problem leads to a feedback system with the property that the components of the state vector $\vec{x}(k)$ are kept near the desired state vector $\vec{x}_D$ without excessive expenditure of control energy [87]. The existence of the control problem is obtained from the solution of the Hamilton–Jacobi equation which is defined everywhere to obtain the maximum conditions for the DLQR control problem. The maximum principles shown in equation (7.132), (7.130b), and (7.130c) are used to obtain the necessary conditions to maximize the solution of this nonlinear problem. In discrete format, the co-state vector and the control vector are equal to,

$$\vec{\lambda}(k) = \frac{\partial \mathcal{H}(k)}{\partial \vec{x}(k)} = \Delta t\, \boldsymbol{Q}(\vec{x}(k) - \vec{x}_D) + \hat{\boldsymbol{A}}^T \lambda(k+1) + \vec{\gamma}(k+1) \tag{7.136a}$$

$$\frac{\partial \mathcal{H}(k)}{\partial \vec{u}(k)} = \Delta t \boldsymbol{R} u(k) + \hat{\boldsymbol{B}}^T \vec{\lambda}(k+1) = 0 \tag{7.136b}$$

The control vector can be defined as,

$$u(k) = -(\Delta t \boldsymbol{R})^{-1} \hat{\boldsymbol{B}}^T \vec{\lambda}(k+1) \tag{7.137}$$

The transversality condition that relates the state vector with the co-state vector is determined in Appendix F. The same transversality condition in discrete format [100] is defined as,

$$\vec{\lambda}(k) = \boldsymbol{P}(k)\vec{x}(k) + \vec{G}(k) \tag{7.138}$$

where the $\boldsymbol{P}(k)$ and $\vec{G}(k)$ matrices, respectively, have dimensions of a $n \times n$ matrix and a $n \times 1$ matrix which depend on the final time. These matrices do not depend on the initial state; for this reason, $\boldsymbol{P}(N_f)$ and $\vec{G}(N_f)$ are set equal to $\boldsymbol{F}$ and $\boldsymbol{F}\vec{x}_D$, respectively. This condition is used to guarantee that the state vector stays near zero after an initial transient interval. This restriction on the $\boldsymbol{P}(k)$ and $\vec{G}(k)$ matrices introduces a process to obtain the solution of the state, co-state, and control vector. Substituting equations (7.137) and (7.138) into equation (7.134a), the state vector is defined as,

$$\vec{x}(k+1)\ [\boldsymbol{I} + \boldsymbol{S}\boldsymbol{P}(k+1)]^{-1}\ [\hat{\boldsymbol{A}}\vec{x}(k) - \boldsymbol{S}\vec{G}(k+1) + \Delta t \vec{\psi}[\vec{x}^L(k)]] \tag{7.139}$$

where,

$$S = \hat{B}(\Delta t R)^{-1}\hat{B}^T$$

The co-state equation is rewritten by substituting the transversality condition (equation (7.138)) and the discrete state equation (equation (7.139)) into equation (7.136a). From this single equation, certain terms can be grouped together to obtain $P(k)$ and $\vec{G}(k)$ as follows,

$$P(k) = \Delta t Q + \hat{A}^T P(k+1)\,[I + SP(k+1)]^{-1}\hat{A} \tag{7.140a}$$

$$\vec{G}(k) = -\Delta t Q\vec{x}_D + \hat{A}^T\vec{G}(k+1) + \vec{\gamma}(k+1) + \hat{A}^T P(k+1)\,[I + SP(k+1)]^{-1}\,[-S(k)\vec{G}(k+1) + \Delta t\vec{\psi}[\vec{x}^L(k)]] \tag{7.140b}$$

$P(k)$ and $\vec{G}(k)$, respectively, is defined as the discrete Riccati and adjoint Riccati equation. The purpose of the hierarchical control scheme is to take advantage of the linear form of the equations of motion to solve the control problem. Equation (7.140a) shows the advantage of the hierarchical control scheme, and equation (7.140b) is used to update the controller with the nonlinear effects of the state equations. These equations are solved in a stepwise manner to obtain the final suboptimal control problem. The solution of the DLQR problem is very similar to that shown in Ref. [100], but this problem becomes different when the error vector is updated in equation (7.140b) if the tolerance value is not satisfied. The following partial derivative is performed to update the error vector:

$$\frac{\partial\mathcal{H}(k)}{\partial\vec{x}^L(k)} = 0 = \Delta t\hat{J}^T\vec{\lambda}(k+1) - \vec{\gamma}(k+1)$$

$$\vec{\gamma}(k+1) = \Delta t\hat{J}^T[P(k+1)\,\vec{x}(k+1) + \vec{G}(k+1)] \tag{7.141}$$

where $\hat{J}$ is the Jacobian matrix of the column vector, $\psi[\vec{x}^L(k)]$. To obtain the value of $\vec{x}^L(k)$ in the updating process, the following partial derivative is performed in the Hamiltonian (equation (7.135)):

$$\frac{\partial\mathcal{H}(k)}{\partial\vec{\gamma}^T(k+1)} = 0 = \vec{x}(k) - \vec{x}^L(k) \tag{7.142}$$

Since $\vec{x}^L(k) \in \vec{x}(k)$, equation (7.142) is used to update the nonlinear terms in equations (7.134a) and (7.140b) for the following iteration $(L + 1)$.

This control scheme contains two levels for the calculation of the suboptimal control scheme. The first level contains the initial conditions and the solution of the Riccati equation since it does not depend on $\vec{x}^L(k)$ or the error vector. The second level contains the adjoint Riccati equation, the state vector, and the control vector. Also, this second level updates the error vector by means of equations (7.141) and (7.142) if the tolerance value is not satisfied.

This procedure at the second level stops when the difference between the previous (L) and following ($L + 1$) solution of the state vector is less than a tolerance value. When the tolerance value is satisfied, the final suboptimal control solution has been achieved. To calculate the error between the previous (L) and subsequent ($L + 1$) runs, the mean square error in discrete format is defined as follows,

$$E_r^2 = \Delta t \sum_{k=0}^{N_f-1} (\vec{x}^{L+1}(k) - \vec{x}^{L}(k))^T (\vec{x}^{L+1}(k) - \vec{x}^{L}(k)) \qquad (7.143)$$

where E_r^2 is the mean square error, and L defines the number of corrections performed in the second level of the hierarchical control scheme. If $E_r^2 < tol$ (tol is the tolerance value), the suboptimal solution has been achieved (i.e. the hierarchical control scheme converges); otherwise, the second level is executed once more to obtain the new state, control, and error vectors.

Figure 7.28 shows the flow chart representing the procedure to solve the hierarchical active control scheme. In Figure 7.28, the first level contains the calculation of the initial conditions and the solution to the Riccati equation. The Riccati equation is solved backwards in time from $\boldsymbol{P}(N_f) = \boldsymbol{F}$ to the initial time. The second level contains the correction of the adjoint Riccati equation, the state vector, the control vector, the error vector, and equation (7.141) to update the states. In the first execution ($L = 1$) of the second level, the error vector (equation (7.141)) and the perturbation vector, $\psi[\vec{x}^L(k)]$, is set equal to zero. In this way, the hierarchical control scheme solves the system of equations in the linear range [96]. The adjoint Riccati equation is also solved backwards in time from $\vec{G}(N_f) = \boldsymbol{F}\vec{x}_D$ to the initial time. The state and control vector are solved forward in time, knowing the initial conditions. Equation (7.143) is used to calculate the mean square error between the previous (L) and subsequent ($L + 1$) state vector to determine if the tolerance value has been achieved. If it is achieved, the solution of the DLQR is suboptimal; otherwise, the error vector is updated with equations (7.141) and (7.142), and L is increased by one. This last step is performed until the tolerance value is achieved.

Figure 7.28 shows that the previous state vector (for the run L) must be stored to be compared with the subsequent run ($L + 1$) of the state vector. The first run ($L = 1$) of the hierarchical control scheme is not sufficient to decide if the solution is suboptimal. For this reason, a second run ($L = 2$) of the second level is required to compare $L = 1$ and $L + 1 = 2$; in this way,

Figure 7.28 Flow chart explaining the hierarchical control scheme

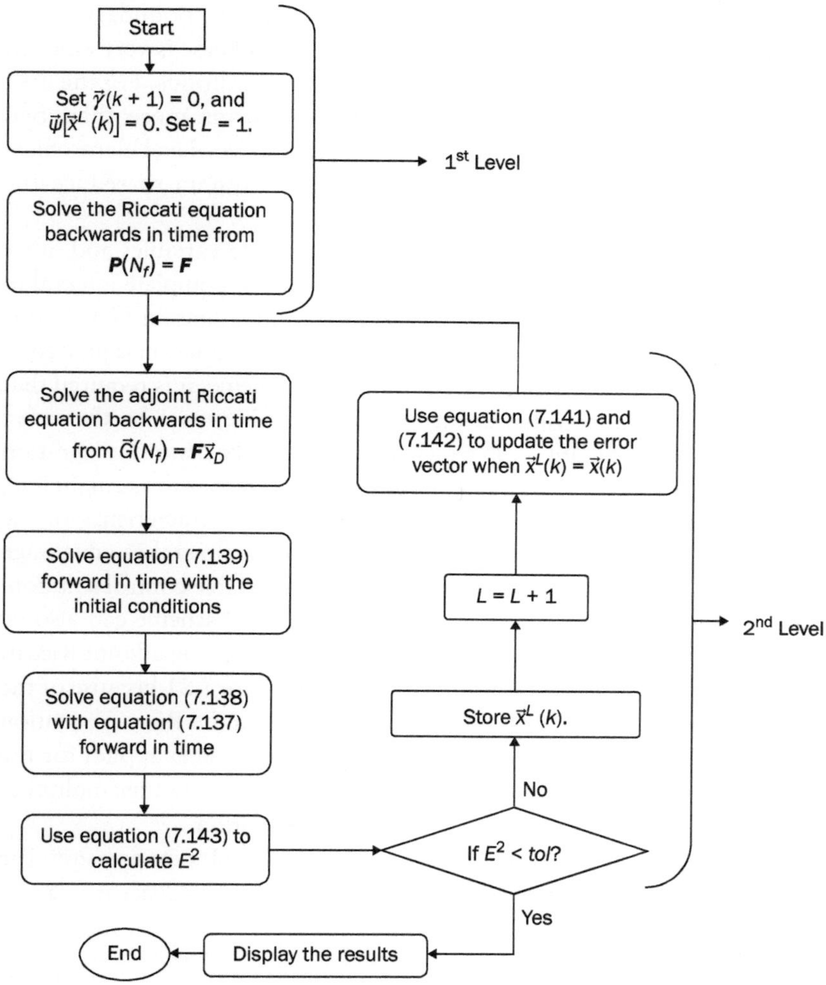

the system can decide if a suboptimal control solution is obtained. Also, L is set to a maximum value of iterations in case the DLQR control scheme fails to obtain a solution for the control problem.

7.4.4.2.2 Steady-state discrete linear quadratic regulator

The two-level hierarchical control scheme requires an iteration procedure to solve the suboptimal control problem as shown in Figure 7.28. The limitation of the two-level hierarchical method is the repetitive calculation of the second level as shown in Figure 7.28. For this reason, the computational method may require a longer period of time before a solution is obtained [103]. For solving two point boundary value problems [104], the two-level hierarchical control scheme is very efficient because of the requirements at the initial and end conditions. For the drift correction, the hierarchical control scheme performs more than two iterations before the solution is obtained; hence, the two-level hierarchical control scheme is re-expressed as a steady-state system to reduce the computational time and process.

A steady-state system is obtained when the co-state variables and other Lagrange multipliers have the same value through the complete interval in time [105]. In discrete format, the Riccati equation (equation (7.140a)) is solved backwards in time. One solution of the Riccati equation is plotted in Figure 7.21. For the discrete steady-state control scheme, it is required that the present and future value of any co-state variable must be the same; hence, the curve in Figure 7.21, which are due to the backward integration process, is eliminated, and the steady-state region, which is the straight line, is extended to infinity to maintain the steady-state characteristics of the system [87]. The steady-state hierarchical control scheme takes advantage of the linear form of the state equations to obtain a stable control solution. The advantage of the steady-state hierarchical control scheme can also be observed in the Riccati equation, or equation (7.140a). The adjoint Riccati equation also has a similar solution as shown in Figure 7.21 because of the backward integration process and the dependence on the Riccati equation to obtain its solution [87]. The steady-state condition also applies for the error vector, $\vec{\gamma}(k+1)$, in equation (7.141) because it is a Lagrange multiplier that is related to the Riccati and adjoint Riccati equation.

Applying the steady-state condition, the variables expressing the Riccati equation, adjoint Riccati equation, and error vector can be written as,

$$\hat{P}_{\infty} = P(k) = P(k+1) \tag{7.144a}$$

$$\hat{\vec{G}}_{\infty} = \vec{G}(k) = \vec{G}(k+1) \tag{7.144b}$$

$$\hat{\vec{\gamma}}_{\infty} = \vec{y}(k) = \vec{y}(k+1) \tag{7.144c}$$

Substituting equation (7.144a) into equation (7.140a), the Riccati equation can be written as,

$$\hat{P}_{\infty} = \Delta t Q + \hat{A}^T \hat{P}_{\infty} [I + S\hat{P}_{\infty}]^{-1} \hat{A} \tag{7.145}$$

In equation (7.145), the solution of the Riccati equation is obtained by solving a system of linear equations and shows the main advantage of the steady-state hierarchical control problem. In practice, this calculation can be performed once on a ground-based computer. Subsequently, these values can be supplied to the onboard computer and are used as part of the control software to calculate the control effort; in this way, the satellite does not require a communication link to the ground station to access the values. Because of this, the calculation process is significantly accelerated to reduce the computational time [**103**].

Using equation (7.144b), the steady-state form of the adjoint Riccati equation can be expressed as,

$$\begin{aligned}\vec{\hat{G}}_\infty &= -\Delta t Q\vec{x}_D + \hat{A}^T\vec{\hat{G}}_\infty + \vec{\hat{\gamma}}_\infty + \hat{A}^T\hat{P}_\infty [I + S\hat{P}_\infty]^{-1} \\ &\quad\times [-S\vec{\hat{G}}_\infty + \Delta t\vec{\psi}[\vec{x}(k)]]\end{aligned} \tag{7.146}$$

With equation (7.144), the steady-state form of the control and error vector equation can be written as,

$$u(k) = -(\Delta t R)^{-1}\hat{B}^T[\hat{P}_\infty\vec{x}(k) + \vec{\hat{G}}_\infty] \tag{7.147a}$$

$$\vec{\hat{\gamma}}_\infty = \Delta t\hat{J}^T[\hat{P}_\infty\vec{x}(k) + \vec{\hat{G}}_\infty] \tag{7.147b}$$

In equations (7.147), $\vec{x}(k)$ is defined in the present value (k) instead of the future value $(k + 1)$ because of the steady-state condition imposed on the problem and the transversality condition defined for the DLQR problem.

Using equation (7.147b), the steady-state adjoint Riccati equation can be expanded as,

$$\begin{aligned}\vec{\hat{G}}_\infty &= -\Delta t(M(k))^{-1}Q\vec{x}_D + \Delta t(M(k))^{-1}\hat{J}^T\hat{P}_\infty\vec{x}(k) \\ &\quad+\Delta t(M(k))^{-1}\hat{A}^T\hat{P}_\infty[I + W\hat{P}_\infty]^{-1}\vec{\psi}[\vec{x}(k)]\end{aligned} \tag{7.148}$$

where,

$$M(k) = [I - \hat{A}^T - \Delta t\hat{J}^T + \hat{A}^T\hat{P}_\infty[I + S\hat{P}_\infty]^{-1}$$

Substituting equation (7.148) into equation (7.147a) and grouping the terms, the control equation can be written as,

$$\vec{u}(k) = -K\vec{x}(k - K_\psi\vec{\psi}[\vec{x}(k)] + K_d\vec{x}_D \tag{7.149a}$$

where,

$$K = (\Delta t R)^{-1}\hat{B}^T\hat{P}_\infty + R^{-1}\hat{B}^T(M(k))^{-1}\hat{J}^T\hat{P}_\infty \tag{7.149b}$$

$$K_\psi = R^{-1}\hat{B}^T(M(k))^{-1}\hat{A}^T\hat{P}_\infty[I + S\hat{P}_\infty]^{-1} \tag{7.149c}$$

$$K_d = R^{-1}\hat{B}^T(M(k))^{-1}Q \tag{7.149d}$$

K is the control gain, K_ψ is the gain due to the nonlinear terms in the perturbation vector, $\psi[\vec{x}(k)]$, and K_d is the gain due to the desired gain. Equation (7.149b) shows that the effects of the nonlinear terms also affect the control gain. The Jacobian matrix is determined from the present states, $\vec{x}(k)$. Also, the gains in equations (7.149b–d) can be obtained depending on the actual state, $\vec{x}(k)$. Substituting equation (7.149a) into the discrete state equations (in equation (7.134a)), the state equation can be expressed as,

$$\vec{x}(k+1) = (\hat{\boldsymbol{A}} - \hat{\boldsymbol{B}}K)\vec{x}(k) + [\Delta t\boldsymbol{I} - \hat{\boldsymbol{B}}K_\psi]\vec{\psi}[\vec{x}^L(k)] + \hat{\boldsymbol{B}}K_d\vec{x}_D \qquad (7.150)$$

Using equation (7.150), Figure 7.29 shows the block diagram that defines the control procedure for the steady-state hierarchical control scheme. In Figure 7.29, all the states are known because they are defined in the present value (k).

Equation (7.100) is used to show a comparison between the continuous and discrete LQR controller. For this case, the Euler approximation is used to transform from the continuous domain to the discrete domain. The sampling time is 0.1 seconds, and the weighting matrices are the following: $Q = 5$ and $R = 0.11$. Figure 7.30 shows the solution for both continuous and discrete LQR. In Figure 7.30, the state shows a very similar solution, but the control input function is completely different. For the continuous LQR, the control input tends to move toward zero while the discrete LQR tends to stabilize near the value of the perturbation function. In addition, the control input function begins at a higher control effort in the discrete LQR than in the continuous LQR. These are the effects of choosing the sampling time and the weighting matrix. In the following chapter, a technique is introduced to help the designer to obtain

Figure 7.29 Block diagram for the discrete steady-state linear quadratic regulator

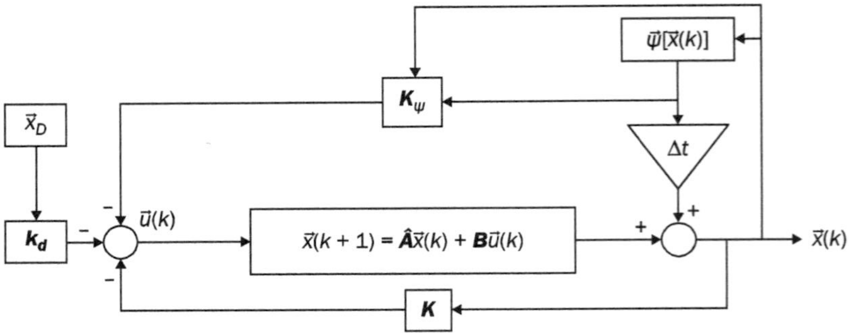

Figure 7.30 **Continuous and discrete linear quadratic regulator solution**

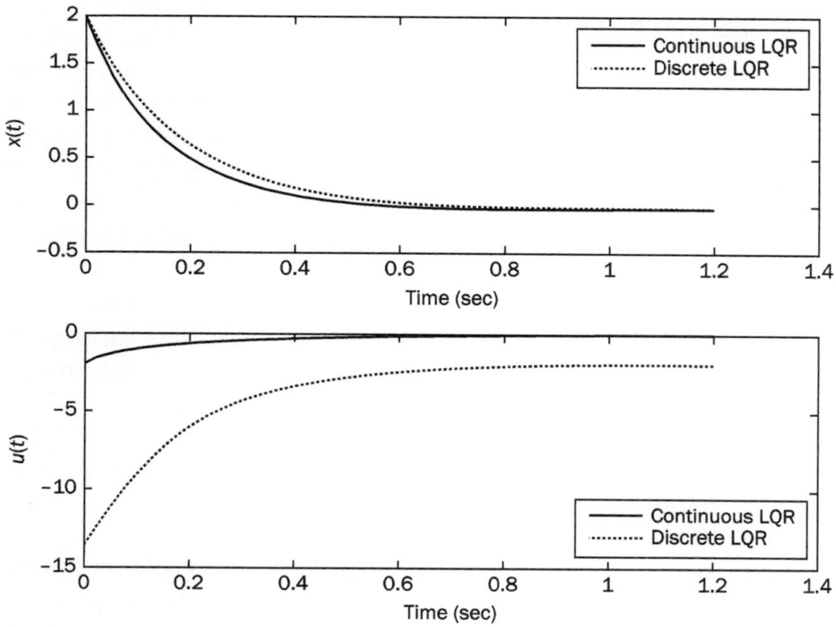

the control weighting matrix based on the desired transient response. With these known matrices, the control input function for the discrete domain can be reduced with respect to the continuous LQR.

7.5 Adaptive and intelligent controls

For launch vehicles and satellite applications, there is a common controller that has been used for a long period of time. This is the PID controller and requires to be thoroughly analyzed before it is used in any space vehicle. As shown in Chapters 3 and 6, the environment can be modeled, but these models are not 100% accurate. There are variations that are not taken into account while any vehicle is flying in Space or within the Earth's atmosphere. Because of these constraints on the model, an additional controller can be added to take into account these model errors and can compensate for those errors while flying the vehicle. These controllers are called adaptive control schemes and intelligent controllers.

The adaptive control schemes are used to either adapt the control gains or the control input functions such that it can take into account effects that are not modeled in the environment. On the other hand, intelligent controllers can be used as knowledge-based and learning systems to learn from the errors and make decisions about the amount of control effort used for a correction. Both adaptive and intelligent controllers use the error in the states as the input functions. For the adaptive control schemes, the formulation is more mathematical and is based on the solution of Lyapunov's stability criteria. In intelligent controllers, there are mathematical formulations but these are based on the human knowledge, human brain, and natural environment effects.

The objective of this section is to identify and provide some information about these controllers. In addition, the following chapters use some of these controllers to perform reconfiguration procedures and attitude corrections for a single satellite and formation flying.

7.5.1 Intelligent controllers

In recent years, the emergence of intelligent systems revolutionizes the area of control systems [**106**]. These intelligent systems adapt, learn, and take different decisions based on the particular problem [**107**]. Among these intelligent systems are the fuzzy logic (FL), neural networks (NN), and genetic algorithms (GA). These systems are widely used in industrial [**108**] [**109**] [**110**], pattern recognition, and image recognition applications [**107**]. In space applications, different intelligent systems have been used to solve the optimal control and data recognition problems, but these systems are commonly used in industrial applications because of the learning capability which leads to an adaptive control scheme.

The neural networks (NN) [**107**] is an intelligent system based on the reasoning of the human brain which uses the basis of machine learning to adapt and learn for the problem that it is solving. In space applications, the system [**111**] has been used to compress different orbital ephemeris data. The NN recognizes certain characteristics in the data such that it can be arranged and compressed; also, the NN eliminates the use of complex mathematical models. The NN requires prior knowledge of the data before its implementation which is implemented in reference [**111**], but its learning capability allows the NN system to adapt to the problem such that it can be used as an adaptive control scheme as shown in Reference [**109**].

A genetic algorithm (GA) is a different intelligent system based on the evolution of the code. The evolution of the GA consists in selection, mutation, and reproduction; also, this intelligent system is based on machine learning to adapt and learn from the problem. In space applications, this intelligent system has been used as an optimal controller [**112**] for station-keeping procedures and as an adaptive control system for interplanetary navigation [**113**]. The NN and GA are widely used through the literature as an adaptive controller because of their machine learning capabilities.

The NN and GA use the basis of machine learning which involves an adaptive mechanism that enables computers to learn by example and by analogy [**107**]. For this reason, NN and GA are mostly used in adaptive control systems [**107**] [**110**], but the disadvantage in both systems is that the data must be known before both systems are used in the actual process; thus, the NN and GA systems involve training. For this reason, the fuzzy logic controller can be used as a nonlinear controller. The fuzzy logic controller is based on the knowledge of the problem that is solved. Bennis et al. [**114**] showed that the fuzzy logic controller can be used for rendezvous and docking procedures with the International Space Station. In this study, the authors show that the fuzzy logic controller can be easily implemented with electronic devices, but, in Reference [**107**], Negnevitsky explained that a simple algorithm can be implemented into C language to obtain the fuzzy logic system. The fuzzy logic controller has not been used previously for reconfiguration procedures in satellite formations. Fuzzy logic systems can be converted into a hybrid system [**107**] in which this intelligent system combines the use of knowledge with a machine learning capability. This hybrid system was used in Reference [**115**] for the rendezvous and docking procedure of an automated vehicle, but this system is widely used in industrial applications [**116**] for the control of industrial motors.

FL can be applied as a nonlinear control because it is an intelligent system that does not require previous knowledge of the data and depends on the conditions and conclusions imposed on the system. NN and GA can be added to the FL controller to provide the learning capabilities to the knowledge based system. Because of these combinations, there are other forms of intelligent controllers such as the neuro-fuzzy controller that are used in some applications. Reference [**107**] explains the combinations that can be performed with the intelligent systems to provide a more robust intelligent controller for the vehicle. To demonstrate these controllers, Chapter 11 shows a fuzzy logic controller for the reconfiguration procedure of a constellation, and

Reference [107] shows the formulation and specification of a neural network system.

7.5.2 Adaptive controllers

In addition to intelligent control systems, there are controllers that can be adapted depending on the states. For intelligent systems, the controller is based on the knowledge and learning parameters. In adaptive controllers, a deep mathematical formulation is required to understand its development. The main reason to use adaptive controllers is due to the inaccuracy of the models for the vehicle motion and disturbances. The perturbations can be very accurately modeled, but there are unpredictable forces acting on the satellite that cannot be modeled. For this reason, the mathematical formulations developed for adaptive control schemes provide the necessary information to design a robust control system.

In order to understand adaptive control schemes, one of the common formulations used to develop the adaptive control scheme [117] is the Lyapunov function. Lyapunov [118] developed a criterion for the stability of certain nonlinear systems. The Lyapunov's direct method does not require a solution to the differential equations. If there exists a homogeneous function (V_L) which is positive definite, the system is asymptotically stable if,

$$\dot{V}_L < 0 \tag{7.151}$$

Using this Lyapunov condition for stability, an adaptive controller can be developed to satisfy the desired control function.

A simple mathematics explanation is provided as a basis to develop adaptive control schemes, but the full characterization of the problem is not shown here. Figure 7.31 shows the most common block diagram used in practice to develop adaptive control schemes. It is known as the model reference adaptive system. In Figure 7.31, an adaptive control function is based on the error obtained from the difference between the reference model and the process model [117], [119]. The process model is the actual motion of the body measured by the sensors. The reference model is the desired motion of the vehicle which is mostly described as a set of linear equations. The difference between the models is the error that has to be reduced. The reference model uses the same control input as the process model, and the common control system is known as the

Figure 7.31 **Model reference adaptive system**

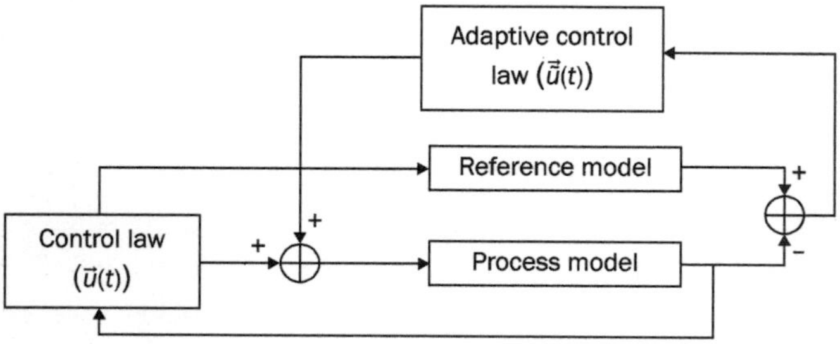

baseline controller. On the other hand, this baseline controller can be augmented with an adaptive controller that is based on the error. If the error is zero, the control input is not adapted; otherwise, the adaptive control scheme keeps varying the control effort. If a PID controller is augmented with an adaptive PD control scheme, the following equation is obtained,

$$\vec{u}_A(t) = \vec{u}(t) + \vec{\tilde{u}}(t) \tag{7.152a}$$

where,

$$\vec{u}(t) = K_p\vec{x} + K_d\dot{\vec{x}} + K_I\int x\,dt \qquad \vec{\tilde{u}}(t) = \tilde{K}_p\vec{e}_x + \tilde{K}_d\vec{e}_{\dot{x}} \tag{7.152b}$$

$\vec{u}(t)$ is the baseline PID control law, and $\vec{\tilde{u}}(t)$ is the adaptive control law. $\tilde{K}_p$ and $\tilde{K}_d$ are the adapted proportional and derivative control gains, respectively. e_x is the error for the actual state and $e_{\dot{x}}$ is the error due to the derivative. If the error is zero, the adaptive scheme becomes zero, and the PID controller is not affected.

These controllers based on equation (7.152) provide the basis for many problems in aerospace systems. Many of this problems are used in flight control as shown in References [**120**] and [**121**]. In a similar manner, Figure 7.31 can be changed to include a neural network instead of an adaptive control scheme. Basically, the neural network is updated with the error function obtained from the difference between the reference and the process models [**122**]. Furthermore, the adaptive models can be extended to fuzzy controllers [**123**].

In summary, adaptive controllers is another technique that can be used to provide additional robustness to the controller. In Chapter 11, it is shown how an adaptive scheme can be added to a reconfiguration procedure for a constellation.

7.6 Suggested problems

Problem 7.1. Assuming the normal force can be written as,

$$N_F = \frac{\partial C_{N_F}}{\partial \alpha} QS\alpha = N'\alpha$$

$$\gamma = \frac{\dot{z}}{|\vec{V}|}$$

Rewrite the linearized attitude for the Ares V rocket (see equations 7.4) to,

$$\ddot{\varphi} = N_{RoCS}/J_1$$

$$\ddot{\theta} = -d_1\theta + d_2\dot{z} + b_1\beta$$

$$\ddot{\psi} = -d_3\psi + b_2\delta$$

where,

$$d_1 = \frac{L_{CP}(A_F + N') - 2gI_{SP}^{SRB}\dot{m}_{SRB}}{J_2} \qquad d_2 = \frac{L_{CP}N'}{J_2|\vec{V}|} \qquad d_3 = \frac{L_{CP}A_F - 2gI_{SP}^{SRB}\dot{m}_{SRB}}{J_3}$$

$$b_1 = \frac{6gI_{SP}^{RS-68b}\dot{m}_{RS-68b}L_{GP}}{J_2} \qquad b_2 = \frac{6gI_{SP}^{RS-68b}\dot{m}_{RS-68b}L_{GP}}{J_3}$$

Problem 7.2. Using the assumption in **Problem 7.1** and the linearization assumptions, show that the linear equations for the translational motion in the inertial frame are written as,

$$m\ddot{x} = F$$

$$m\ddot{y} = e_1\psi + c_1\delta$$

$$m\ddot{z} = e_2\theta + e_3\dot{z} - c_1\beta$$

where

$$e_1 = A_F - 2gI_{SP}^{SRB}\,\dot{m}_{SRB} + mg$$

$$e_2 = -A_F + 2gI_{SP}^{SRB}\,\dot{m}_{SRB} - mg - N'$$

$$e_3 = \frac{N\varepsilon'}{|\vec{V}|} \qquad c_1 = 6gI_{SP}^{RS-68b}\dot{m}_{RS-68b}$$

$$F = -A_F + 2gI_{SP}^{SRB}\,\dot{m}_{SRB} + 6gI_{SP}^{RS-68b}\,\dot{m}_{RS-68b} - mg$$

Problem 7.3. The pitch motion and the Z translational motion for the Ares V rocket are decoupled in the linear equations. Write the pitch and Z translational equation of motion in state vector format as,

$$\dot{\vec{Y}} = A\vec{Y} + \boldsymbol{B}u$$

where $\vec{Y} = [\theta\ \dot{\theta}\ \dot{z}]^T$ and $u = \beta$. Determine the state matrix $\boldsymbol{A}$ and control matrix $\boldsymbol{B}$. Assuming $y \approx 0$, what are the conditions such that the system is stable?

Problem 7.4. When the Shuttle is launched from Kennedy Space Center, the complete space transportation system (STS) does not have a roll control system (RoCS) because it is controlled with the solid rocket boosters. Let's assume that it is possible to do the same roll control maneuver with SRBs from Ares V while the RS-68bs are not gimbaled during the first stage flight. The equations of motion can be written as,

$$\ddot{\varphi} = b_1(\beta_R + \beta_L)$$

$$\ddot{\theta} = -d_1\theta + b_2(\beta_L - \beta_R)$$

$$\ddot{\psi} = -d_3\psi - b_3\delta$$

where,

$$d_1 = \frac{L_{CP}(A_F + N') - 6gI_{SP}^{RS\text{-}68b}\dot{m}_{RS\text{-}68b}}{J_2} \qquad d_2 = \frac{L_{CP}N'}{J_2|\vec{V}|} \qquad d_3 = \frac{L_{CP}A_F - 6gI_{SP}^{RS\text{-}68b}\dot{m}_{RS\text{-}68b}}{J_3}$$

$$b_1 = \frac{2gI_{SP}^{SRB}\dot{m}_{SRB}L_{GP}}{J_1} \qquad b_2 = \frac{2gI_{SP}^{SRS}\dot{m}_{SRB}L_{GP}}{J_2} \qquad b_3 = \frac{2gI_{SP}^{SRS}\dot{m}_{SRB}L_{GP}}{J_3}$$

β_L and β_R is the left and right gimbal angle for the left and right SRB, respectively.

a. What equations are couple and decouple?

b. Write the state based representation for the couple equation.

c. Write the state based representation for the decouple equation. Under what conditions the decouple equation is stable?

Problem 7.5. Using the yaw equation,

$$\ddot{\psi} = -d_3\psi + b_2\delta$$

A closed-loop system can be drawn with an actuator ($G_a(s)$) and control ($G_c(s)$) transfer function as shown in Figure 7.32. In Figure 7.32, an

Figure 7.32 Problem 7.5

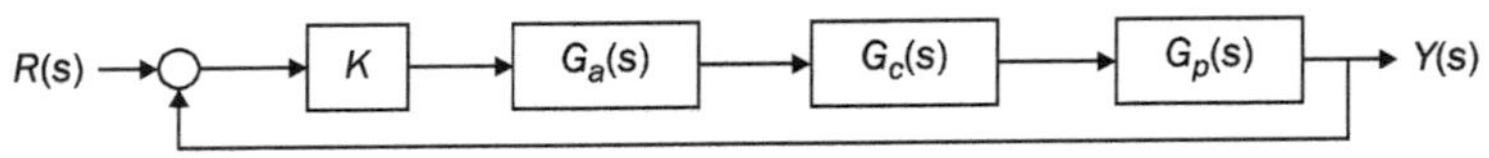

additional gain is multiplied in the same control loop system to compensate for the known actuator in the system.

a. Obtain the open-loop transfer function ($Y(s)/R(s)$) and develop a PD controller when K and $G_a(s)$ are unity. Assume that the desired damping and natural frequency for the system is 0.6 and 0.4 rad/sec. Also $d_3 = 0.4$. and $b_2 = 0.2$.

b. Assume that,

$$Ga(s) = \frac{z_1}{s + z_1}\left(\frac{\omega_n^2}{s^2 + 2\xi\omega_n s + \omega_n^2}\right)$$

where $z_1 = 7$, $\xi = 0.5$, and $\omega_n = 56$. It is not desired to calculate a new set of PD gains for the controller; instead, the gain K is used to determine the compensation for the controller in Figure 7.32. Using the Routh–Hurwitz method, determine the minimum value of K that provides the necessary compensation for the actuator model knowing the PD gains.

c. Plot the results without the actuator model. Compare the results with the actuator model and the gain K.

Problem 7.6. Use the state based form of the couple set of equations in **Problem 7.4** to determine the control gain matrix, **K** when

$$\mathbf{K} = \begin{bmatrix} k_1 & k_2 & k_3 & k_4 \\ k_1 & -k_2 & k_3 & -k_4 \end{bmatrix}$$

and the desired second order responses for the roll and pitch can be written as,

$$F(s) = (s^2 + 2\xi_\phi\omega_{n,\phi}s + \omega_{n,\phi}^2)(s^2 + 2\xi_\theta\omega_{n,\theta}s + \omega_{n,\theta}^2)$$

where $\xi_\phi = 0.6$, $\xi_\theta = 0.65$, and $\omega_{n,\phi} = \omega_{n,\theta} = 0.4$. Plot the results of the state based equation and the control input function, $\vec{u} = -\mathbf{K}\vec{x}$, when $\vec{x}(0) = [1\ {-2}\ 0\ 0.1]^T$. Write the state vector as $\vec{x} = [\phi\ \phi\ \dot{\theta}]^T$. Also assumed $b_1 = 0.1$, $b_2 = 0.2$, and $d_1 = 0.4$.

Problem 7.7. It is known that the shortest distance between two points is a straight line. For $x(t_0) = a$ and $x(t_1) = b$, the cost function describing this particular path can be written as,

$$\mathcal{F} = \int_{t_0}^{t_1} \sqrt{1 + \dot{x}^2}$$

Solve the two point boundary value problem using the Euler–Lagrange equation.

Problem 7.8. Show that the Euler–Lagrange equation,

$$\frac{\partial F}{\partial x} - \frac{d}{dt}\left(\frac{\partial F}{\partial \dot{x}}\right) = 0$$

can be written in the alternate form as,

$$\frac{d}{dt}\left(F - \dot{x}\frac{\partial F}{\partial \dot{x}}\right) = \frac{\partial F}{\partial x}$$

Problem 7.9. Using the alternate form of the Euler–Lagrange equations, solve the two point boundary value problem for the following cost function,

$$\mathcal{F} = \int_{t_0}^{t_1} (m\dot{x}^2 - cx^2)dt$$

The conditions for the problem are the following: $x(t_0) = a$ and $x(t_1) = b$. This cost function defines the optimal path to move a vehicle from altitude a to altitude b in a period of time. Plot some results when $x(t_0 = 0) = 0$ and $x(t_1 = 100) = 1000$ meters for $m = 100$ kg and $c = 9.81$ (1/sec).

Problem 7.10. Develop an expression for the locus of the first switching point in the fourth quadrant for the minimum gas consumption problem presented in Section 7.3.6.2.6. Assume that the initial states is defined by $x^0{}_1$, $x^0{}_2$ for the sequence {–1,0,1}. Note that in general $x_1^0 \neq 0$, $x_2^0 \neq 0$.

Problem 7.11. Consider the single axis attitude control of a rigid body under the influence of a single actuator which provides torque only about that axis. Denote the deviation in attitude between a desired pointing direction and a reference axis on the vehicle by α.

a. In the absence of other disturbing torques write down the equation of motion and recast this in the form: $\dot{\vec{x}} = \boldsymbol{A}\vec{x} + \boldsymbol{B}\vec{u}$. Indicate clearly the dimensionality of every matrix in this equation.

b. The optimal control is selected such as to minimize the following cost functional:

$$J = \frac{1}{2}\int_{t_0}^{\infty}(\vec{x}^T Q\vec{x} + \vec{u}^T R\vec{u})dt$$

where the elements of the Q matrix are selected such that $q_{ii} = q$ for the penalty weights on the coordinates and $q_{jj} = q_1$ for the penalty weight on the angular velocities and $q_{ij} = 0$ for $i \neq j$, and the R is selected so that the penalty on the control, $\vec{u}$ is r. For this application write out the form of the matrix Riccati differential equation and obtain the steady state solution for the matrix, P_∞. Be sure to obtain the solution which results in P_∞ being positive definite.

c. Use the results from (b) to obtain the optimal control law of the form $\vec{u}(t) = -K_\infty \vec{x}(t)$. Show that the position gain element(s) in P_∞ are equal to $\sqrt{qr/r}$ and that the rate gain element(s) also depend on q_1.

Problem 7.12. Using the pitch equation in Problem 7.1 when $\gamma \approx 0$,

$$\ddot{\theta} = -d_1\theta + b_1\beta$$

a. Obtain the Z-transform of the pitch equation when $d_1 = -0.2$ and $b_1 = 0.4$. Assume that the sampling time is 0.1 seconds.

b. Write the Z-transfer function for a PD controller knowing that $\beta(t) = K_p\theta + K_d\dot{\theta}$.

c. Obtain the open-loop Z-transfer function for the closed-loop system shown in equation (7.119).

d. Use Routh–Hurwitz to determine the minimum condition to obtain a stable condition for the discrete control problem.

e. Write the state-based formulation for the discrete equation in part (a) and use the Ackermann's formula to determine the control gain matrix when the desired second order function has $\xi_\theta = 0.65$ and $\omega_{n,\theta} = 0.4$. Assume that the sampling time is 0.1 seconds.

f. Show some results and compare part (d) and (e).

Problem 7.13. Using the minimum time consumption problem when the final time is unknown, develop the problem in the discrete domain using equations (7.130). Also, show the procedure to solve the problem using the boundary conditions expressed in Section 7.3.6.2.5. Use the same assumed values for the state equation, and assume a sampling time of 0.1 seconds.

8

Example

Abstract: The control schemes can be applied to correct the attitude motion of a satellite. In this chapter, two common methods used in small satellites are presented. The B-dot controller and the linear quadratic regulator (LQR) are explained to control a small satellite with only magnetic torquers; in addition, some sensors are mentioned because the control scheme depends on the type of sensors in the satellite. In summary, the purpose of this chapter is to demonstrate different schemes to maintain the attitude motion of a small satellite.

Key words: attitude control scheme, frame rotation scheme, b-dot controller, periodic linear quadratic regulator, weight design for the linear quadratic regulator, small satellite attitude control and motion.

8.1 Introduction to examples in spacecraft attitude dynamics and control

In the previous chapters, the Ares V rocket is used as the main example used for the definition of the attitude dynamics and control. The purpose of this chapter is to apply the same attitude formulation for nanosatellites. A nanosatellite does not refer to the size of the satellite; it refers to the weight of the satellite. Table 8.1 shows the categories of small satellites classified by weight. As an example, Figure 8.1 shows a minisatellite called the Fast Affordable Science and Technology Satellite (FASTSAT) developed by NASA George C. Marshall Space Flight Center. These small satellites have the same attitude hardware as a larger satellite without

Table 8.1 Satellite classification

Satellite classification	Weight (kg)
Minisatellite	100 to 500
Microsatellite	10 to 100
Nanosatellite	1 to 10
Picosatellite	0.1 to 1

Figure 8.1 Fast Affordable Science and Technology Satellite (FASTSAT) structural diagram

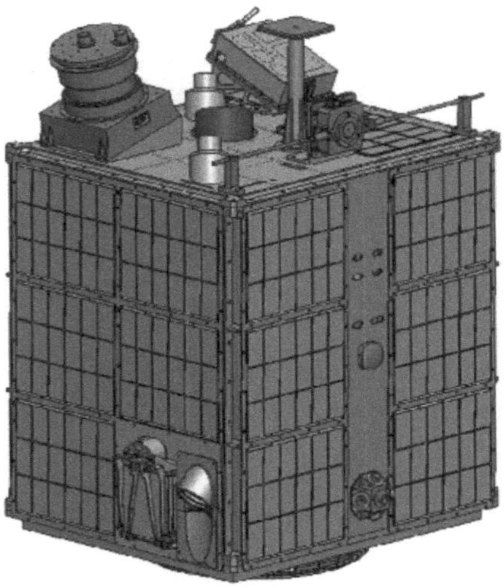

occupying a large space. In addition, these satellites are used in formation flying because they can be stacked together and launched once the rocket reaches the orbit. In summary, this chapter shows some of the attitude control techniques used to maintain the orientation of the satellite in a circular orbit based on the attitude hardware available for the mission.

8.2 Nanosatellite problem definition

The first step in working with nanosatellites is to understand the attitude dynamics. From these equations of motion, the stability conditions can be obtained to maintain a local vertical orientation with respect to the gravity of the Earth. The steps to describe the attitude motion of a satellite are the same as explained in Section 5.7.

First, assume a small satellite with mass, m, orbiting about the Earth in a circular orbit. Also, the satellite is assumed to have three-axes magnetic torquers to maintain the orientation of the satellite. The attitude motion of the satellite is sensed with sun sensors, magnetometers, and star trackers [124]; but the explanation of these sensors is beyond the scope of this book. The satellite has a reference body axis expressed in the local vertical/local horizontal frame, and the body frame is described with the aerospace rotation. The reference (X^+, Y^+, Z^+) and body (x^+, y^+, z^+) frame is shown in Figure 8.2. Two environmental torques are added to the formulation – gravity-gradient and magnetic torques. The gravity-gradient can be used to stabilize the satellite such that the satellite is always pointing toward the Earth; the magnetic torquer is the actuator that interacts with the environment to cause a rotation of the body.

Knowing the specifications of the problem, the first step is to describe the rotational matrix with the aerospace rotation as,

$$\mathbf{R} = \begin{bmatrix} \cos\psi\cos\theta & \sin\psi\cos\varphi & -\sin\theta \\ \cos\psi\sin\theta\sin\varphi - \sin\psi\cos\varphi & \sin\psi\sin\theta\sin\varphi + \cos\psi\cos\varphi & \cos\theta\sin\varphi \\ \cos\psi\sin\theta\cos\varphi + \sin\psi\sin\varphi & \sin\psi\sin\theta\cos\varphi - \cos\psi\sin\varphi & \cos\theta\cos\varphi \end{bmatrix}$$

Figure 8.2 Satellite reference and body frames

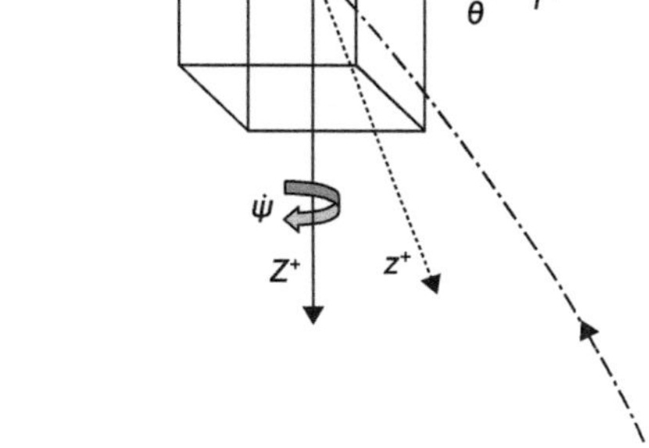

Using small angle assumptions, the rotational matrix is approximated by [11],

$$\mathbf{R} \approx \begin{bmatrix} 1 & \psi & -\theta \\ -\psi & 1 & \varphi \\ \theta & -\varphi & 1 \end{bmatrix} \tag{8.1}$$

The angular velocity of the vehicle ($\vec{\varpi}_R$) in the reference frame using small angle assumptions can be described as,

$$\vec{\varpi}_R = \begin{bmatrix} \dot{\varphi} - \dot{\psi}\sin\theta \\ \dot{\theta}\cos\varphi + \dot{\psi}\sin\varphi\cos\theta \\ -\dot{\theta}\sin\varphi + \dot{\psi}\cos\varphi\cos\theta \end{bmatrix} \approx \begin{bmatrix} \dot{\varphi} \\ \dot{\theta} \\ \dot{\psi} \end{bmatrix} \tag{8.2}$$

To cause a rotation of the satellite such that it points toward the Earth, the angular rate of the satellite is equal to its mean motion; for this reason, the mean motion of the vehicle can be written as,

$$\vec{\varpi}_n = \begin{bmatrix} 1 & \psi & -\theta \\ -\psi & 1 & \varphi \\ \theta & -\varphi & 1 \end{bmatrix} \begin{bmatrix} 0 \\ -n \\ 0 \end{bmatrix} = \begin{bmatrix} -n\psi \\ -n \\ n\varphi \end{bmatrix} \tag{8.3}$$

Then, the angular rate of the satellite in the orbit is equal to,

$$\vec{\varpi} = \vec{\varpi}_R + \vec{\varpi}_n = \begin{bmatrix} \dot{\varphi} - n\psi \\ \dot{\theta} - n \\ \dot{\psi} + n\varphi \end{bmatrix} \tag{8.4}$$

And,

$$\dot{\vec{\varpi}} = \begin{bmatrix} \ddot{\varphi} \\ \ddot{\theta} \\ \ddot{\psi} \end{bmatrix} + \begin{bmatrix} 0 & 0 & -n \\ 0 & 0 & 0 \\ +n & 0 & 0 \end{bmatrix} \begin{bmatrix} \dot{\varphi} \\ \dot{\theta} \\ \dot{\psi} \end{bmatrix} \tag{8.5}$$

Euler's equation of motion in the principal moments of inertia is expressed as follows,

$$\dot{\vec{\varpi}} = -J^{-1}\vec{\varpi} \times J\vec{\varpi} + J^{-1}\vec{N}_E + J^{-1}\vec{N}_m \tag{8.6}$$

where J is the principal moment of inertia, $\vec{N}_E$ is the gravity-gradient torque, and $\vec{N}_m$ is the magnetic torque.

The first term on the right hand side (RHS) of equation (8.6) can be written as,

$$-J^{-1}\vec{\varpi}\times J\vec{\varpi} = \begin{bmatrix} -n^2\alpha_1 & 0 & 0 \\ 0 & 0 & 0 \\ 0 & 0 & n^2\alpha_3 \end{bmatrix}\begin{bmatrix} \varphi \\ \theta \\ \psi \end{bmatrix} + \begin{bmatrix} 0 & 0 & -n\alpha_1 \\ 0 & 0 & 0 \\ -n\alpha_3 & 0 & 0 \end{bmatrix}\begin{bmatrix} \dot{\varphi} \\ \dot{\theta} \\ \dot{\psi} \end{bmatrix} \tag{8.7}$$

where,

$$\alpha_1 = \frac{J_2 - J_3}{J_1} \qquad \alpha_3 = \frac{J_1 - J_2}{J_3}$$

Second, the gravity-gradient torque is described from equations (6.13) in Section 6.2.1 as,

$$N_{E,x} = 3n^2(J_3 - J_2)\hat{R}_2\hat{R}_3 \tag{8.7a}$$

$$N_{E,y} = 3n^2(J_1 - J_3)\hat{R}_1\hat{R}_3 \tag{8.7b}$$

$$N_{E,z} = 3n^2(J_2 - J_1)\hat{R}_1\hat{R}_2 \tag{8.7c}$$

where $\hat{\hat{R}}$ are the direction cosines of the position vector from the center of the Earth to the satellite location in the orbit; thus,

$$\hat{\hat{R}} = \mathbf{R}\begin{bmatrix} 0 \\ 0 \\ -1 \end{bmatrix} = \begin{bmatrix} +\theta \\ -\varphi \\ -1 \end{bmatrix} \tag{8.8}$$

Substituting equation (8.8) into equation (8.7), the gravity-gradient torque is equal to,

$$J^{-1}\vec{N}_E = \begin{bmatrix} -3n^2\alpha_1 & 0 & 0 \\ 0 & -3n^2\alpha_2 & 0 \\ 0 & 0 & 0 \end{bmatrix} = \begin{bmatrix} \varphi \\ \theta \\ \psi \end{bmatrix} \tag{8.9}$$

where,

$$\alpha_2 = \frac{J_1 - J_3}{J_2}$$

Finally, the magnetic torques are written as follows,

$$\vec{N}_m = -\vec{B}_m \times \vec{D} = \begin{bmatrix} 0 & B_{m,3} & -B_{m,2} \\ -B_{m,3} & 0 & B_{m,1} \\ B_{m,2} & -B_{m,1} & 0 \end{bmatrix} \begin{bmatrix} m_1 \\ m_2 \\ m_3 \end{bmatrix} \tag{8.10a}$$

where,

$$\vec{B}_m = \mathbf{R}\vec{b} \tag{8.10b}$$

$\vec{b}$ is the Earth's magnetic field that can be described by a dipole model shown in Section 6.2.4. For this case, the simple magnetic dipole of the Earth is used,

$$\vec{b}_d = -\frac{D_E \sin\phi}{\vec{R}^3}\hat{\imath} - \frac{2D_E \cos\phi}{\vec{R}^3}\hat{k} \tag{8.11}$$

Performing a rotation about the Z axis and multiplying the transformation matrix to equation (8.11), the Earth's magnetic field in the orbital plane can be described as,

$$\vec{b} = -\frac{D_E}{\vec{R}^3}\begin{bmatrix} -\cos\lambda \sin i \\ \sin\lambda \sin i \\ -2\cos i \end{bmatrix} \tag{8.12}$$

where ϕ is also known as the inclination angle, and λ is the longitude angle equal to nt. Substituting equation (8.12) into equation (8.10b), $\vec{B}_m = \vec{b}$ for small angle assumptions [**68**], and the magnetic torques are equal to,

$$\vec{N}_m = \begin{bmatrix} 0 & \frac{b_3}{J_1} & -\frac{b_2}{J_1} \\ -\frac{b_3}{J_2} & 0 & \frac{b_1}{J_2} \\ \frac{b_2}{J_3} & -\frac{b_1}{J_3} & 0 \end{bmatrix} \begin{bmatrix} m_1 \\ m_2 \\ m_3 \end{bmatrix} \tag{8.13}$$

Finally, Euler's equation in linear form can be written in state-based format from equations (8.6) to (8.13) as,

$$\dot{\vec{x}} = \boldsymbol{A}\vec{x} + \boldsymbol{B}\vec{u} \tag{8.14a}$$

where,

$$A = \begin{bmatrix} \mathbf{0} & I \\ A_q & A_v \end{bmatrix} \qquad B = \begin{bmatrix} \mathbf{0} \\ B_c \end{bmatrix} \tag{8.14b}$$

$$A_q = \begin{bmatrix} -4n^2\alpha_1 & 0 & 0 \\ 0 & -3n^2\alpha_2 & 0 \\ 0 & 0 & n^2\alpha_3 \end{bmatrix} A_v = \begin{bmatrix} 0 & 0 & +n(1-\alpha_1) \\ 0 & 0 & 0 \\ -n(1+\alpha_3) & 0 & 0 \end{bmatrix} \tag{8.14c}$$

$$B_c = \begin{bmatrix} 0 & \frac{b_3}{J_1} & -\frac{b_2}{J_1} \\ -\frac{b_3}{J_2} & 0 & \frac{b_1}{J_2} \\ \frac{b_2}{J_3} & -\frac{b_1}{J_3} & 0 \end{bmatrix} \tag{8.14d}$$

$$\vec{x}(t) = [\varphi(t) \quad \theta(t) \quad \psi(t) \quad \dot{\varphi}(t) \quad \dot{\theta}(t) \quad \dot{\psi}(t)]^T$$
$$\vec{u}(t) = [m_1(t) \quad m_2(t) \quad m_3(t)]^T$$

0 is a 3×3 matrix of zeros, and I is a 3×3 identity matrix. The linearized system can be analyzed with either the Routh–Hurwitz criteria or the eigenvalue problem to determine the stability conditions. In References [43] and [44], the stability conditions are determined such that an additional mass can be located on the body of the satellite to provide gravity-gradient stabilization [53]; this stabilization results in less use of control torque to maintain the attitude motion of the vehicle. Normally, this set of equations can be used to describe the attitude motion for any nanosatellite. Depending on the attitude determination scheme [124], the satellite can sense some of the states providing different attitude control schemes.

8.3 B-dot controller for fast corrections

The designer has to know the available sensors to determine the attitude control system. Depending on the sensor, the error in the attitude knowledge can be reduced [10]. For cube satellites, the attitude knowledge is based on two basic sensors – the sun sensor [125] and the magnetometer [126]. The sun sensor is used to determine the location of the Sun with respect to the satellite. Using the sun sensor only, the attitude motion can be determined with an error of nearly 10 degrees [**10**]. On the other hand, the magnetometer measures the magnetic field of the Earth. Using the magnetometer and the sun sensor, the error in the orientation of the vehicle can be reduced to approximately 5 degrees [**10**].

Using these sensors, there is a method that reduces the angular momentum of the vehicle without the knowledge of the body rates [**127**]. This method is known as the B-dot controller. The commanded magnetic dipole moment is defined to be directly proportional to the time derivative of the Earth's magnetic field. The magnetic dipole command is equal to,

$$\vec{m} = -k_d \dot{\vec{b}} \tag{8.15}$$

where,

$$k_d = \begin{bmatrix} k_{d,x} & 0 & 0 \\ 0 & k_{d,y} & 0 \\ 0 & 0 & k_{d,z} \end{bmatrix}$$

k_d is the gain matrix for the b-dot controller. The time derivative of the Earth's magnetic field is determined by the measurements obtained from the magnetometer. By using Euler's theorem in the present time or filtering the signal with a high-pass filter, the time derivative of the magnetic field is written as,

$$\dot{\vec{b}} = \frac{\vec{b}(k) - \vec{b}(k-1)}{\Delta t} \tag{8.16}$$

$\vec{b}(k)$ is the present magnetic field of the Earth, and $\vec{b}(k-1)$ is the previous magnetic field. $\vec{b}$ should be sampled at a higher rate to have stability and efficiency for the attitude correction. The angular momentum decreases over an orbit as the magnetic field changes direction. Usually, the damping of the angular rates is performing at least in one orbit. The b-dot controller is less effective for low inclination and equatorial orbits. Once the rates are damped, the satellite begins to rotate at 2 revolutions per orbit (see Chapter 6).

Take as an example a microsatellite with the following principal moments of inertia: $J_1 = 15\ \mathrm{kg-m^2}$, $J_2 = 16\ \mathrm{kg-m^2}$, and $J_3 = 12\ \mathrm{kg-m^2}$. Also, the microsatellite is in a circular polar orbit ($i = 90°$) with an altitude of 800 km. The gains of the b-dot control scheme are $k_{d,x} = k_{d,z} = 0$, and $k_{d,y} = 5 \times 10^{11}$. These gains are sufficiently large so that they can produce at least $1\ A - m^2$ of dipole moment. Figure 8.3 shows the solution for the

Figure 8.3 B-dot controller solution for the pitch motion

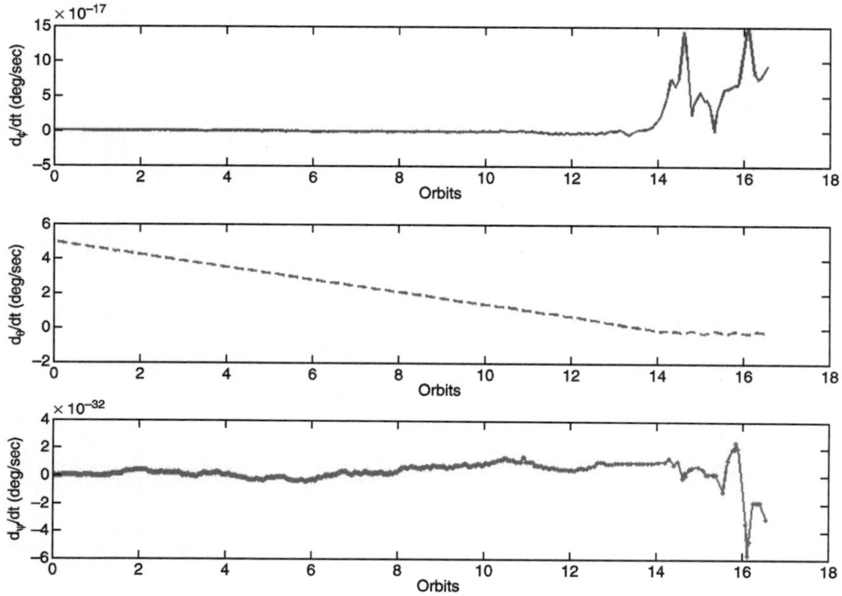

controller when the yaw magnetic torquer is used. For this microsatellite, the pitch angle has been gradually reduced in time. Initially, the satellite has an angular rate about pitch of 5 (deg/sec). Figure 8.4 shows the solution in which the pitch angular rate has been reduced in approximately 14 orbits. If a larger gain is used, the angular rate can be reduced in a shorter period of time. The other two rates (roll and yaw) are not highly affected by using this type of control scheme.

8.4 Linear quadratic regulator for attitude correction

It is possible to know the attitude rates if there are geographical positioning system (GPS), accelerometers, and gyroscopes [128]. By knowing this information, other attitude control schemes can be implemented such as the linear quadratic regulator (LQR). To perform science and technology experiments, the satellite is required to have various orientations at certain locations in the orbit; for this reason, the satellite's pointing requirements may vary between attitude orientations. Many of the experiments may require a slew maneuver which makes the satellite

Figure 8.4 B-dot control solution for the body rates

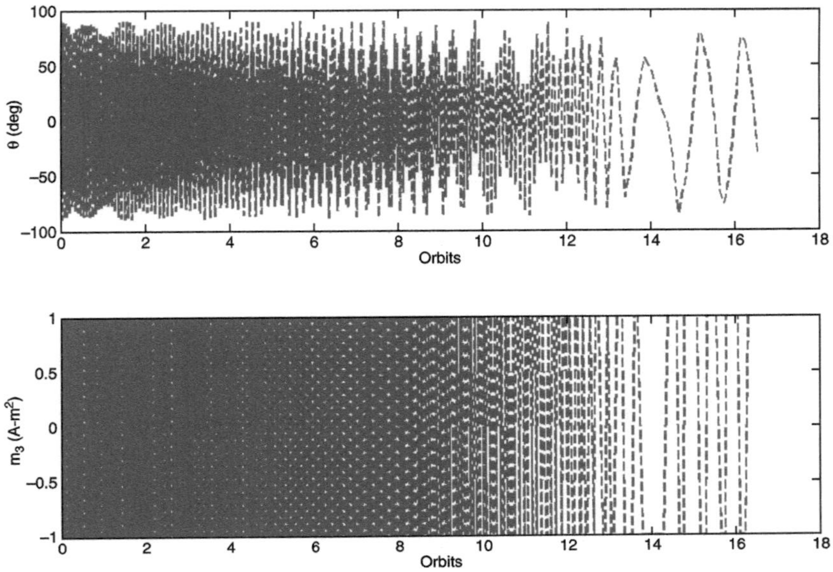

gyroscopically unstable. A triad of magnetic torque rods can be used as the actuator which provides instantaneous torques about two body axes. These axes are not necessarily lined up with the body reference axes and results in one uncontrolled axis at all time. There are other actuators that can be used such as the ion thrusters [**129**] and the reaction wheels [**130**], but these actuators (so far) have not been implemented for cube satellites.

This section shows two different techniques that can be applied to maintain the attitude motion of a satellite with a LQR control scheme. One technique is the steady-state formulation for the Riccati equation based on the periodicity of the magnetic field; and the second technique is the rotational frame technique for the analysis of the attitude motion. These analyses provide insight information about the dynamic equations of the satellite.

8.4.1 Linear quadratic regulator attitude correction via periodic earth's magnetic field

The LQR controller is an optimal controller that minimizes the correction of the states without excessive use of control effort. The LQR uses two weighting matrices for the state and control vectors. The cost function for the discrete LQR controller can be written as follows:

$$\mathcal{J}(k)=\frac{\Delta t}{2}\sum_{k=0}^{N_f-1}\left(\vec{x}(k)-\vec{x}_D\right)^T \boldsymbol{Q}\left(\vec{x}(k)-\vec{x}_D\right)+\vec{u}^T(k)\boldsymbol{R}\vec{u}(k)$$

$\boldsymbol{Q}$ is a $n \times n$ positive semidefinite matrix, and $\boldsymbol{R}$ is a $m \times m$ positive definite matrix. n is the number of state variables, and m is the number of control inputs. The minimum conditions are applied to the cost function and are constrained to the state vector in equation (8.14); the solution is given by the Riccati equation and is described as follows,

$$\hat{P}_\infty = \Delta t\boldsymbol{Q} + \hat{\boldsymbol{A}}^T\hat{P}_\infty [\boldsymbol{I} + \boldsymbol{S}\hat{P}_\infty]^{-1}\hat{\boldsymbol{A}} \tag{8.17}$$

where $\boldsymbol{P}_\infty$ is the Riccati matrix, $\hat{\boldsymbol{A}} = \boldsymbol{I} + \boldsymbol{A}\Delta t$, $\hat{\boldsymbol{B}} = \Delta t\boldsymbol{B}$, and $\boldsymbol{S} = \boldsymbol{B}\boldsymbol{R}^{-1}\boldsymbol{B}^T$. The Riccati equation enhances the development of the control system. One of the disadvantages of the LQR controller is the selection of the weighting matrices. This selection is not intuitive and it is required to guess the weighting matrices.

Because of the definition of the Earth's magnetic field, the Riccati matrix should be calculated for every location in the orbit. This calculation creates a new control gain which causes a larger computational time in the computer onboard the satellite. On the contrary, Psiaki [68] proposed a different scheme on the periodicity of the Earth's magnetic field. In equation (8.12), the magnetic field of the Earth is a periodic function, i.e. $\boldsymbol{B}(t) = \boldsymbol{B}(t + \mathrm{T})$ where T is the period of the orbit. If it is assumed that the Earth's magnetic field is periodic, the solution of the Riccati equation is also periodic. By taking the average over one period of the square of the $\boldsymbol{B}$ matrix,

$$\tilde{\boldsymbol{B}}\tilde{\boldsymbol{B}}^T = \frac{1}{T}\int_0^T \boldsymbol{S}(\tau)\,d\tau = \frac{\Delta t}{T}\sum_{k=0}^{N-1}\boldsymbol{S}(k) \tag{8.18}$$

A steady state solution of the Riccati equation can be obtained and is written as,

$$\hat{P}_{SP} = \Delta t\boldsymbol{Q} + \hat{\boldsymbol{A}}^T\hat{P}_{SP} [\boldsymbol{I} + \tilde{\boldsymbol{B}}\tilde{\boldsymbol{B}}^T \hat{P}_{SP}]^{-1}\hat{\boldsymbol{A}} \tag{8.19}$$

where $\hat{P}_{SP}$ is the steady-state periodic solution of the Riccati equation. Then, the control input function, $\vec{u}(t)$, becomes as,

$$\vec{u}(k) = -(\Delta t\boldsymbol{R})^{-1}\hat{\boldsymbol{B}}^T\hat{P}_{SP}\,\vec{x}(k) = -\boldsymbol{K}_{SP}\,\vec{x}(k) \tag{8.20}$$

$\hat{P}_{SP}$ is a constant matrix. Substituting equation (8.14d) into equation (8.18) and taking $\boldsymbol{R}$ as an identity matrix, the average over one period of the Earth's magnetic field is expressed as,

$$\tilde{\boldsymbol{B}}\tilde{\boldsymbol{B}}^T = \frac{1}{T}\int_0^T \begin{bmatrix} \mathbf{0} & \mathbf{0} \\ \mathbf{0} & \boldsymbol{B}_C\boldsymbol{B}_C^T \end{bmatrix} d\tau \Rightarrow \tilde{\boldsymbol{B}}_C\tilde{\boldsymbol{B}}_C^T = \frac{1}{T}\int_0^T \boldsymbol{B}_C\boldsymbol{B}_C^T\, d\tau \tag{8.22}$$

A unit control weighting matrix maintains a minimum use of control effort. As shown by Battin [**16**], the average over one period can be determined from the orbital elements and is shown in Reference [**131**]. On the other hand, the average over one period can be performed numerically [**14**]. Solving equation (8.22) numerically, the average over one period for the square of the magnetic field in the LVLH frame can be approximated to,

$$\tilde{\boldsymbol{B}}_C\tilde{\boldsymbol{B}}_C^T = \frac{\Delta t}{T}\sum_{k=0}^{N_f-1} \boldsymbol{B}_C(k)\boldsymbol{B}_C^T(k) \approx \begin{bmatrix} c_1^2 & 0 & 0 \\ 0 & c_2^2 & 0 \\ 0 & 0 & c_3^2 \end{bmatrix} \tag{8.23}$$

Equation (8.23) simplifies the solution of the Riccati equation. The main objective of using equation (8.23) is to decouple equations (8.14); in addition, the analysis of the LQR controller can be easily simplified and solved in the discrete domain.

Using the same orbit as in the previous example but reducing the principal inertia matrix to J_1 = 0.15 kg – m^2, J_2, = 0.165 kg – m^2, and J_3 = 0.12 kg – m^2, Figure 8.5 shows the solution for LQR control scheme when $Q = diag(2,1,2,10,15,20)$. In approximately 5 orbits, the attitude angle and rates have been reduced for the satellite. In Figure 8.6, the dipole used has been maintained to a minimum in which the magnetic torquer for the yaw axis shows the maximum (initial) dipole moment.

8.4.2 Frame transformation attitude control scheme

All the linearizations of the equations of motion are performed about the origin of the satellite body axis. If a commanded attitude angle is larger than 10 degrees, the equations of motion are not the same. In addition, the LQR control scheme does not provide an appropriate compensation for the attitude motion of the satellite. For this reason, the attitude dynamics can be studied to determine how the magnetic field, the gyroscopic terms, and gravity gradient terms affect the satellite motion. The frame rotation scheme can be implemented to determine the new equations of motion such that the LQR optimal control can be applied.

Figure 8.5 Angular motion and body rates for the periodic control solution

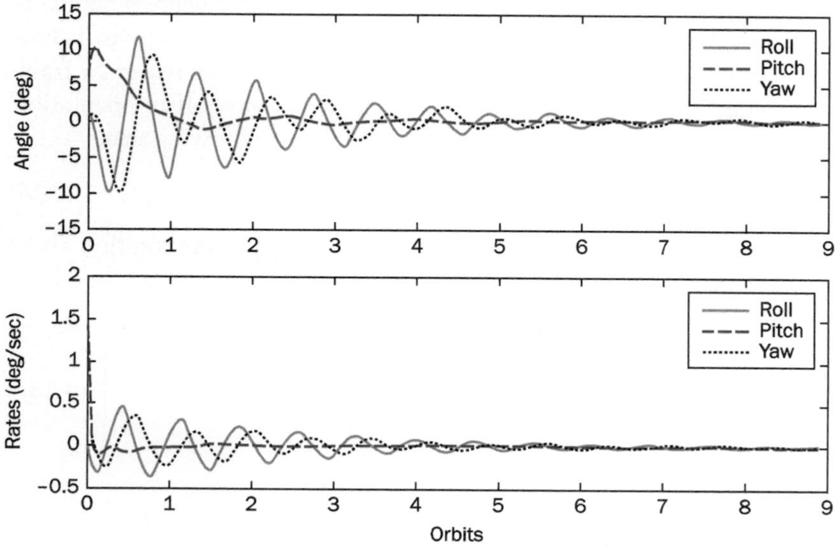

Figure 8.6 Dipole moment for the periodic solution

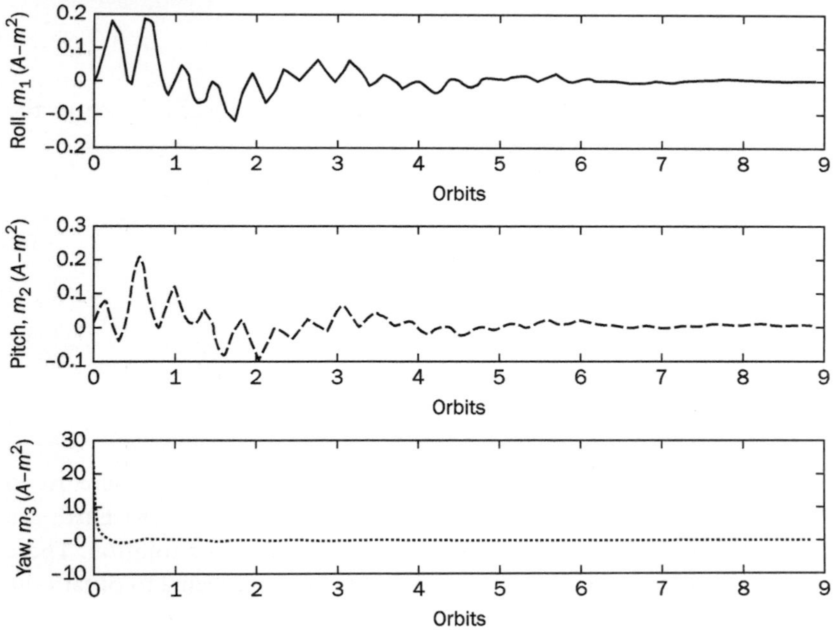

The discrete time invariant LQR optimal controller [**103**] can be used but should be modified depending on the desired orientation of the satellite. As an example, a satellite is required to be rotated 90 degrees about the yaw axis. Because of this rotation, the dynamical equations behave differently leading to an unstable attitude point. For this reason, a frame rotation is required to express the dynamics in the yaw offset orientation. Following the same procedure as in Section 8.2 but knowing that,

$$\vec{\omega}_{yaw} = \boldsymbol{R}_{LVLH2YAW}\vec{\omega} \tag{8.17}$$

The linearized equations of motion for a 90 degree frame rotation about the yaw axis from the LVLH frame are given by,

$$\boldsymbol{A_q} = \begin{bmatrix} -4n^2\alpha_2 & 0 & 0 \\ 0 & 3n^2\alpha_1 & 0 \\ 0 & 0 & n^2\alpha_3 \end{bmatrix} \boldsymbol{A_v} = \begin{bmatrix} 0 & 0 & n(1-\alpha_2) \\ 0 & 0 & 0 \\ -n(1+\alpha_3) & 0 & 0 \end{bmatrix} \tag{8.18a}$$

$$\boldsymbol{B_c} = \begin{bmatrix} \frac{b_3}{J_2} & 0 & \frac{-b_2}{J_2} \\ 0 & \frac{b_3}{J_1} & \frac{b_1}{J_1} \\ -\frac{b_1}{J_3} & 0 & \frac{-b_2}{J_3} \end{bmatrix} \tag{8.18b}$$

where $\boldsymbol{R}_{LVLH2YAW}$ is the rotational matrix from the LVLH frame to the yaw offset frame and is written as,

$$\boldsymbol{R}_{LVLH2YAW} = \boldsymbol{R}_{yaw}\boldsymbol{R} \approx \begin{bmatrix} -\psi & 1 & \varphi \\ -1 & -\psi & \theta \\ \theta & \varphi & 1 \end{bmatrix}$$

$$\boldsymbol{R} = \begin{bmatrix} 0 & 1 & 0 \\ -1 & 0 & 0 \\ 0 & 0 & 1 \end{bmatrix}$$

From equation (8.18a), the roll equation is decoupled if there is no control effort. Also, the elements of the magnetic field of the Earth are rearranged to take into account the 90 degree yaw offset rotation. These equations can be used in the previous LQR control scheme to obtain the new periodic steady-state solution for the Riccati equation.

The control algorithm can be described in terms of the quaternions and angular velocities. In the computer onboard the satellite, the attitude determination system outputs the quaternions and angular rates in the reference frame. A transformation is performed to convert the quaternions into the body frame and is be expressed as,

$$q_B = q q_{R2B} \tag{8.19}$$

q and q_B, respectively, are the quaternions specifying the orientation of the satellite in the LVLH and body frame. q_{R2B} is the quaternion defining the rotation from the LVLH frame to the body frame and is defined depending on the desired orientation. The angular velocities in the body frame are transformed as follows,

$$\vec{\varpi}_B = \boldsymbol{R}_{R2B}\, \vec{\varpi} \tag{8.20}$$

where $\boldsymbol{R}_{R2B}$ is the rotational matrix from the LVLH to the body frame. For any of the rotating frames, the attitude maneuvers are performed about the center of mass of the satellite. With this information, the controller commands the magnetic torquers to change the orientation of the satellite.

On the ground, one steady-state Riccati matrix is stored in the computer onboard the satellite per desired orientation. In orbit, the attitude control scheme calls the Riccati matrix associated with any other commanded orientation. The attitude control algorithm works as follows:

1. Initialize the quaternions and angular rates in the reference frame.
2. Check the desired orientation for the satellite.
3. Transform the quaternion and angular rates from the reference frame to the desired body frame with equations (8.19) and (8.20).
4. Look up the $\hat{P}_\infty$ for the desired new frame.
5. Calculate the control input in equation (8.20).
6. Send the control input solution to the magnetic torquers to actuate the vehicle.

After commanding the satellite, the attitude controller is not based on one orientation but is defined according to the desired attitude pointing. Every attitude pointing has unique characteristics which are specified in the attitude controller. This is the main advantage of the attitude controller presented here.

The attitude controller uses the principal axis moments of inertia to simplify the formulation. For a fully populated inertia matrix, the principal axis moments of inertia and the direction cosines are determined [43]; then a transformation from the reference frame into the principal axis frame is performed for the quaternions, angular rates, and control input functions. The control input functions are then used to control the magnetic torquers.

8.5 Linear quadratic regulator control weight design

With the properties of the LQR, any system can be stabilized if the system has at least one pole in the negative left hand side (LHS) of the imaginary axis [87]. This solution of the LQR control problem depends on the chosen and/or guessed state and control weighting matrices. Therefore, the selection of the weighting matrices is an important design variable. Typically, a good initial guess and various iterations to update the weighting matrices are required in order to obtain the desired second order response for the system.

The literature contains several methods for obtaining the LQR weighting matrices that move beyond the typical trial-and-error method. Kawasaki and Shimemura [132] determine the weighting matrices by performing successive iterations of their algorithm. The weighting matrices are not ensured to be positive definite but can be used to locate the poles in a specified region. Broussard [133] defines a cost function based on the quadratic weighting of the control gain matrix to calculate the weighting matrices in a minimization constraint problem. Broussard [133] ensures the positive definiteness of the weighting matrices by expressing the cost function as a quadratic multiplication; but the computational scheme does not provide the desired output in a short period of time because it requires the selection of a desired control gain. On the other hand, Becus and Sonmez [134] use a different method based on the eigenstructure assignment [88]. By manipulating the variables symbolically and using a minimization scheme, they can obtain a positive definite weighting matrix. They define two constraints based on the positive definiteness of the solution of the Riccati equation and the desired negative roots of the system. In this method, the control weighting matrix is defined as an identity matrix to provide a minimum consumption of control energy.

The eigenstructure assignment is a very efficient method to obtain the control matrices. Choi and Seo [135] use the eigenstructure assignment to obtain the weighting matrix for a desired set of closed-loop control poles. This method requires that the designer indentify the closed-loop eigenvalues and eigenvectors to determine the weighting matrices for the LQR controller. This method does not ensure that the weighting matrix is positive definite because the designer is forcing the LQR controller to have a certain set of eigenvalues and eigenvectors. Luo and Lan [136] use another minimization procedure by selecting certain values for the weighting matrices. This method uses the desired eigenvalues to determine the weighting matrices. Once more, the weighting matrices are not ensured to be positive definite. In Reference [137], the authors explained that a negative definite weighting matrix can be used to obtain a stable problem. This contradicts the solution shown in Athans and Falb [87] that verifies and explains why the weighting matrices are positive definite.

More recent papers use genetic algorithms [138] and multi-objective evolution algorithms [139] to obtain the weighting matrices for the LQR. The authors of these papers define the state weighting matrix as a principal diagonal matrix with the off-diagonal terms equal to zero. In their work, the weighting matrix becomes positive definite when the principal diagonal terms are larger than zero. Also, they use a control weighting matrix different than the identity matrix. In common practice, the control weighting matrix is set to an identity matrix to maintain a minimum control effort; but, the state weighting matrix is set to any value to stabilize the system in minimum time.

The objective of this section is to present a similar minimization and symbolic procedure to determine the state weighting matrix as in Reference [134]. The difference between this section and Reference [134] is the constraints used to minimize the equation which are based on the positive definite form of the state weighting matrix. The control weighting matrix is always equal to the identity matrix to maintain a minimum control effort. In addition, the method in this section uses a desired set of poles, but the eigenvectors are expressed as a combination of known and unknown variables. Depending on the selection of the eigenvector, the state weighting matrix increases the emphasis on the translational and/or rotational motion of a system. After solving the minimization procedure, the state weighting matrix is always positive definite and provides the desired poles for the system.

8.5.1 Weighting matrix optimization

The formulation of the optimization scheme is based on the steady-state formulation of the LQR, but the weighting matrices determined in this scheme can be used to solve the discrete LQR. Using the steady-state system, the Riccati equation reduces to,

$$P_\infty(t)A + A^T P_\infty(t) - P_\infty(t)SP_\infty(t) + Q = 0 \tag{8.21}$$

And, the optimal control problem solution [87] can be described as,

$$\vec{u}\ (t) = -R^{-1}B^T\ (t)P_\infty\ (t)\vec{x}(t) \tag{8.22a}$$

$$\dot{\vec{x}}(t) = A_C\vec{x}(t) \qquad A_C = A - B(t)R^{-1}B^T\ (t)P_\infty(t) \tag{8.22b}$$

A_C is known as the closed-loop state matrix. The solution of equation (8.21) is obtained by solving a system of linear equations. The current disadvantage of the LQR is the somewhat ambiguous selection of the weighting matrices to determine the best response. The selection of these matrices must be performed in an iterative manner to obtain the desired response and roots of the closed-loop system. In many references [74] [138] [139], the Q and R matrices are assumed to be principal diagonal matrices with positive values, although, mathematically, a positive definite matrix is not required to be principal diagonal and can have off diagonal terms. An eigenstructure assignment is used here to select the weighting matrices. The control weighting matrix is set equal to the identity matrix to maintain a minimum consumption of control effort.

The eigenstructure assignment [88] is used to obtain the control gain for a system. This problem is also known as a pole placement technique. Also, the eigenstructure assignment requires the definition of a desired eigenvector system. The eigenstructure assignment is expressed in terms of an eigenvalue problem [134] for a closed-loop system defined as,

$$A_C y = y\Lambda \tag{8.23}$$

where Λ is a $2n \times 2n$ matrix containing the eigenvalues of A_C in the principal diagonal, and y is the eigenvector matrix associated with A_C. By substituting the definition for A_C in equation (8.22b), equation (8.23) can be written as,

$$M = (A - B(t)B^T(t)P_\infty(t))y_D - y_D\Lambda_D = 0 \tag{8.24}$$

M is the closed loop equation for the LQR problem. For this development, R is an identity matrix. Choi and Seo [135] solved a similar eigenstructure assignment problem to determine the weighting matrices which are not guarantee to be positive definite. The main reason is that they are forcing the eigenvectors and eigenvalues of a desired closed-loop system to the actual system. Instead, the method presented by Becus and Sonmez [134] enforces the positive definiteness of the weighting matrices by minimizing an equation. In this method, the desired real negative roots for the closed-loop system are defined, but the eigenvectors are expressed as a combination of unknown and known variables. Their method requires the solution of the eigenvalue problem at every iteration to determine if the roots are negative. The disadvantage of this method [134] is that the state weighting matrix is not used as a constraint. The only constraint to ensure a positive definite Q matrix is the Riccati equation.

The present problem is solved for the proportional-derivative (PD) controller because the state vector formulation is commonly used in control problems. The general formulation for the state vectors can be written in the form of equation (8.14b), but the control matrix is expressed as,

$$B_C = \begin{bmatrix} d_1 & 0 & 0 \\ 0 & d_2 & 0 \\ 0 & 0 & d_3 \end{bmatrix} \tag{8.25}$$

B_C is a $m \times m$ principal diagonal control matrix with constant terms d_1, d_2, and d_3. In many cases, the B_C matrix is not a principal diagonal matrix but can be used for the general formulation. The Riccati equation, the desired eigenvectors, and the desired eigenvalues can be written as,

$$P_\infty = \begin{bmatrix} P_{11} & P_{12} \\ P_{12}^T & P_{22} \end{bmatrix} \quad y_D = \begin{bmatrix} y_{11} & y_{12} \\ y_{21} & y_{22} \end{bmatrix} \quad \Lambda_D = \begin{bmatrix} \Lambda_{11} & 0 \\ 0 & \Lambda_{22} \end{bmatrix} \tag{8.26}$$

All the submatrices in equation (8.26) have $n \times n$ dimensions. The Riccati matrix, P_∞, is assumed to be a symmetric matrix because of the symmetry of Q and the quadratic multiplication of P_∞ in the left hand side (LHS) of equation (8.21). Solving equation (8.24) with the general form of equations (8.14b), (8.25) and (8.26), the eigenvector assignment in equation (8.24) can be rewritten as,

$$M_{11} = y_{21} - y_{11}\Lambda_{11} \quad M_{12} = y_{22} - y_{12}\Lambda_{22} \tag{8.27a}$$

$$M_{21} = (A_q - B_C(t)B_C^T(t)P_{12}^T)y_{11} + (A_v - B_C(t)B_C^T(t)P_{22})y_{21} - y_{21}\Lambda_{11} \tag{8.27b}$$

$$M_{22} = (A_q - B_C(t)B_C^T(t)P_{12}^T)y_{12} + (A_v - B_C(t)B_C^T(t)P_{22})y_{22} - y_{22}\Lambda_{22} \tag{8.27c}$$

Equations (8.27a) can be solved to define the eigenvectors in equations (8.27b) and (8.27c). The solution of equations (8.27a) depends on the desired eigenvectors, Λ_D. After the substitution of equations (8.27a) into equations (8.27b) and (8.27c), a constraint minimum problem is solved to obtain the values for the Riccati equation and eigenvectors. The minimization equation can be written as follows,

$$G = \min \Sigma_{i=1}^{n} \Sigma_{j=1}^{n} [M_{21}]_{ij}^{2} + [M_{22}]_{ij}^{2} \tag{8.28}$$

where $[\quad]_{ij}^{2}$ refers to the square of the $i^{\underline{th}}$ and $j^{\underline{th}}$ element of the matrix between the brackets. There are more unknown variables in equations (8.28) than known variables. For this reason, the constraint equations are obtained from the steady-state Riccati equation.

Substituting equations (8.14b), (8.25), and (8.26) into equation (8.21), the state weighting matrix, $\mathbf{Q}$, can be written as,

$$\mathbf{Q}_{11} = \mathbf{P}_{12}\mathbf{B}_C\mathbf{B}_C^T\mathbf{P}_{12}^T - \mathbf{P}_{12}\mathbf{A}_q - \mathbf{A}_q^T\mathbf{P}_{12}^T \tag{8.29a}$$

$$\mathbf{Q}_{12} = \mathbf{P}_{12}\mathbf{B}_C\mathbf{B}_C^T\mathbf{P}_{22}^T - \mathbf{P}_{11} - \mathbf{A}_q^T\mathbf{P}_{22} - \mathbf{P}_{12}\mathbf{A}_v \tag{8.29b}$$

$$\mathbf{Q}_{21} = \mathbf{P}_{22}\mathbf{B}_C\mathbf{B}_C^T\mathbf{P}_{12}^T - \mathbf{P}_{11} - \mathbf{A}_v^T\mathbf{P}_{12}^T - \mathbf{P}_{22}\mathbf{A}_q \tag{8.29c}$$

$$\mathbf{Q}_{22} = \mathbf{P}_{22}\mathbf{B}_C\mathbf{B}_C^T\mathbf{P}_{22}^T - \mathbf{P}_{12} - \mathbf{A}_v^T\mathbf{P}_{22} - \mathbf{P}_{22}\mathbf{A}_v - \mathbf{P}_{12}^T \tag{8.29d}$$

These are the constraint equations used to minimize the problem with equation (8.28). To ensure a positive definite matrix, it is known that the principal diagonal of the $\mathbf{Q}$ matrix should be larger and/or equal to zero, and the off-diagonal terms of $\mathbf{Q}$ should be zero. For equation (8.29a) and (8.29d), the following constraints can be established:

$$diag(\mathbf{Q}_{11}) \geq 0 \text{ and } diag\,(\mathbf{Q}_{22}) \geq 0 \tag{8.30a}$$

$$off\ diag(\mathbf{Q}_{11}) = 0 \text{ and } off\ diag\,(\mathbf{Q}_{22}) = 0 \tag{8.30b}$$

where *diag* ( ) and *off diag* ( ) refers to the principal diagonal and off-diagonal terms of a matrix, respectively. Knowing that $\mathbf{Q}$ must be symmetric ($\mathbf{Q} = \mathbf{Q}^T$), equations (8.29b) and (8.29c) are set to zero ($\mathbf{Q}_{12} = \mathbf{Q}_{21} = 0$); then,

$$0 = -\mathbf{P}_{12}\mathbf{B}_C\mathbf{B}_C^T\mathbf{P}_{22}^T + \mathbf{P}_{22}\mathbf{B}_C\mathbf{B}_C^T\mathbf{P}_{22}^T + \mathbf{A}_q^T\mathbf{P}_{22} + \mathbf{P}_{12}\mathbf{A}_v - \mathbf{A}_v^T\mathbf{P}_{12}^T - \mathbf{P}_{22}\mathbf{A}_q \tag{8.30c}$$

Hence, equation (8.28) is minimized using the constraint equations (8.30) to ensure a positive definite matrix for $\mathbf{Q}$. In addition, the negative real roots for the closed-loop problem are obtained. After the minimization problem is solved, P_{11} can be calculated as follows,

$$\mathbf{P}_{11} = \mathbf{P}_{12}\,\mathbf{B}_C\mathbf{B}_C^T\mathbf{P}_{22}^T - \mathbf{A}_q^T\mathbf{P}_{22} - \mathbf{P}_{12}\mathbf{A}_v \tag{8.31}$$

Solving equations (8.29) after the minimization of the constraint problem, the state weighting matrix can be reconstructed as follows,

$$Q = \begin{bmatrix} Q_{11} & 0 \\ 0 & Q_{22} \end{bmatrix} \quad (8.32)$$

Q is a principal diagonal matrix if equation (8.28) is minimized. If a local minimum is obtained, Q_{11} and Q_{22} could be a $n \times n$ matrix but still guarantees a positive definite Q matrix.

There are other forms of this problem that can be solved. If complex conjugate poles are used, the problem in equation (8.24) can be expressed as [134],

$$M = (A - B(t)B^T(t)P_\infty(t))y_D L - y_D \Lambda_D L = 0 \quad (8.33)$$

L is a $n \times n$ transformation matrix used to make M real. As an example, L can be written as,

$$L = \begin{bmatrix} 0.5 & -0.5j & 0 & 0 & 0 \\ 0.5 & 0.5j & 0 & 0 & 0 \\ 0 & 0 & 0.5 & -0.5j & 0 \\ 0 & 0 & 0.5 & 0.5j & 0 \\ 0 & 0 & 0 & 0 & 1 \end{bmatrix} \quad (8.34)$$

L can have any size and can be rearranged as desired because L depends on the location of the complex conjugate poles in the Λ_D matrix.

8.5.2. Example of the optimization scheme

One case is studied for the numerical and symbolic gain selection technique: PD LQR controller with real roots for a set of coupled equations. The linearized state equations are represented by the attitude dynamics of a satellite in equation (8.14b) in which the control input matrix is given by equation (8.25). The decoupled set of linearized equations of motion for a satellite can be expressed as follows:

Pitch Motion,

$$A_q = 3n^2\alpha_2 \quad A_v = 0 \quad B_C = d_2 \quad \vec{x}(t) = [\theta(t) \quad \dot{\theta}(t)]^T \; \vec{u}(t) = m_2(t) \quad (8.35)$$

Roll and Yaw Motion,

$$A_q = \begin{bmatrix} -4n^2\alpha_1 & 0 \\ 0 & n^2\alpha_3 \end{bmatrix} A_v = \begin{bmatrix} 0 & n(1-\alpha_1) \\ n(1+\alpha_3) & 0 \end{bmatrix} B_c = \begin{bmatrix} d_1 & 0 \\ 0 & d_3 \end{bmatrix} \quad (8.36a)$$

$$\vec{x}(t) = [\varphi(t) \quad \psi(t) \quad \dot{\varphi}(t) \quad \dot{\psi}(t)]^T \quad \vec{u}(t) = [m_1(t) \quad m_3(t)]^T \tag{8.36b}$$

The satellite is assumed to be in a circular orbit with an altitude of 650 km. The principal moments of inertia of the satellite for this example are: I_{xx} = 15 kg – m^2, I_{yy} = 16.5 kg – m^2, and I_{zz} = 12 kg – m^2. The example is only solved for the roll and yaw motion of the vehicle to show the computational method.

For this problem, assumed $d_1 = 2/15$ and $d_3 = 1/6$. This case shows the capability of this calculation to obtain a positive definite matrix for the state weighting matrix. In this example, equations (8.36) are solved with the minimization procedure to obtain the solution of equation (8.28) using the constraint equations (8.30). The following desired eigenvalues and eigenvectors are selected for this coupled set of equations:

$$\Lambda_{11} = \begin{bmatrix} -1 & 0 \\ 0 & -0.5 \end{bmatrix} \quad \Lambda_{22} = \begin{bmatrix} -2 & 0 \\ 0 & -2.5 \end{bmatrix} \tag{8.37a}$$

$$y_{21} = \begin{bmatrix} 1 & a \\ b & 1 \end{bmatrix} \quad \Lambda_{22} = \begin{bmatrix} 1 & c \\ d & 1 \end{bmatrix} \tag{8.37b}$$

The eigenvectors in equation (8.37b) are related to the coupling of the roll and yaw angular velocities. Equations (8.27) are equal to,

$$\boldsymbol{M}_{11} = 0 \Rightarrow y_{11} = \begin{bmatrix} -1 & -2a \\ -b & -2 \end{bmatrix} \tag{8.38a}$$

$$\boldsymbol{M}_{12} = 0 \Rightarrow y_{12} = \begin{bmatrix} -0.5 & -\frac{2}{3}c \\ -0.5d & -\frac{2}{3} \end{bmatrix} \tag{8.38b}$$

$$\boldsymbol{M}_{21} = (\boldsymbol{A}_q - \boldsymbol{B}_C(t)\boldsymbol{B}_C^T(t)\boldsymbol{P}_{12}^T)y_{21} + (\boldsymbol{A}_v - \boldsymbol{B}_C(t)\boldsymbol{B}_C^T(t)\boldsymbol{P}_{22})y_{21}\Lambda_{11} - y_{21}\Lambda_{11}^2 \tag{8.38c}$$

$$\boldsymbol{M}_{22} = (\boldsymbol{A}_q - \boldsymbol{B}_C(t)\boldsymbol{B}_C^T(t)\boldsymbol{P}_{12}^T)y_{22} + (\boldsymbol{A}_v - \boldsymbol{B}_C(t)\boldsymbol{B}_C^T(t)\boldsymbol{P}_{22})y_{22}\Lambda_{22} - y_{22}\Lambda_{22}^2 \tag{8.38d}$$

Equations (8.38c) and (8.38d) have 11 unknowns because the entries in $\boldsymbol{P}_{11}$ are calculated after the minimization procedure is performed. To reduce the number of unknowns from 11 to 7, the terms in equations (8.38c) and (8.38d) can be manipulated. It is left to the reader to show these terms as an exercise. Using the reduce set of equations, the constraint optimization problem is solved. There are quasi-Newton methods [14] that can be used, but the reduction of unknowns aid in the computational time for this problem. Solving the minimization problem, the Riccati matrix becomes,

$$\boldsymbol{P}_{11} = \begin{bmatrix} 337.482 & -0.015 \\ -0.015 & 135.008 \end{bmatrix} \quad \boldsymbol{P}_{12} = \begin{bmatrix} 112.493 & -0.158 \\ 0.098 & 45.001 \end{bmatrix} \tag{8.39a}$$

$$\boldsymbol{P}_{22} = \begin{bmatrix} 168.746 & -0.036 \\ -0.036 & 108.003 \end{bmatrix} \tag{8.39b}$$

And the eigenvectors are equal to,

$$y_{21} = \begin{bmatrix} 1 & 0.0004 \\ 0.0034 & 1 \end{bmatrix} \quad y_{22} = \begin{bmatrix} 1 & 0.0112 \\ 0.0126 & 1 \end{bmatrix} \tag{8.40a}$$

$$y_{11} = \begin{bmatrix} -1 & -0.0008 \\ -0.0034 & -2 \end{bmatrix} \quad y_{12} = \begin{bmatrix} -0.5 & -0.0045 \\ -0.0036 & -0.4000 \end{bmatrix} \tag{8.40b}$$

The state weighting matrix is equal to $\boldsymbol{Q} = diag$ (224.986, 56.254, 281.239, 234.013). Since the principal diagonal of the $\boldsymbol{Q}$ matrix is positive and the off-diagonal terms are zero, $\boldsymbol{Q}$ is positive definite. Because of the selection of the eigenvectors, the angular velocities are weighted more than the angular position.

A second execution of the minimization algorithm is performed, but the following values are selected: $d_1 = 2.0938 \times 10^{-6}$ and $d_3 = 1.4036 \times 10^{-6}$. Solving the same minimization problem, the Riccati matrix becomes,

$$\boldsymbol{P}_{11} = \begin{bmatrix} 1.365 \times 10^{12} & 1.775 \times 10^{9} \\ -3.1473 \times 10^{9} & 1.903 \times 10^{12} \end{bmatrix} \quad \boldsymbol{P}_{12} = \begin{bmatrix} 4.554 \times 10^{11} & -2.190 \times 10^{8} \\ 1.371 \times 10^{8} & 6.343 \times 10^{11} \end{bmatrix} \tag{8.41a}$$

$$\boldsymbol{P}_{22} = \begin{bmatrix} 6.839 \times 10^{11} & 1.429 \times 10^{8} \\ 1.429 \times 10^{8} & 1.522 \times 10^{12} \end{bmatrix} \tag{8.41b}$$

And the eigenvectors are,

$$y_{21} = \begin{bmatrix} 1 & -0.905 \\ -2.038 & 1 \end{bmatrix} \quad y_{22} = \begin{bmatrix} 1 & -0.667 \\ 0.129 & 1 \end{bmatrix} \tag{8.42a}$$

$$y_{11} = \begin{bmatrix} -1 & 1.8103 \\ 2.038 & -2 \end{bmatrix} \quad y_{12} = \begin{bmatrix} -0.5 & -0.267 \\ -0.064 & -0.4000 \end{bmatrix} \tag{8.42b}$$

The state weighting matrix is expressed as,

$$\boldsymbol{Q}_{11} = \begin{bmatrix} 9.093\times10^{11} & 6.050\times10^{3} \\ 6.050\times10^{3} & 7.927\times10^{11} \end{bmatrix} \tag{8.43a}$$

$$\boldsymbol{Q}_{12} = \boldsymbol{Q}_{21} = \begin{bmatrix} 0 & 0 \\ 0 & 0 \end{bmatrix} \tag{8.43b}$$

$$\boldsymbol{Q}_{22} = \begin{bmatrix} 1.139\times10^{12} & -4.808\times10^{9} \\ -4.808\times10^{9} & 3.298\times10^{12} \end{bmatrix} \tag{8.43c}$$

To show that $\boldsymbol{Q}$ is positive definite, the eigenvalues of $\boldsymbol{Q}$ are calculated. If the eigenvalues are larger than zero, $\boldsymbol{Q}$ is positive definite [81]. The eigenvalues of $\boldsymbol{Q}$ in equation (8.43) are equal to $\lambda_Q = 10^{12}$ $[0.909 \quad 0.793 \quad 1.139 \quad 3.298]^T$; hence, $\boldsymbol{Q}$ is positive definite. This state weighting matrix is substituted back into the Riccati equation to determine the closed-loop eigenvalues. The closed-loop eigenvalues using equations (8.43) are:

$$\boldsymbol{\Lambda}_{11} = \begin{bmatrix} -0.9984 & 0 \\ 0 & -0.5 \end{bmatrix} \quad \boldsymbol{\Lambda}_{22} = \begin{bmatrix} -1.9997 & 0 \\ 0 & -2.4995 \end{bmatrix} \tag{8.44}$$

This example demonstrates the capability of the numerical method to find a local minimum that can satisfy the constraints. There are other solutions for the state weighting matrix that could lead to a positive definite state weighting matrix. Also, the solutions of the state weighting matrix depends on the initial condition for the minimization problem. To decrease the computational process, a state weighting matrix can be guessed to determine a Riccati matrix and the eigenvectors; in this way, these guessed values can be used as the initial conditions for the minimization procedure. Once the weighting matrices are determined, the Riccati solution can be used again to determine the steady-state control gain for the desired response.

8.6 Suggested problems

Problem 8.1. The equations of rotational motion for small oscillation of a gravity-gradient satellite in a circular orbit can be written:

$$\dot{\vec{x}} = A\vec{x} + \vec{N}$$

where,

$$A = \begin{bmatrix} \mathbf{0} & I \\ A_q & A_v \end{bmatrix} \quad \vec{N} = \begin{bmatrix} N_{d1} \\ N_{d2} \\ N_{d3} \end{bmatrix}$$

$$A_q = \begin{bmatrix} -4n^2\alpha_1 & 0 & 0 \\ 0 & 3n^2\alpha_2 & 0 \\ 0 & 0 & n^2\alpha_3 \end{bmatrix} \qquad A_v = \begin{bmatrix} 0 & 0 & +n(1-\alpha_1) \\ 0 & 0 & 0 \\ -n(1+\alpha_3) & 0 & 0 \end{bmatrix}$$

$$\vec{x}(t) = [\varphi(t) \quad \theta(t) \quad \psi(t) \quad \dot{\varphi}(t) \quad \dot{\theta}(t) \quad \dot{\psi}(t)]^T$$

where J_1, J_2, and J_3 are the principal moments of inertia, and

$$\alpha_1 = \frac{J_2 - J_3}{J_1} \quad \alpha_2 = \frac{J_1 - J_3}{J_2} \quad \alpha_3 = \frac{J_1 - J_2}{J_3}$$

a. For the case where the disturbance and control torques $\vec{N}$ are zero, develop the system characteristic equation. Analyze the stability of this system about the motion $\varphi(t) = \theta(t) = \psi(t) = 0$, both from an examination of the roots of the characteristic equation and an application of the Routh–Hurwitz criterion. Be sure to carefully distinguish between the definition(s) of stability that you are considering.

b. For the particular gravity-gradient satellite shown in Figure 8.7 apply the results in part (a). (Note: The satellite pitch axis is

Figure 8.7 **Problem 8.1**

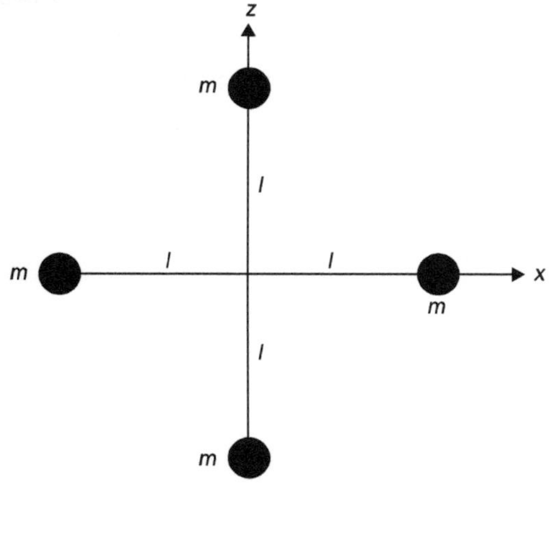

perpendicular to the plane of the paper. When in equilibrium the satellite z axis is aligned along the vertical, and the x axis along the local horizontal.

Problem 8.2. The interesting fact about the Frame Rotations scheme is to rotate the reference frame to another attitude pointing so that the stability of the equations can be determined. One unstable attitude point is a rotation of 90 degrees of the reference frame about yaw.

a. Show that the linearized equations of motion for a 90 degree frame rotation about the yaw axis from LVLH are given by,

$$A_q = \begin{bmatrix} -4n^2\alpha_2 & 0 & 0 \\ 0 & 3n^2\alpha_1 & 0 \\ 0 & 0 & n^2\alpha_3 \end{bmatrix} \quad A_v = \begin{bmatrix} 0 & 0 & n(1-\alpha_2) \\ 0 & 0 & 0 \\ -n(1+\alpha_3) & 0 & 0 \end{bmatrix}$$

$$B_c = \begin{bmatrix} \frac{b_3}{J_2} & 0 & -\frac{b_2}{J_2} \\ 0 & \frac{b_3}{J_1} & \frac{b_1}{J_1} \\ -\frac{b_1}{J_3} & 0 & -\frac{b_2}{J_3} \end{bmatrix}$$

b. What equations are coupled and decoupled?

c. From the decoupled equation, what is the condition to obtain stability?

d. What is the stability condition for the coupled equations?

Problem 8.3. As shown previously the weighting matrix can be optimized for any type of state vector system. Using equations (8.35), taking the same assumption of the satellite, and assuming $d_2 = 0.12$, solve the following:

a. Obtain the ***Q*** and ***P*** matrix when,

$$\Lambda_D = \begin{bmatrix} -0.5 & 0 \\ 0 & -1 \end{bmatrix} \quad y_D = \begin{bmatrix} 1 & 1 \\ a & b \end{bmatrix}$$

b. Obtain the ***Q*** and ***P*** matrix and the condition that ensures ***Q*** to be positive definite when,

$$\Lambda_D = \begin{bmatrix} -L_1 + L_2 i & 0 \\ 0 & -L_1 - L_2 i \end{bmatrix} \quad y_D = \begin{bmatrix} 1 & 1 \\ a+bi & a-bi \end{bmatrix}$$

Assume that $L_1 = 1$ and $L_2 = 7$ to show the $\boldsymbol{Q}$ and $\boldsymbol{P}$ matrix.

Problem 8.4. In Section 8.5.2, equations (8.38c) and (8.38d) have 11 unknowns and can be reduced to 7 unknowns. Using the following equations,

$$\boldsymbol{\Lambda}_{11} = \begin{bmatrix} -L_1 & 0 \\ 0 & -L_2 \end{bmatrix} \quad \boldsymbol{\Lambda}_{22} = \begin{bmatrix} -L_3 & 0 \\ 0 & -L_4 \end{bmatrix}$$

$$\boldsymbol{P}_{11} = \begin{bmatrix} p_1 & p_2 \\ p_2 & p_5 \end{bmatrix} \quad \boldsymbol{P}_{12} = \boldsymbol{P}_{12}^T = \begin{bmatrix} p_3 & p_6 \\ p_4 & p_7 \end{bmatrix} \quad \boldsymbol{P}_{22} = \begin{bmatrix} p_8 & p_9 \\ p_9 & p_{10} \end{bmatrix}$$

Manipulate symbolically equations (8.38c) and (8.38d) to obtain the solutions for the variables p_3, p_7, p_8, and p_{10}.

9

Formation flying

Abstract: In recent years, formation flying has been a wide research area involving multiple satellites which provide different measurements from various points at the same time. There are proposed constellations to be launched in the following years. In this chapter, the Tschauner–Hempel and Clohessy–Wiltshire equations are explained to define a pair of satellites in an elliptical and circular orbit, respectively. In addition, different perturbations due to the Earth and Sun are added to the equations of motion. The hierarchical control scheme is used to demonstrate one of the thrust arcs explained by Lawden since the disturbance effects caused additional nonlinear terms. Finally, the controller is implemented in the steady-state condition to demonstrate its application in the computer onboard the satellite. Thus, the purpose of this chapter is to describe the motion of a constellation, the active control scheme for formation flying, and the implementation of the controller in the satellite.

Key words: Tschauner–Hempel equations, Clohessy–Wiltshire equations, hierarchical control scheme, steady-state controller, controller implementation formation flying, disturbances in the constellation, Lawden thrust arcs, two point boundary value problems, drift correction.

9.1 Introduction to formation flying

From the beginning of the space age, monolithic satellites have been used for many years for communications and experimentations on the Earth and in Space [10]. This has been the discussion of this book in the last chapters. In 1945, Arthur C. Clarke [140] published the first paper discussing the use of more than one satellite for relay communications; in this way, a group of three satellites in a geosynchronous orbit can provide continuous coverage to ground stations on Earth, and also to satellites in lower Earth orbits. In 1960, Clohessy and Wiltshire [**141**] developed a set of equations of motion that describe the motion of a pair of satellites in a circular orbit. Five years later, Tschauner–Hempel [**142**] expanded the Clohessy–Wiltshire equations to explained the relative motion in elliptical orbits. These ideas were expanded by Walker [**143**] in 1977. To this date, the constellation is known as the Walker patterns because it uses the orbital elements to create a complete coverage of the Earth with a certain number of satellites [**10**].

Recently, the desire to perform multiple measurements at the same time takes a different perspective, and these objectives are achieved by the use of formation flying or constellations. Formation flying or constellations describe a group of satellites in a certain configuration that, simultaneously, perform different experiments and analysis. A single constellation can provide better and continuous coverage over the Earth and has a higher reliability than a series of individual satellites; in addition, the satellites within the constellation are smaller and simpler satellites because these satellites contain fewer instruments than a single more complex satellite. Because of this, there have been different proposed constellations such as the string of pearls [**97**], flower constellations [**144**], coulomb spacecraft formation control [**145**], and tetrahedron constellation [**146**].

This chapter develops the Tschauner–Hempel (TH) equations based on the Carter–Humi formulation [**5**]. The TH equations provide the most general form for any elliptical and circular orbits. After developing these formulations, the perturbation due to the Earth and Sun are added. In order to correct the separation distance between a pair of satellites, the hierarchical controller is used to maintain the separation distance between satellites. This controller demonstrates one of the thrust arcs explained by Lawden [**147**]. At the end of the chapter, it is shown how the hierarchical controller can be implemented in the satellite. In summary, the main objective of this chapter is to show the formation flying formulation and control based on the TH equations. This provides the reader with the necessary information to work with formation flying and constellations.

9.2 Tschauner–Hempel formulation

The linearized Tschauner–Hempel (TH) equations are developed from the gravitational potential function when the Earth is assumed as a homogeneous sphere. The formulation presented here contains additional disturbing forces acting on the satellite which are explained in the following sections. The gravitational potential function defined by Brouwer [17] containing the J2 perturbation is defined as,

$$U = \frac{\mu m}{|\vec{R}|} + U_{J_2} \tag{9.1a}$$

where U_{J2} is the potential due to the J2 perturbation and written as,

$$U_{J_2} = -B_2 \frac{\mu m}{2|\vec{R}|^3} + B_2 \frac{3\mu m Z^2}{2|\vec{R}|^5} \tag{9.1b}$$

where $\vec{R} = X\hat{\imath} + Y\hat{\jmath} + Z\hat{k}$, $\mu = GM_E$, m is the mass of the satellite,

$$B_2 = -\frac{2}{3} J_2 R_E \tag{9.2}$$

and R_E is the radius of the Earth. J_2 [10] is approximated to 1.08263×10^{-3}. The symbol $|\quad|$ means the norm of a vector, and $\vec{R}$ is the position of the satellite measured from the center of the Earth.

The equations of motion are derived for a maneuvering satellite and a reference (or target) satellite which is orbiting about the Earth in an elliptical orbit as shown in Figure 9.1. The maneuvering satellite is assumed to have a scalar point mass $m(t)$ and an applied thrust vector $T(t)$ projected along the reference-axis system. The target satellite is acted on by a force derived from a gravitational potential function (defined in equation (9.1)) and directed toward the center of the Earth. $\vec{R}(t)$ is the vector measured from the center of the Earth to the reference (or target) satellite, and $\vec{r}(t)$ is the vector from the center of the Earth to the maneuvering satellite. $\vec{\rho}(t)$ (dashed line in Figure 9.1) is the vector measured from the reference satellite to the maneuvering satellite and describes their relative separation distance. Throughout this work, the Euclidian space [52] is used to define the coordinates of the maneuvering and target spacecraft. The coordinate system is given in the following manner: 1) X_1 is opposed to the motion of the maneuvering satellite and perpendicular to the X_2 axis, whose positive direction is along $\vec{R}(t)$; 2) X_3 is positive when the right handed system is completed. υ_s is the relative angular velocity of the satellites about the Earth.

The equation of motion for the maneuvering satellite can be written as,

$$\ddot{\vec{r}}(t) = \frac{-\mu(\vec{R} + \vec{\rho})}{|\vec{R} + \vec{\rho}|^3} + \frac{\vec{f}_{J_2,M}}{m} + \frac{\vec{f}_{P,M}}{m} + \frac{\vec{T}}{m} \tag{9.3a}$$

Figure 9.1 **Reference and maneuvering satellite motion about the Earth**

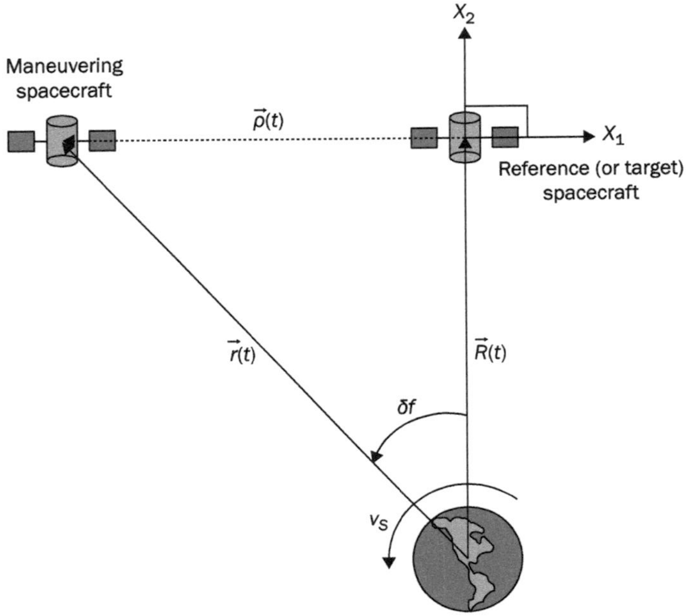

where $\vec{f}_{J_2,M}$ is the force due to the J2 perturbation for the maneuvering satellite, $\vec{T}$ is the thrust applied over some reference-axis in the maneuvering satellite, and $\vec{f}_{P,M}$ is the force due to other disturbing forces acting on the body of the satellite. For the reference satellite, a similar expression can be written as,

$$\ddot{\vec{R}}(t) = \frac{-\mu(\vec{R})}{|\vec{R}|^3} + \frac{\vec{f}_{J_2,R}}{m} + \frac{\vec{f}_{P,R}}{m} \tag{9.3b}$$

where $\vec{f}_{J_2,R}$ is the force due to the J2 perturbation for the reference satellite, and $\vec{f}_{P,R}$ is the force due to other disturbing forces acting on the body of the reference satellite. The relative motion of a pair of satellites about the Earth can be defined as,

$$\ddot{\vec{\rho}} = \ddot{\vec{r}} - \ddot{\vec{R}}$$

Substituting equations (9.3) into the last equation, the motion of a pair of satellites about the Earth for a perturbed system can be written as,

$$\ddot{\vec{\rho}} = \frac{\mu(\vec{R})}{|\vec{R}|^3} - \frac{\mu(\vec{R}+\vec{\rho})}{|\vec{R}+\vec{\rho}|^3} + \frac{\vec{F}_D}{m} + \frac{\vec{T}}{m} \tag{9.4}$$

where $\vec{F}_D = \vec{F}_{J_2} + \vec{F}_P$, $\vec{F}_{J_2} = \vec{f}_{J_2,M} - \vec{f}_{J_2,R}$, and $\vec{F}_s = \vec{f}_{S,M} - \vec{f}_{S,R}$. The denominator of the first two terms on the right hand side (RHS) of equation (9.4) is approximated with the binomial theorem [**13**]. The first two terms of the expansion is used to define the linearized system of equations. The study of the equations with higher order terms of this series is beyond the scope of this book. The second term on the RHS in equation (9.4) can be approximated within the linear range using the binomial expansion as,

$$\left|\vec{R}+\vec{\rho}\right|^{-n} \approx \frac{1}{\left|\vec{R}\right|^{n}}\left[1-\frac{n}{2}\left(\frac{\left|\vec{\rho}\right|^{2}}{\left|\vec{R}\right|^{2}}+2\frac{\vec{R}\cdot\vec{\rho}}{\left|\vec{R}\right|^{2}}\right)\right] \tag{9.5}$$

Equation (9.4) is expanded using equation (9.5) as,

$$\ddot{\vec{\rho}} = -\frac{\mu}{\left|\vec{R}\right|^{3}}\left[\vec{\rho}-\frac{3}{2}\left(\frac{\left|\vec{\rho}\right|^{2}}{\left|\vec{R}\right|^{2}}+2\frac{\vec{R}\cdot\vec{\rho}}{\left|\vec{R}\right|^{2}}\right)\left(\vec{R}+\vec{\rho}\right)\right]+\frac{\vec{F}_D}{m}+\frac{\vec{T}}{m} \tag{9.6}$$

Since $|\vec{\rho}| << |\vec{R}|$, equation (9.6) can be approximated by,

$$\ddot{\vec{\rho}} = -\frac{\mu}{\left|\vec{R}\right|^{3}}\left[\vec{\rho}-3\frac{\vec{R}\cdot\vec{\rho}}{\left|\vec{R}\right|^{2}}\vec{R}\right]+\frac{\vec{F}_D}{m}+\frac{\vec{T}}{m} \tag{9.7}$$

$\ddot{\vec{\rho}}$ (in equation (9.7)) is expressed in the inertial frame, but the relative separation distance vector between the two satellites is rotating about the Earth as shown in Chapter 5. The motion of the separation distance vector between a pair of satellites can be represented as:

$$\ddot{\vec{\rho}}|_{Space} = \ddot{\vec{\rho}}|_{Body} + \dot{\vec{\Omega}}_E \times \vec{\rho} + 2\vec{\Omega}_E \times \dot{\vec{\rho}} + \vec{\Omega}_E \times \vec{\Omega}_E \times \vec{\rho} \tag{9.8}$$

where $\vec{\Omega}_E = \dot{f}(t)\hat{k} = \upsilon_s\hat{k}$, and $\times$ is the vector cross product. If $\vec{R}$ is chosen to be along the radial direction and is pointing to the reference satellite; then, $\vec{R} = |\vec{R}|\hat{j} = R\hat{j}$. $\vec{\rho}$ is the relative measurement (or separation distance) vector from the target (or reference) satellite to the maneuvering satellite and can be written as:

$$\vec{\rho} = x_1\hat{\imath} + x_2\hat{j} + x_3\hat{k} \tag{9.9}$$

where,

$$x_1 = X_{1,M} - X_{1,R} \quad x_2 = X_{2,M} - X_{2,R} \quad x_3 = X_{3,M} - X_{3,R}$$

Solving equation (9.7) with equations (9.8) and (9.9), the equations of motion expressing the separation distance between the maneuvering and the target spacecraft in component form are,

$$\ddot{x}_1 - \dot{\upsilon}_S x_2 - 2\upsilon_S \dot{x}_2 - \upsilon_S^2 x_1 = -\frac{\mu}{R^3} x_1 + \frac{F_{1D}}{m} + \frac{T_1}{m}$$

$$\ddot{x}_2 + \dot{\upsilon}_S x_2 + 2\upsilon_S \dot{x}_1 - \upsilon_S^2 x_1 = \frac{2\mu}{R^3} x_2 + \frac{F_{2D}}{m} + \frac{T_2}{m} \qquad (9.10)$$

$$\ddot{x}_3 = -\frac{\mu}{R^3} x_3 + \frac{F_{3D}}{m} + \frac{T_3}{m}$$

Equations (9.10) contain the linear terms relating the motion of a pair of satellites about the Earth in a Keplerian motion and the terms F_{jD} (where j = 1,2,3) expressing the disturbance forces. This set of equations of motion is called the Tschauner–Hempel equations for highly elliptical orbits. Equations (9.10), when F_{jD} and T_i equals zero, are confined to a plane which is defined by the well known Keplerian orbit (or equation of a conic section):

$$R = \frac{\alpha\left(1-e^2\right)}{1+e\cos f} = \frac{h^2}{\mu}\left(1+e\cos f\right)^{-1} \qquad (9.11)$$

where α is the semimajor axis, e is the eccentricity, f is the true anomaly angle, and h is the angular momentum. The angular momentum magnitude is defined as:

$$h = R^2 \upsilon_S \qquad (9.12)$$

and $\upsilon_S = \dot{f}$. From equation (9.12), the true anomaly angle (f) is an increasing function of time. Because of this correspondence, it is convenient to change from the time domain (t) into the true anomaly angle domain (f); then,

$$\frac{d\varsigma}{dt} = \upsilon_S \frac{d\varsigma}{df} = \upsilon_S \varsigma' \qquad (9.13)$$

$$\frac{d}{dt}\left(\frac{d\varsigma}{dt}\right) = \upsilon_S^2 \varsigma'' + \upsilon_S \; \upsilon_S' \varsigma' \qquad (9.14)$$

where ς is a function of the true anomaly angle and is twice differentiable. The prime symbol denotes differentiation with respect to f where,

$$\varsigma' = \frac{d\varsigma}{df} \quad \varsigma'' = \frac{d^2\varsigma}{d^2 f} \qquad (9.15)$$

Using equations (9.13) and (9.14), equations (9.10) can be rewritten as,

$$\upsilon_S x_1'' + \upsilon_S' x_1' - \upsilon_S' x_2 - 2\upsilon_S x_2' - \upsilon_S x_1 = -\upsilon_S^{1/2} k_h x_1 + \frac{F_{1D}}{m\upsilon_S} + \frac{T_1}{m\upsilon_S}$$

$$\upsilon_S x_2'' + \upsilon_S' x_2' + \upsilon_S' x_1 + 2\upsilon_S x_1' - \upsilon_S x_2 = 2\upsilon_S^{1/2} k_h x_2 + \frac{F_{2D}}{m\upsilon_S} + \frac{T_2}{m\upsilon_S} \quad (9.16)$$

$$\upsilon_S x_3'' + \upsilon_S' x_3' = -\upsilon_S^{1/2} k_h x_3 + \frac{F_{3D}}{m\upsilon_S} + \frac{T_3}{m\upsilon_S}$$

where,

$$\frac{\mu}{R^3} = \upsilon_S^{3/2} k_h \quad (9.17a)$$

$$k_h = \frac{\mu}{h^{3/2}} \frac{1}{R^2} = \frac{\upsilon_S}{h} \quad (9.17b)$$

Equations (9.17) can be obtained from equations (9.11) and (9.12). Using equations (9.11) and (9.12), the angular velocity (υ_s) and its derivative can be written as,

$$\upsilon_S = k_h^2(1 + e\cos f)^2 \quad \upsilon_S' = -2k_h^2(e\sin f)(1 + e\cos f) \quad (9.17c)$$

And, from equation (9.11),

$$\frac{1}{R} = \frac{\mu}{h^2}\left(1 + e\cos f\right) \quad (9.17d)$$

By means of equations (9.17c) and (9.17d), equations (9.16) can be rewritten in terms of the true anomaly angle in component format as,

$$(1 + e\cos f)x_1'' - 2(e\sin f)x_1' + 2(e\sin f)x_2 - 2(1 + e\cos f)x_2' - (e\cos f)x_1 = d_1 + c_1$$

$$(1 + e\cos f)x_2'' - 2(e\sin f)x_2' - 2(e\sin f)x_1 + 2(1 + e\cos f)x_1' - (e\cos f)x_2 = 3x_2 + d_2 + c_2 \quad (9.18a)$$

$$(1 + e\cos f)x_3'' - 2(e\sin f)x_3' = -x_3 + d_3 + c_3$$

where,

$$d_j = \frac{F_{jD}}{m\upsilon_S k_h^2\left(1 + e\cos f\right)} \quad (j = 1, 2, 3) \quad (9.18b)$$

$$c_j = \frac{T_j}{m\upsilon_S k_h^2\left(1 + e\cos f\right)} \quad (j = 1, 2, 3) \quad (9.18c)$$

Using the transformations,

$$y_j = (1 + e\cos f)x_j$$

$$y_j' = (1 + e\cos f)x_j' - (e\sin f)x_j \tag{9.19a}$$

$$y_j'' = (1 + e\cos f)x_j'' - (e\sin f)x_j' - (e\cos f)x_j$$

and,

$$\{(1 + e\cos f)x_j\}' = (1 + e\cos f)x_j' = (e\sin f)x_j \tag{9.19b}$$

equations (9.18a) are reduced to,

$$y_1'' = 2y_2' + d_1 + c_1$$

$$y_2'' = -2y_1' + 3\kappa y_2 + d_2 + c_2 \tag{9.20}$$

$$y_3'' = -y_3 + d_3 + c_3$$

where,

$$\kappa = \frac{1}{(1 + e\cos f)}$$

These equations are the Tschauner–Hempel equations for a Keplerian motion (or without perturbations) when $F_{jD} = 0$. In reference [5], there is a closed-form solution for the linearized TH equation in an environment without perturbations.

In Chapter 6, a thrust condition can be defined for this set of equations. A thrust capability can be developed for equations (9.18c) as,

$$c_j = \frac{h^6}{\mu^4} \frac{T_m}{(1 + e\cos f)^3 m_0} u_j(f) = bu_j(f)\kappa^3 \tag{9.21}$$

where T_m is the maximum thrust, and,

$$b = \frac{h^6}{\mu^4} \frac{T_m}{m_0} u_j(f) = \frac{T_j(f)}{T_m} \tag{9.22}$$

Equation (9.21) is used to define the transformed thrust acceleration in equations (9.20). It is assumed that the initial and final mass of the spacecraft are the same because of the small consumption of thrust for the correction of the separation distance between a pair of satellites. For this reason, equations (9.20) and (9.21) are implemented with an active controller to determine the thrust consumption. As shown in Chapter 7, the digital form of equations (9.20) can be represented as:

$$\vec{y}(k+1) = \hat{\mathbf{A}}(k)\vec{y}(k) + \hat{\mathbf{B}}(k)\,\vec{u}(k) + \Delta f \vec{\psi}(k) \tag{9.23a}$$

where,

$$\hat{A}(k) = \begin{bmatrix} 1 & 0 & 0 & \Delta f & 0 & 0 \\ 0 & 1 & 0 & 0 & \Delta f & 0 \\ 0 & \phi & 1 & 0 & 0 & \Delta f \\ 0 & 0 & 0 & 1 & 2\Delta f & 0 \\ 0 & 3\kappa\Delta f & 0 & -2\Delta f & 1 & 0 \\ 0 & 0 & -\Delta f & 0 & 0 & 1 \end{bmatrix} \tag{9.23b}$$

$$\hat{B}(k) = b\kappa^3\Delta f \begin{bmatrix} 0 & 0 & 0 \\ 0 & 0 & 0 \\ 0 & 0 & 0 \\ 1 & 0 & 0 \\ 0 & 1 & 0 \\ 0 & 0 & 1 \end{bmatrix} \tag{9.23c}$$

$$\kappa = \frac{1}{1 + e\cos f(k)} \qquad f(k) = f_0 + k\Delta f \tag{9.23d}$$

$$\vec{y}(k) = [y_1(k) \quad y_2(k) \quad y_3(k) \quad y_1'(k) \quad y_2'(k) \quad y_3'(k)]^T \tag{9.23e}$$

$$\vec{u}(k) = [u_1(k) \quad u_2(k) \quad u_3(k)]^T \tag{9.23f}$$

$$\vec{\psi}(k) = [0\ 0\ 0\ d_1\ d_2\ d_3]^T \tag{9.23g}$$

where f_0 is the initial true anomaly angle, and Δf is the sampling interval in the true anomaly angle. This Δf works very similar to the sampling interval in the time domain, but in the following topics, there is a relation between the sampling interval in time and in the true anomaly angle.

9.3 Clohessy–Wiltshire formulation

The TH equations are considered as a general form for any type of orbit, and, from these equations, the Clohessy–Wiltshire (CW) equations can be developed. The Clohessy–Wiltshire equations [141] defined the motion of a pair of satellites in a circular orbit about the Earth. For a circular orbit ($e = 0$), the following relationships can be obtained,

$$\upsilon_S = n \qquad \dot{\upsilon}_S = 0 \qquad \frac{\mu}{R^3} = n^2 \tag{9.24}$$

where n is the mean motion. Substituting equations (9.24) into equations (9.10), the TH equations are reduced to the CW equations in the following manner:

$$\begin{aligned} \ddot{x}_1 - 2n\dot{x}_2 &= \frac{F_{1D}}{m} + \frac{T_1}{m} \\ \ddot{x}_2 + 2n\dot{x}_1 - 3n^2 x_2 &= \frac{F_{2D}}{m} + \frac{T_2}{m} \\ \ddot{x}_3 + n x_3 &= \frac{F_{3D}}{m} + \frac{T_3}{m} \end{aligned} \tag{9.25}$$

There are closed form solutions for the CW equations but it is left to the reader as an exercise to obtain this solution without any forces and with a constant applied force.

9.4 Earth oblateness and solar effects in formation flying

The disturbance effects due to the Earth and the Sun can cause changes in the orbit and is demonstrated in the development shown in Chapter 3. This formulation for a single satellite can be expanded to a constellation. For highly elliptical orbits and circular orbits, the perturbations due to the Earth such as the drag force and J2 perturbation affects the separation distance constraints between a pair of satellites within a certain number of orbits [148]. The solar pressure effects depend on how long the satellite remains within the shadow of the Earth [149]. In the formulation, the perturbation vector is divided into the perturbation due to the Earth's oblateness and solar pressure effects. The objective of this section is to include the effects due to the J2 perturbation and the solar pressure into the linearized TH equations.

9.4.1 Earth oblateness in formation flying

The $\vec{F}_{J_2}$ in equation (9.10) is shown in this section, but more details can be found in Reference [104]. As shown before, the potential function is divided between the effects of the spherical Earth and the Earth's oblateness. From Reference [17], it is known that,

$$X = |\vec{R}| \cos f \cos i \tag{9.26a}$$

$$Y = |\vec{R}| \sin f \cos i \tag{9.26b}$$

$$Z = |\vec{R}| \sin i \tag{9.26c}$$

where i is the inclination angle. From equation (9.1b), the effects due to the Earth oblateness in equation (9.4) can be written as,

$$\begin{aligned} \vec{F}_{J_2} &= -\frac{3\mu B_2}{|\vec{R}|^5}\left(\frac{\vec{R}}{2} - \frac{5}{2}\frac{\vec{R}X_{3,R}^2}{|\vec{R}|^2} + X_{3,R}\hat{k}\right) \\ &+ \frac{3\mu B_2}{|\vec{R}+\vec{\rho}|^5}\left(\frac{\vec{R}+\vec{\rho}}{2} - \frac{5}{2}\frac{\left(\vec{R}+\vec{\rho}\right)X_{3,R}^2}{|\vec{R}+\vec{\rho}|^2} + X_{3,M}\hat{k}\right) \end{aligned} \tag{9.27}$$

Expanding equation (9.27) with the linear approximation in equation (9.5) and assuming $|\vec{\rho}| << |\vec{R}|$, equation (9.27) can be reduced to,

$$\begin{aligned} \vec{F}_{J_2} &= -\frac{3\mu B_2}{|\vec{R}|^5}\left(\frac{5\vec{R}}{2|\vec{R}|^2}\left(X_{3,M}^2 - X_{3,R}^2\right) - \left(X_{3,M} - X_{3,R}\right)\hat{k}\right) \\ &+ \frac{3\mu B_2}{|\vec{R}|^5}\left[\frac{1}{2}\left(\vec{\rho} - 5\frac{\vec{R}\cdot\vec{\rho}}{|\vec{R}|^2}\vec{R}\right) - \frac{5X_{3,M}^2}{2|\vec{R}|^2}\left(\vec{\rho} - 7\frac{\vec{R}\cdot\vec{\rho}}{|\vec{R}|^2}\vec{R}\right) - 5X_{3,M}^2\frac{\vec{R}\cdot\vec{\rho}}{|\vec{R}|^2}\hat{k}\right] \end{aligned} \tag{9.28}$$

The following relations are defined for the motion of the maneuvering and reference spacecraft in the X_3 direction:

$$X_{3,M} = |\vec{R} + \vec{\rho}| \sin i_M \quad X_{3,R} = |\vec{R}| \sin i_R \tag{9.29}$$

i_M and i_R is the inclination angle for the maneuvering and reference spacecraft, respectively. Some components of equation (9.28) are approximated with equations (9.5) and (9.29) and are written as,

$$X_{3,M}^2 - X_{3,R}^2 = (X_{3,M} - X_{3,R})^2 + 2X_{3,M}\,X_{3,R} \tag{9.30a}$$

$$\frac{X_{3,M}X_{3,R}}{|\vec{R}|^2} \approx \left(1+2\frac{\vec{R}\cdot\vec{\rho}}{|\vec{R}|^2}\right)^2 \sin i_M \ \sin i_R \tag{9.30b}$$

$$\frac{X_{3,M}^2}{|\vec{R}|^7} \approx \frac{1}{|\vec{R}|^5}\left(1+2\frac{\vec{R}\cdot\vec{\rho}}{|\vec{R}|^2}\right)^2 \sin^2 i_M \tag{9.30c}$$

$$\frac{X_{3,M}}{|\vec{R}|^5} \approx \frac{1}{|\vec{R}|^4}\left(1+2\frac{\vec{R}\cdot\vec{\rho}}{|\vec{R}|^2}\right)^{1/2} \sin\ i_M \tag{9.30d}$$

Substituting equations (9.30) into equation (9.28), the relative motion of a pair of satellites are described as,

$$\begin{aligned}\vec{F}_{J_2} = &-\frac{3\mu B_2}{|\vec{R}|^5}\left(\frac{5\vec{R}}{2|\vec{R}|^2}\left(X_{3,M}-X_{3,R}\right)^2+5\vec{R}\left(1+2\frac{\vec{R}\cdot\vec{\rho}}{|\vec{R}|^2}\right)^{1/2}\sin i_M \sin i_R-\left(X_{3,M}-X_{3,R}\right)\hat{k}\right)\\ &+\frac{3\mu B_2}{|\vec{R}|^5}\left[\frac{1}{2}\left(\vec{\rho}-5\frac{\vec{R}\cdot\vec{\rho}}{|\vec{R}|^2}\vec{R}\right)-\frac{5}{2}\left(\vec{\rho}-7\frac{\vec{R}\cdot\vec{\rho}}{|\vec{R}|^2}\vec{R}\right)\left(1+2\frac{\vec{R}\cdot\vec{\rho}}{|\vec{R}|^2}\right)\sin^2 i_M\right]\\ &-\frac{3\mu B_2}{|\vec{R}|^5}\left[|\vec{R}|\left(5\frac{\vec{R}\cdot\vec{\rho}}{|\vec{R}|^2}\right)\left(1+2\frac{\vec{R}\cdot\vec{\rho}}{|\vec{R}|^2}\right)^{1/2}\sin\ i_M\right]\hat{k}\end{aligned} \tag{9.31}$$

The inclination angles for the maneuvering and the reference satellites are the same if the pair of satellites is in the same plane of motion. The inclination angle for a satellite in the out-of-plane motion is only slightly different for larger orbital dimensions such as in the NASA Benchmark Tetrahedron Constellation problem [**146**]. Because of this small difference in the inclination angle, this angle is assumed here to be the same for every pair of satellites ($i = i_M = i_R$). Solving equation (9.31) in component format, assuming the same inclination angle, and taking in account that $|\vec{\rho}| << |\vec{R}|$, the disturbance force for the J2 perturbation can be written as,

$$\begin{aligned}F_{J_2,1} &= \frac{3\mu B_2}{R^5}x_1\left(1-5\sin^2 i\right)\\ F_{J_2,2} &= -\frac{15\mu B_2}{2R^6}x_3^2-\frac{15\mu B_2}{2R^4}x_1\ \sin^2 i-\frac{3\mu B_2}{R^5}x_2\left(2-15\sin^2 i\right)\\ F_{J_3,3} &= \frac{3\mu B_2}{2R^5}x_3\left(3-5\sin^2 i\right)-\frac{15\mu B_2}{2R^5}x_2\ \sin\ i\end{aligned} \tag{9.32}$$

Performing the same procedure as in equations (9.16) to (9.20) and knowing that,

$$\frac{\mu}{R^4}=\frac{\mu \upsilon_S^{3/2} k_h}{h^2}\left(1+e\cos f\right) \quad \frac{\mu}{R^5}=\frac{\upsilon_S^{5/2} k_h}{h} \quad \frac{\mu}{R^6}=\frac{\upsilon_S^3 k_h^2}{\mu} \tag{9.33}$$

Equations (9.32) can be reduced to,

$$\begin{aligned} d_1 &= \frac{3B_2 k_h^2}{2h\kappa} y_1\left(1-5\sin^2 i\right) \\ d_2 &= -\frac{15B_2 k_h^4}{2\mu\kappa} y_3^2 - \frac{15B_2 k_h^4 h}{2\mu\kappa}\sin^2 i - \frac{3B_2 k_h^2}{h\kappa} y_2\left(2-15\sin^2 i\right) \\ d_3 &= -\frac{3B_2 k_h^2}{2h\kappa} y_3\left(1-5\sin^2 i\right) + \frac{15B_2 k_h^2}{h\kappa} y_2 \sin i \end{aligned} \tag{9.34}$$

Equations (9.34) can be substituted into equation (9.23g) to explain the motion of the vehicle with the J2 perturbation; in addition, these nonlinear terms only affect the translational motion of the vehicle.

9.4.2 Solar pressure in formation flying

A column matrix $\vec{\Gamma}(f(k))$ can be obtained from the disturbance matrix $\vec{\psi}(k)$. This matrix is used to explain the disturbance due the incidence of light on the surface of a satellite. This incidence of light creates outside forces or torques that can change the orbital and attitude dynamics of the satellite. As shown in Chapter 3, Karymov [32] developed a formulation to define the forces and torques due to this perturbation. This formulation depends on the shape of the satellite, but the direction of the solar rays is defined in terms of the position of the satellite.

As shown in Chapter 3, the force due to the incidence of light on the surface of a satellite is given by,

$$\vec{f}_s = (1-\varepsilon)\,\vec{F}^{+} + \varepsilon\vec{F}^{-} \tag{9.35}$$

where ε is the reflectivity coefficient, $\vec{F}^{+}$ is the force due to an absorbing surface, and $\vec{F}^{-}$ is the force due to a reflective surface. Equation (9.35) is used to represent the forces due to the solar pressure for the maneuvering and target spacecraft.

The direction of the incident light ($\hat{\sigma}$) over the satellite should be established to determine the solar force on the body of the satellite. The direction of the incidence of light can vary depending on the position of the satellite along the orbit and on the inclination angle of the Sun with respect to the Earth. In Reference [35], it is assumed that the direction of the light can be determined with an axis transformation defining the position of the satellites and the inclination angle of the Sun with respect to the Earth. Also, this relation is established as the difference between the maneuvering and target spacecraft as shown in equation (9.4). Using this assumption [35], the direction of the light can be written as:

$$\begin{aligned} a_0 &= \cos f \cos i_S \\ b_0 &= \sin f \cos i_S \\ c_0 &= \sin i_S \end{aligned} \tag{9.36a}$$

where i_S is the inclination angle of the Sun with respect to the Earth. Equations (9.36a) are used to define the direction of the incident light acting on the target (or reference) satellite. For the maneuvering spacecraft, the direction of the light can be represented in the following form:

$$\begin{aligned} a_{0,M} &= \cos(f + \delta f) \cos i_S \\ b_{0,M} &= \sin(f + \delta f) \cos i_S \\ c_{0,M} &= \sin i_S \end{aligned} \tag{9.36b}$$

where δf is the difference in the true anomaly angle (Figure 9.1). This difference can vary with every correction performed by the controller to maintain the separation distance between any pair of satellites within a constellation. δf can be determined with the most general triangle using the law of cosines as follows,

$$|\vec{\rho}(t)|^2 = |\vec{R}(t)|^2 + |\vec{r}(t)|^2 - 2|\vec{R}(t)||\vec{r}(t)| \cos \delta f \tag{9.37}$$

Using equation (9.5) and

$$|\vec{r}(t)| = |\vec{R}(t)| \left(1 + \frac{|\vec{\rho}(t)|^2}{|\vec{R}(t)|^2} + 2\frac{\vec{R} \cdot \vec{\rho}}{|\vec{R}|^2} \right)^{1/2}$$

equation (9.37) can be expanded in the following way,

$$\frac{|\vec{\rho}(t)|^2}{2|\vec{R}(t)|^2}\left(1-\frac{|\vec{\rho}(t)|^2}{|\vec{R}(t)|^2}+2\frac{\vec{R}\cdot\vec{\rho}}{|\vec{R}|^2}\right)=\frac{1}{2}\left[1-\frac{1}{2}\left(\frac{|\vec{\rho}(t)|^2}{|\vec{R}(t)|^2}+2\frac{\vec{R}\cdot\vec{\rho}}{|\vec{R}|^2}\right)\right]$$
$$+\frac{1}{2}\left(1+\frac{|\vec{\rho}(t)|^2}{|\vec{R}(t)|^2}+2\frac{\vec{R}\cdot\vec{\rho}}{|\vec{R}|^2}\right)^{1/2}-\cos\delta f \quad (9.38)$$

Knowing that $|\vec{\rho}| << |\vec{R}|$, and $\vec{R} = R\hat{j}$, the second and third terms on the left-hand side and the second, third, fifth and sixth terms on the right-hand side of equation (9.38) are considered as higher-order terms; thus, equation (9.38) is approximated as,

$$\cos\delta f \approx 1 \Rightarrow \delta f = 0 \quad (9.39)$$

From equation (9.39), the approximate directions of the light for the maneuvering ($\hat{\sigma}_M$) and reference ($\hat{\sigma}$) satellite are the same.

There are two ways to analyze the solar pressure on a pair of satellites: 1) Assume the same satellite dimensions with a difference in the reflectivity coefficient **[150]**, or 2) assume the same reflectivity coefficient with a difference in the satellite dimensions **[151]**. Both cases can demonstrate variations in the controller effects. To explain both formulations, the solar pressure force for a right circular cylinder is used. This formulation is developed in Section 3.5 and is shown here. The solar pressure force for a total absorbing ($\vec{F}^{+}$) and total reflective ($\vec{F}^{-}$) right circular cylinder is written as,

$$\vec{F}^{+} = -R_C\bar{h}_0(4l_C b_0 + \pi R_C c_0)\hat{\sigma} \quad (9.40a)$$

$$\vec{F}^{-} = R_C\bar{h}_0\begin{bmatrix} -\frac{16}{3}l_C a_0 b_0 \\ \frac{4l_C}{3}\left(2a_0^2+8b_0^3\right) \\ -\pi R_C c_0^2 \end{bmatrix} \quad (9.40b)$$

Using equations (9.35), the force due to the solar pressure can be determined for the maneuvering and target spacecraft when the reflectivity coefficient is different and the satellite dimensions are the same. In the formulation of the TH equations for a perturbed motion, the force due to solar pressure is the difference in the force between the reference and the maneuvering satellite; thus, the total force due to the solar pressure for an arbitrary surface can be written as,

$$\vec{F}_S = \vec{f}_{S,M} - \vec{f}_{S,R} \tag{9.41}$$

where,

$$\vec{f}_{S,R} = (1 - \varepsilon_R)\vec{f}^{+}_R + \varepsilon_R \vec{F}^{-}_R \quad \vec{f}_{S,M} = (1 - \varepsilon_M)\vec{F}^{+}_M + \varepsilon_M \vec{F}^{-}_M$$

Since $\hat{\sigma} = \hat{\sigma}_M$, $\vec{F}^{+} = \vec{F}^{+}_R = \vec{F}^{+}_M$, and $\vec{F}^{-} = \vec{F}^{-}_R = \vec{F}^{-}_M$, consequently, the total force due to the solar pressure for an arbitrary surface reduces to,

$$\vec{F}_S = \Delta\varepsilon(\vec{F}^{-} - \vec{F}^{+}) \tag{9.42}$$

where $\Delta\varepsilon = \varepsilon_R - \varepsilon_M \cdot \varepsilon_R$ and ε_M are the reflective coefficients for the reference and maneuvering satellite, respectively. If $\varepsilon_R = \varepsilon_M$, the solar pressure is not disturbing the pair of satellites in the elliptical orbit (same reflective coefficient properties). It could happen that the reflective coefficients for the maneuvering and reference satellites are not the same due to the construction of the outside panels of the satellite. For this reason, equation (9.42) is used to define this small difference between the reflective coefficients for the maneuvering and reference spacecraft. A column matrix $\vec{\Gamma}(k)$ can be separated from the disturbance matrix to contain these coefficients. The column vector is expressed as,

$$\vec{\Gamma}(k) = [0 \quad 0 \quad 0 \quad f_1 \quad f_2 \quad f_3] \tag{9.43}$$

where,

$$f_j = \frac{F_{jS}}{m v_S k_h^2 \left(1 + e\cos f(k)\right)} \qquad (j = 1, 2, 3)$$

On the other hand, the reflectivity coefficient between both spacecraft can be the same ($\varepsilon = \varepsilon_R = \varepsilon_M$), but the dimensions can be different. Using this assumption and equations (9.41), the difference in solar pressure between the reference and maneuvering spacecraft is,

$$\vec{F}_S = (1 - \varepsilon)(\vec{F}^{+}_R - \vec{F}^{+}_M) + \varepsilon(\vec{F}^{-}_R - \vec{F}^{-}_M) \tag{9.44}$$

Using, the right circular cylinder example, equation (9.44) can be written as,

$$\vec{F}^{+}_R - \vec{F}^{+}_M = 4b_0\bar{h}_0\Delta(l_c R_C)\hat{\sigma} \tag{9.45a}$$

$$\vec{F}^{-}_R - \vec{F}^{-}_M = -\frac{4}{3}\bar{h}_0\Delta\left(l_C R_C\right)\left[4a_0 b_0\hat{\imath} - \left(2a_0^2 + 8b_0^3\right)\hat{\jmath}\right] - 2\pi\bar{h}_0\Delta R_C^2\hat{k} \tag{9.45b}$$

and,

$$\Delta(l_C R_C) = l_{C,R} R_{C,R}(1 - r_l r_{RC}) \quad \Delta R_C^2 = R_{C,R}^2(1 - r_{RC}^2) \tag{9.46a}$$

$$r_l = \frac{l_{C,M}}{l_{C,R}} \quad r_{R_C} = \frac{R_{C,M}}{R_{C,R}} \tag{9.46b}$$

$l_{C,M}$ and $l_{C,R}$ are the length of the maneuvering and reference spacecraft, respectively. $R_{C,M}$ and $R_{C,R}$, respectively, is the radius of the cylinder (R_C) for the maneuvering and reference satellite. Equations (9.45) depend on the ratio of the length (r_l) and radius of the cylinder (r_{RC}) between the maneuvering and reference satellite. If both ratios are the same, the solar pressure does not affect the motion of a pair of satellites in an elliptical orbit. Once more, these formulations can be substituted in equation (9.43) to describe the solar pressure force in the discrete TH equations.

9.5 Lawden solution

The active control scheme can be understood by examining the solution of the TH equations. Since the CW equations are a special form of the TH equations, the following solutions can be applied for constellations in circular orbits. There are two types of problems: 1) two point boundary value problems (TPBVP) and 2) drift correction.

In TPBVP, the desired states are known and must be satisfied at the final true anomaly angle. As shown in reference [5], the TPBVP for the linearized TH equation in an environment without perturbations can be solved with primer vector relations. The primer vector was proposed by Lawden [147] in 1963. Reference [152] defines the primer vector as a byproduct of applying Calculus of Variations techniques to the problem of minimizing fuel usage of impulsive trajectories, and, in the primer vector theory, the optimal number of impulses and their location depend on the trajectory or path of a satellite. Reference [152] provides the conditions to prove that the state equations can be used with the primer vector theory, and, in reference [5], Carter and Humi showed that the linearized TH equations for an unperturbed motion satisfied the primer vector relations. In primer vector theory, there exist three types of arcs that can be obtained in the solution of the problem. These arcs are the following: 1) NT – Null-Thrust Arc; 2) IT – Intermediate-Thrust Arc; 3) MT – Maximum-Thrust Arc. As explained in reference [152], the NT arcs happen when a single impulse is applied at the beginning and at the end of the maneuver such that the satellite can travel in a predefined arc without applying any type of thrust between full thrusts. The IT and MT thrust arcs are applied along the path or trajectory of the satellite. Lawden [147] explained different conditions for the primer vector relation in terms of the thrust magnitude switching (TMS) function. The TMS function relates the expenditure of mass and the co-state variables. In an NT arc, the TMS function is positive and shows a direct proportional relation between the TMS function and the co-state variables. For an MT arc, the TMS is negative and is defined by a first order linear differential equation which is in terms of the mass expenditure and the co-state variable. Over an IT arc, the TMS function vanishes identically. This classification of the thrust arcs is used in the solution of the TPBVP to determine the arc that is used in the hierarchical control scheme.

Using the linear quadratic regulator (LQR) control scheme, the solution does not provide an NT arc because its solution has linear control laws. For fuel-optimal problems, IT arcs are singular controls, but, for the LQR problem, they are not; for linear differential equations, they are linear without MT or NT.

If it is not desired to solve the TPBVP, the drift correction [**100**] can be used. The drift is defined as the difference between the desired coordinates and the initial coordinates of any pair of satellites within a constellation. For the drift correction, the desired state is set to zero such that the drift can be reduced in a short period of time. In reference [**153**], the drift correction is performed with the LQR. The drift correction is performed faster when the varying coefficients for the position of the satellite in the Q matrix are weighted more than the constant coefficients for the velocity of the satellite. This weighting for the Q matrix is used because the coefficients in the position are multiplied by the varying term, κ shown in equation (9.23).

9.6 Discrete optimal control problem for formation flying

References [**154**] and [**155**] solve the optimal control problem of the TH equations by solving a bang-bang controller. On the other hand, Carter and Humi [**5**] solve the same problem with a bang-off-bang controller. In any of these solutions, the TH equations lead to either a null thrust or a full thrust arc. This section uses the linear quadratic cost function to solve the TH equations to determine another thrust arc.

The LQR cost function for the TH equations can be written as follows,

$$\mathcal{J}(k) = \frac{\Delta t}{2} \sum_{k=0}^{N_{\hat{f}-1}} \left(\vec{y}(k) - \vec{y}_D\right)^T \boldsymbol{Q} \left(\vec{y}(k) - \vec{y}_D\right) + \vec{u}^T(k) \boldsymbol{R} \vec{u}(k) \tag{9.47}$$

$\vec{y}_D$ is the desired state vector, and $\vec{y}(k)$ is the state vector described in equation (9.23). $\boldsymbol{Q}$ and $\boldsymbol{R}$ is the state and control weighting matrix, respectively. Reference [153] shows that this cost function can be applied to low eccentric orbits (between 0 and 0.6). In addition, the varying term in the TH equation is updated in a piecewise manner as shown in Reference [156]. The update scheme is based on the location of the satellite in the elliptical orbit, and the state matrices are maintained constant for short periods of times.

References [5] and [157] used the same linearized TH equations, but the equations of motion are expressed in terms of the eccentricity and true anomaly angle. Marec [158] established that the cost function, the state matrix, and the control matrix must be in terms of the true anomaly angle to obtain a minimum-time problem. Marec says that:

> *The optimization problem then appears as a 'minimum-time' problem: the question is to transfer from orbit $\vec{q}_0$ to orbit $\vec{q}_f$ with a minimum consumption C_0. In this problem, not only the thrust direction $\vec{D}$ but* also the true anomaly v must be considered *as* controls: *in effect, the duration* being *free and the osculating orbits 0 being* elliptic, *the choice of point v of 0 where the thrust is applied is entirely free: it suffices to wait on 0 till the suitable position is reached.*

In addition, Inalhan, Tillerson, and How [159] say that, 'for an eccentric orbit, the shape of the closed form solution is not a perfect ellipse, but is actually skewed and scaled', but they said that the eccentricity term must be part of the equations of motion.

Carter–Humi [5] used Pontryagin minimum principles to obtain an admissible control to rendezvous in an optimal way between the target and the reference satellite. They define the following cost function for the optimal control as:

$$\mathcal{J} = \int_{f_0}^{f_L} \frac{\left|\vec{u}(f)\right|}{\left(1+e\cos f\right)^2}\, df \tag{9.48}$$

Through this development of the discrete LQR, the cost function is defined in terms of the varying terms in the $\hat{\boldsymbol{A}}$ and $\hat{\boldsymbol{B}}$ matrix as:

$$\mathcal{J}(k) = \frac{\Delta f}{2} \sum_{k=0}^{N_{f-1}} \left(\vec{y}(k)-\vec{y}_D\right)^T \tilde{\boldsymbol{Q}}\left(\vec{y}(k)-\vec{y}_D\right) + \vec{u}^T(k)\tilde{\boldsymbol{R}}\vec{u}(k) \tag{9.49}$$

where,

$$\tilde{\boldsymbol{Q}} = \frac{\boldsymbol{Q}}{\left(1+e\cos f(k)\right)^2} \qquad \tilde{\boldsymbol{R}} = \frac{\boldsymbol{R}}{\left(1+e\cos f(k)\right)^2}$$

and,

$$N_f - 1 = \frac{f_L - f_0}{\Delta f}$$

f_L is the final true anomaly angle, and Δf is the sampling interval in the true anomaly angle. This proposed cost function is in accordance with Marec's statement for a minimum-time problem in an elliptical orbit and is based on the cost function defined by Carter–Humi.

Using the following state vector representation for the hierarchical control scheme,

$$\vec{y}(k+1) = \hat{\boldsymbol{A}}\vec{y}(k) + \hat{\boldsymbol{B}}\vec{u}(k) + \Delta f\vec{\psi}[\vec{y}^L(k)] + \Delta f\vec{\Gamma}(\mathrm{k}) \quad (9.50\mathrm{a})$$

$$\vec{y}^L(k) \in \vec{y}(k) \quad (9.50\mathrm{b})$$

The Hamiltonian (H) for a hierarchical control system in discrete format can be written as,

$$\begin{aligned}\hat{A}(k) = {} & \frac{\Delta \mathrm{f}}{2}\left(\vec{y}(k) - \vec{y}_D\right)^T \boldsymbol{Q}\left(\vec{y}(k) - \vec{y}_D\right) + \frac{\Delta f}{2} \boldsymbol{u}^T(k)\boldsymbol{R}\boldsymbol{u}(k) \\ & + \vec{\lambda}^T(k+1)\left(\hat{\boldsymbol{A}}\vec{y}(k) + \hat{\boldsymbol{B}}\vec{u}(k) + \Delta f\vec{\psi}\left[\vec{y}^L(k)\right]\right) \\ & + \vec{\gamma}^T(k+1)\left(\vec{y}(k) - \vec{y}^L(k)\right)\end{aligned} \quad (9.51)$$

where $\vec{\lambda}(k+1)$ is the co-state variable. The minimum principle is used to obtain the necessary conditions to determine the suboptimal control problem for the nonlinear system. For the hierarchical control scheme, the solution of the DLQR problem becomes as,

$$\boldsymbol{P}(k) = \Delta f\boldsymbol{Q} + \hat{\boldsymbol{A}}^T\boldsymbol{P}(k+1)[\boldsymbol{I} + \boldsymbol{SP}(k+1)]^{-1}\hat{\boldsymbol{A}} \quad (9.52\mathrm{a})$$

$$\begin{aligned}\vec{G}(k) = {} & -\Delta f\boldsymbol{Q}\vec{y}_D + \hat{\boldsymbol{A}}^T\,\vec{G}(k+1) + \vec{\gamma}(k+1) \\ & + \hat{\boldsymbol{A}}^T\,\boldsymbol{P}(k+1)[\boldsymbol{I} + \boldsymbol{SP}(k+1)]^{-1}[-\boldsymbol{S}(k)\vec{G}(k+1) \\ & + \Delta f\vec{\psi}[\vec{y}^L(k)] + \Delta f\vec{\Gamma}(k)]\end{aligned} \quad (9.52\mathrm{b})$$

$$\begin{aligned}\vec{y}(k+1) = {} & [\boldsymbol{I} + \boldsymbol{SP}(k+1)]^{-1}[\hat{\boldsymbol{A}}\vec{y}(k) - \boldsymbol{S}\vec{G}(k+1) \\ & + \Delta f\vec{\psi}[\vec{y}^L(k)] + \Delta f\vec{\Gamma}(k)]\end{aligned} \quad (9.52\mathrm{c})$$

$$\vec{u}(k) = -(\Delta f\boldsymbol{R})^{-1}\,\hat{\boldsymbol{B}}^T\,(\boldsymbol{P}(k)\vec{y}(k) + \vec{G}(k)) \quad (9.52\mathrm{d})$$

$$\vec{\gamma}(k+1) = \Delta f\hat{\boldsymbol{J}}^T[\boldsymbol{P}(k+1)\vec{y}(k+1) + \vec{G}(k+1)] \quad (9.52\mathrm{e})$$

$$0 = \vec{y}(k) - \vec{y}^L(k) \quad (9.52\mathrm{f})$$

$$\hat{J} = \frac{\partial \vec{\psi}\left[\vec{y}^L(k)\right]}{\partial \vec{y}^L(k)} \tag{9.52g}$$

$P(k+1)$ and $\vec{G}(k+1)$ are the Riccati and adjoint Riccati equation, respectively. In equation (9.52e), $\hat{J}$ is the Jacobian matrix of the column vector, $\vec{\psi}[\vec{y}^L(k)]$. Equations (9.52b–e) are related to the error vector and $\vec{y}^L(k)$, and, for the first run, $\vec{y}^L(k) = \vec{\gamma}(k+1) = 0$; in this way, the hierarchical control scheme is solving equation (9.50a) in the linear range. The second run updates the error vector and $\vec{y}^L(k)$ with equations (9.52e) and (9.52g) to take into account the nonlinear terms of the J2 perturbation. For the J2 perturbation, the Jacobian column vector can be written as,

$$\hat{J} = \frac{\partial \vec{\psi}\left[\vec{y}^L(k)\right]}{\partial \vec{y}^L(k)} = \begin{bmatrix} 0 & 0 & 0 & 0 & 0 & 0 \\ 0 & 0 & 0 & 0 & 0 & 0 \\ 0 & 0 & 0 & 0 & 0 & 0 \\ \frac{\partial \psi_4}{\partial y_1^L} & 0 & 0 & 0 & 0 & 0 \\ 0 & \frac{\partial \psi_5}{\partial y_2^L} & \frac{\partial \psi_5}{\partial y_3^L} & 0 & 0 & 0 \\ 0 & \frac{\partial \psi_6}{\partial y_2^L} & \frac{\partial \psi_6}{\partial y_3^L} & 0 & 0 & 0 \end{bmatrix} \tag{9.53a}$$

where,

$$\frac{\partial \psi_4}{\partial y_1^L} = \frac{3B_2k_h^2}{2h\kappa}\left(1-5\sin^2 i\right) \quad \frac{\partial \psi_5}{\partial y_2^L} = -\frac{3B_2k_h^2}{h\kappa}\left(2-15\sin^2 i\right) \tag{9.53b}$$

$$\frac{\partial \psi_5}{\partial y_3^L} = -\frac{15B_2k_h^4}{\mu\kappa}y_3^L \qquad \frac{\partial \psi_6}{\partial y_2^L} = -\frac{15B_2k_h^2}{h\kappa}\sin\, i \tag{9.53c}$$

$$\frac{\partial \psi_6}{\partial y_3^L} = \frac{3B_2k_h^2}{2h\kappa}\left(3-5\sin^2\, i\right) \tag{9.53d}$$

Using equations (9.52) and (9.53), the hierarchical control scheme can be solved (Figure 7.28 shows the algorithm). In the solution, it is important that the transformation for the $\vec{x}$ domain to the $\vec{y}$ domain is performed with equations (9.19a). In the transformation performed with equations (9.13) and (9.19a), the dimensions for the y_1, y_2, y_3 system are the same as for the x_1, x_2, x_3 system.

9.6.1 Two point boundary value problem solution

One specific size of the NASA Benchmark Tetrahedron Constellation is used to show the control solution [**146**] (see Appendix G for explanation). The NASA Benchmark Tetrahedron Constellation requires that the separation distance between any pair of satellites is 10 km at the apogee point. At any other point in the HEO, this separation distance should not be less than one kilometer. The orbital elements for one specific size (or phase) of this proposed constellation [**146**] are shown in Table 9.1.

To include an ion thruster capability into the simulation, the maximum thrust [10] (T_m) is assumed to be 0.5 N. The $\boldsymbol{Q}$ and $\boldsymbol{R}$ matrix, respectively, is a 6 × 6 and 3 × 3 diagonal matrix. The weights in the $\boldsymbol{Q}$ matrix are *diag*([20, 20, 20, 1, 1, 1]), and the weights of the $\boldsymbol{R}$ matrix are *diag*([1, 1, 1]). The mass of the satellite at the initial true anomaly angle is 90 kg (assumed constant), and the inclination angle of the satellites is defined in Table 9.1. The sampling in the true anomaly angle is set to 0.05 radians to obtain a better approximation because of the nonlinearities in the TH equations for a perturbed motion. To implement this control scheme in the computer onboard the satellite, it is required to transform from the time domain into the true anomaly angle domain with Kepler's transcendental equation, but this transformation is not pursued at this point. In the hierarchical control scheme, the maximum number of iterations in the second level is 60 (L = 60), and the tolerance value is equal to 1e-10. In addition, the satellites are assumed to have the same dimensions with a difference in the reflectivity coefficient equal to 0.01.

A station keeping process is performed to maintain the separation distance conditions of a pair of satellites. The desired and initial

Table 9.1 **Orbital elements for one specific size of the proposed constellation**

Dimensions	First phase
Radius of perigee *(r_p)*	1.2 R_E
Radius of apogee *(r_a)*	12 R_E
Simimajor axis *(a)*	42,095.7 km
Eccentricity *(e)*	0.818
Inclination angle *(i)*	18.5°
Period *(days)*	1

conditions for a pair of satellites within the proposed tetrahedron constellation are:

$$\vec{x}_D = [-8.7 \text{ km} \quad 4.7 \text{ km} \quad 1.6 \text{ km} \quad 3.5 \times 10^{-4} \text{ km/sec} \quad 0 \quad 0]^T \quad (9.54a)$$

$$\vec{x}(f_0) = [-9.9 \text{ km} \quad 4.7 \text{ km} \quad 1.5 \text{ km} \quad 3.5 \times 10^{-4} \text{ km/sec} \quad -1.4 \times 10^{-5} \text{ km/sec} \quad -9 \times 10^{-6} \text{ km/sec}] \quad (9.54b)$$

These desired and initial conditions are obtained from previous simulations performed in an orbit propagator shown in Reference [**148**]. These conditions are obtained when a pair of satellites within the proposed constellation first violates the separation distance conditions [**148**].

Figure 9.2 shows the solution of the TPBVP for a station keeping procedure. To obtain the solution, the hierarchical control scheme requires 3 iterations to obtain the suboptimal control solution. The separation distance forms different arcs to satisfy the suboptimal control problem. Near the perigee point, the velocities cross through zero. Also, the separation distance and the suboptimal control show minimum and/or maximum values near the perigee point. One interesting result is the thrust arcs shown in the suboptimal control graph. The thrust arcs

Figure 9.2 **Two point boundary value problem solution for the hierarchical active control scheme**

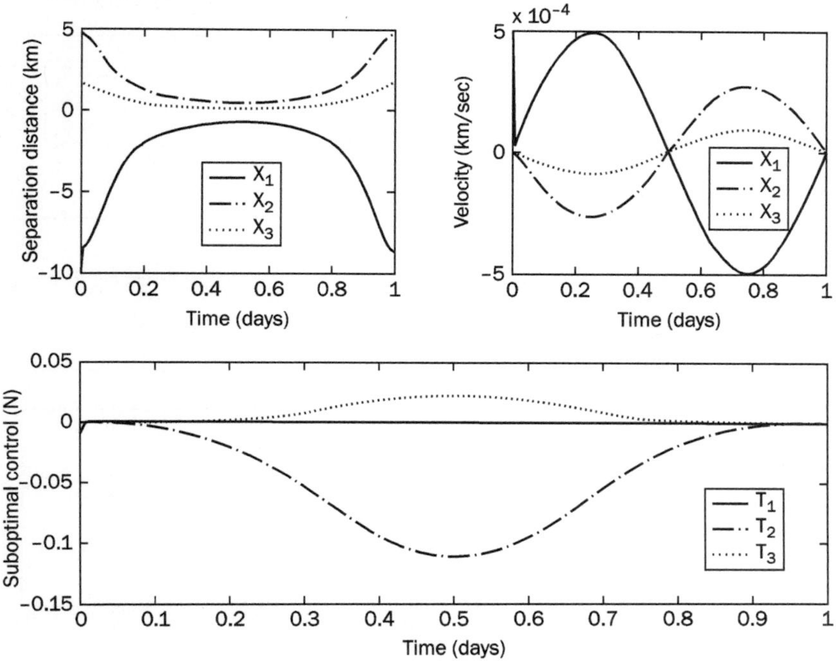

vanish identically at the apogee point, and the maximum value of control happens at the perigee point. Also, this discrete control scheme applies small constant impulses along the path of the orbit to solve the TPBVP. From the definitions provided by Lawden [**147**], the thrust arcs shown in Figure 9.2 can be classified as intermediate thrust arcs (IT). With this cost function, the IT arcs are proven and demonstrated the third thrust arc written first by Lawden [**147**]. In addition, this demonstration is performed by using discrete control schemes.

9.6.2 Drift correction

The same conditions used in the TPBVP are used for the drift correction of the proposed constellation. Taking the difference between the initial and the desired conditions (equation (9.54)), the drift between the pair of satellites is the following:

$$\vec{x}(f_0) = [-1.2 \text{ km} \quad 0.3 \text{ km} \quad 0.01 \text{ km} \quad 2. \times 10^{-4} \text{ km/sec} \quad -1.4 \times 10^{-5} \text{ km/sec} \quad -9 \times 10^{-6} \text{ km/sec}] \tag{9.55}$$

Figure 9.3 Drift correction with the hierarchical active control scheme

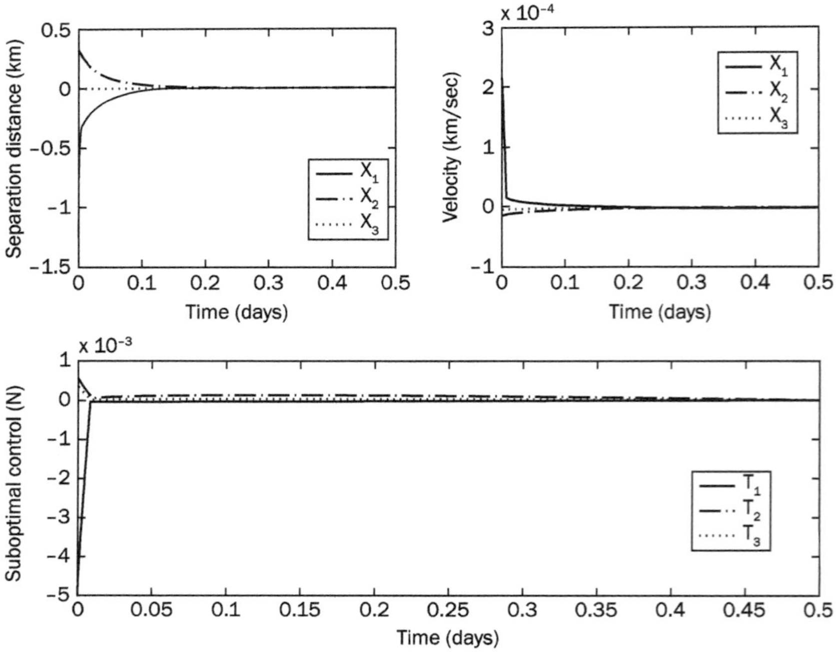

The purpose of the drift correction is to minimize the drifts as fast as possible.

Figure 9.3 shows the results for the drift correction for a station keeping procedure. The hierarchical control scheme also takes three iterations at the second level to obtain the suboptimal solution. The correction of the separation distance and velocity drifts is performed before the satellites reach the perigee point. The maximum magnitude of the control effort is less in Figure 9.3 than in Figure 9.2. At the initial apogee point, the suboptimal control applies a higher thrust to cause a variation in the velocity of the satellite. Comparing Figures 9.2 and 9.3, the drift correction shows a better consumption of thrust in comparison to the TPBVP.

9.7 Formation flying controller implementation

The implementation of the discrete controller is complex because of the relationship between the time and true anomaly angle. As mentioned before, the hierarchical controller is developed to correct the separation distance and velocity drifts [103] between a pair of satellites within a constellation in HEO. In HEO, the satellites may not have direct communication to ground stations or to global positioning systems (GPS) at the apogee point [10]. Once the satellites reach the perigee point, the ground stations can obtain the information about the location of the satellites in the orbit. With this information transmitted to the ground stations, the satellites can be updated to determine their exact locations. An orbit propagator [10] imbedded in the computer onboard the satellite can be used to track the location of the satellites at the apogee point. This information obtained from the orbit propagator is used to correct the separation distance and velocity drifts.

To perform this correction, Figure 9.4 shows a schematic for the implementation of the discrete hierarchical controller. The navigation system for the maneuvering and reference satellites provides information about their translational motion when the perturbations due to the Earth and the Sun are added. In a similar manner, the trajectory system provides information about the desired location of this pair of satellites along the HEO with a spherical Earth. The navigation and trajectory systems can contain a highly precise orbit propagator and/or communication system between the satellites that supply accurate information about the location of the satellites along the orbit.

Figure 9.4 **Implementation of the discrete controller**

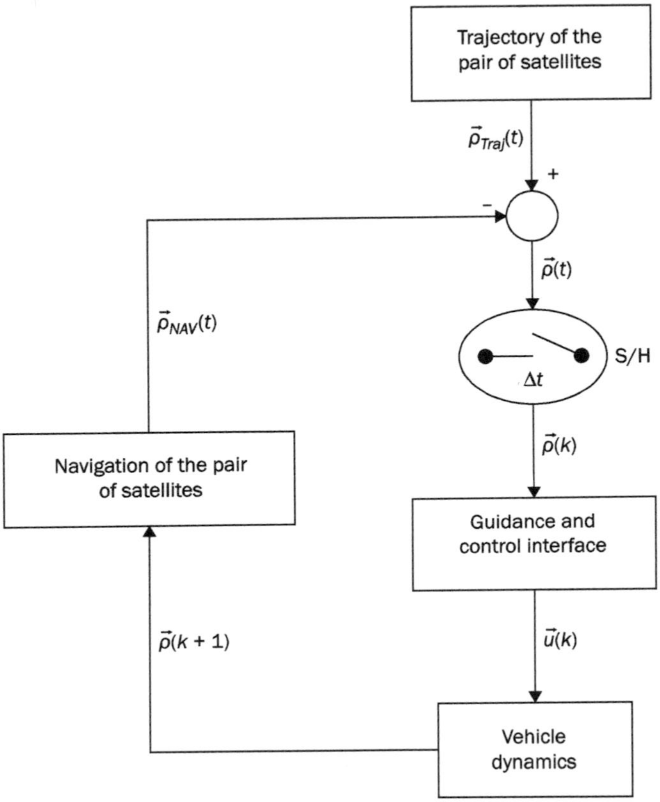

Using this information, the separation distances and velocities of the pair of satellites obtained from the navigation and trajectory systems are subtracted to determine the separation distance and velocity drifts, $\vec{\rho}(t)$. These drifts can be continuously supplied to the computer onboard of one of the satellites, but a sample of this information is used. For this reason, a sample and holder (SH) [78] is used to obtain the separation distance of the pair of satellites at certain locations in the HEO. The sampling time of the S/H is varying along the elliptical orbit because the vehicle dynamics are described by the Tschauner–Hempel (TH) equations. These equations are defined in terms of the eccentricity and the true anomaly angle, and a relationship is required to transform from the true anomaly angle to the time domain.

The separation distance and velocity drifts are sent to the guidance and control interface (GCI) to obtain the control inputs. In the GCI, the separation distance and velocity drifts are transformed, and the control input forces are determined. This information is sent to the vehicle dynamics to determine the following output for the separation distance and velocity drifts. These new separation distance and velocity drifts are sent to the navigation systems to propagate the pair of satellites to determine their new positions and velocities. As shown in Figure 9.4, this control scheme becomes a closed loop system for the correction of the separation distance and velocity drifts. In the following sections, the components of this discrete controller are explained in detail.

9.7.1 Vehicle dynamics

The vehicle dynamics is explained first to understand the information that the other systems receive. The TH equations [5] explain the motion of a pair of satellites about the Earth in a HEO. The vehicle dynamics includes a model based on the perturbations due to the oblateness of the Earth and the solar pressure force. As shown previously, the controller is expressed in the discrete domain to reduce the amount of data [**100**]. For this reason, the discrete form of the TH equations for a perturbed motion can be represented as shown in equation (9.50a),

$$\vec{y}(k+1) = \hat{\boldsymbol{A}}\vec{y}(k) + \hat{\boldsymbol{B}}\vec{u}(k) + \Delta f\,\vec{\psi}[\vec{y}(k)] + \Delta f\,\vec{\Gamma}(k)$$

$\vec{\psi}[\vec{y}(k)]$ describes the J2 perturbation, and $\vec{\Gamma}(k)$ describes the solar pressure force. When the satellites are manufactured, they can have differences in the dimensions or in the type of surface [**150**] [**151**]. To take into account the difference in the reflectivity coefficients between the pair of satellites, the satellites are assumed to have the same dimensions but contain small differences in their surface properties. In addition, the satellites are assumed to be opaque right circular cylinders in which the incident solar rays can affect the motion of the satellite; thus, the solar pressure force in the Karymov theory [32] is expressed in equation (9.42).

9.7.2 True anomaly angle to time transformation

As shown in Figure 9.4, the S/H is an important component of the discrete controller. This S/H provides the necessary information at specific time intervals for use in the controller. If the sampling interval in the true anomaly is constant, the sampling time is varying; therefore, a relationship between the time and the true anomaly angle domain is required in the control procedure.

In the discrete controller, the system is solved in terms of the sample (k) obtained at a certain period of time or location in the orbit (TH equations). To determine the time in terms of the true anomaly angle, Kepler's transcendental equation can be used [9]. With this equation, the system is required to transform from the true anomaly to the eccentric anomaly and, then, from the eccentric anomaly to the time. To determine the sampling time (Δt), the difference between the actual and future time of flight is required; but, this calculation of Δt is a slow process.

On the other hand, the angular momentum equation is considered to obtain Δt in terms of the sampling interval in the true anomaly, Δf, and the sample, k. The angular momentum magnitude [5] is defined in equation (9.12). Using the definition of the derivative [81], equation (9.12) can be written as,

$$\dot{f} = \lim_{\Delta t \to 0} \frac{\Delta f}{\Delta t} = \frac{h}{R^2} \Rightarrow \Delta t(k) = \frac{h^3}{\mu}\left(1 + e\cos f(k)\right)^{-2} \Delta f \tag{9.56}$$

In the discrete TH equations if the sampling interval in the true anomaly is constant, then, the sampling time is varying with respect to the sample, k. The sampling time is expressed as the difference between the present and future time; then, the time in which a measurement is sampled can be written as,

$$t(k+1) = t(k) + \Delta t(k) = t(k) + \frac{h^3}{\mu}\left(1 + e\cos f(k)\right)^{-2} \Delta f \tag{9.57}$$

where $t(0) = t_0$, and t_0 is the initial time. Figure 9.5 shows the variation of time for a = 42095.7 km and e = 0.818 (Table 9.1) when Δf = 0.05 and 0.1 rad. When Δf = 0.05 rad, the controller is using less data in comparison to Δf = 0.1 rad. In both sampling intervals in the true anomaly, the controller is sampling more near the perigee point than at the apogee point. Near the perigee point, the time decreases as the sampling interval in the true anomaly angle decreases because of the number of samples used to discretize the orbit.

9.7.3 Integration process

For the navigation and trajectory systems, an integration process is required to determine the location of the satellites in the HEO. For

Figure 9.5 Time vs. true anomaly angle

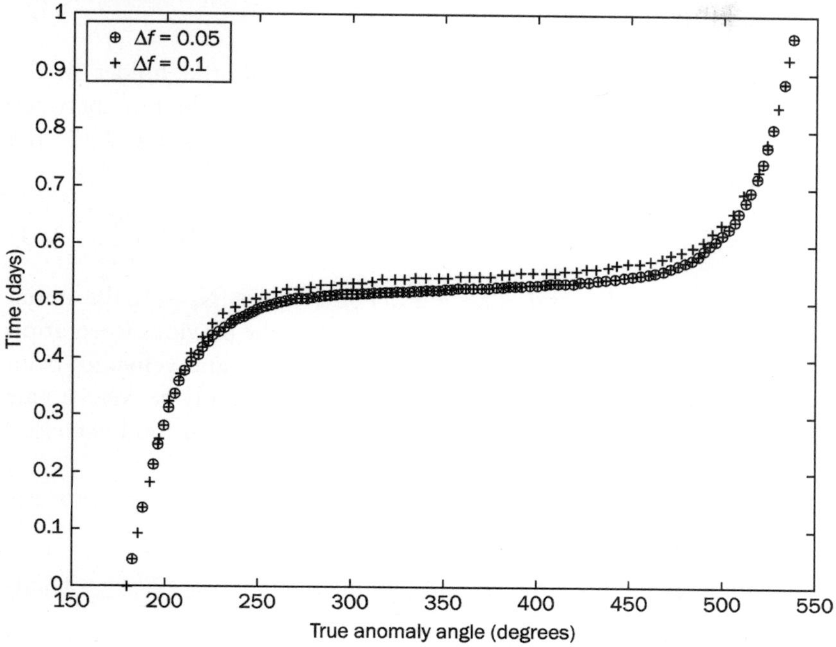

this integration procedure, Cowell's method [9] is used to specify the motion of a satellite in an environment with or without perturbations. The motion of a single satellite [9] about a spherical Earth can be written as,

$$\ddot{\vec{R}}(t)+\mu\frac{\vec{R}}{\left|\vec{R}\right|^{3}}=0 \tag{9.58}$$

where,

$$\vec{R}=[X_1\ X_2\ X_3]^T\ (\text{km}) \tag{9.59a}$$

$$\dot{\vec{R}}=[\dot{X}_1\ \dot{X}_2\ \dot{X}_3]^T\ (\text{km/sec}) \tag{9.59b}$$

X_1, X_2, X_3 and $\dot{X}_1$, $\dot{X}_2$, $\dot{X}_3$ are the positions and velocities of the satellite in the LVLH frame of the Earth for the X, Y, and Z directions, respectively. The initial conditions are expressed as,

$$\vec{R}(0)=[X_1(0)\quad X_2(0)\quad X_3(0)]^T\ (\text{km}) \tag{9.59a}$$

$$\dot{\vec{R}}(0)=[\dot{X}_1(0)\quad \dot{X}_2(0)\quad \dot{X}_3(0)]^T\ (\text{km/sec}) \tag{9.59b}$$

The integration of equation (9.59) can be written as,

$$\vec{R}(t) = \int_{t_0}^{t_1} \dot{\vec{R}}(t)\, dt \quad \dot{\vec{R}}(t) = \int_{t_0}^{t_1} \ddot{\vec{R}}(t)\, dt \tag{9.60}$$

where t_1 is the final time. To include the sample and holder in the integration process, equation (9.57) is used to specify the time between the samples. Using the definition of the integral and equation (9.57), the integration procedure can be written as follows,

$$\vec{R}(t) = \sum_{k=0}^{N_f-1} \int_{t(k)}^{t(k)+\Delta t(k)} \dot{\vec{R}}(t)\, dt \; \dot{\vec{R}}(t) = \sum_{k=0}^{N_f-1} \int_{t(k)}^{t(k)+\Delta t(k)} \ddot{\vec{R}}(t)\, dt \tag{9.61}$$

At $k = 0$, $\vec{R}(t_0) = \vec{R}_0$, and $\dot{\vec{R}}(t_0) = \dot{\vec{R}}_0$. For $1 \leq k \leq N_f - 1$, the initial condition is given by the last values of $\vec{R}$ and $\dot{\vec{R}}$ of the previous integration at $t(k)$. At $t(k + 1)$, the controller uses the positions and velocities from the trajectory and navigation systems to correct the drifts between a pair of satellites within the constellation. These new positions and velocities describe one location in the HEO.

Using the following initial conditions [**149**] in Cartesian form for the HEO in Figure 9.5,

$$\vec{R}(0) = [-8.7 \; -72582.5 \; -24285.8]^T \text{ (km)} \tag{9.62a}$$

$$\dot{\vec{R}}(0) = [0.97308 \quad 0 \quad 0]^T \text{ (km/sec)} \tag{9.63b}$$

and solving numerically equations (9.61) for $\Delta f = 0.05$, Figure 9.6 shows the HEO for this satellite. In Figure 9.6, the circle markers show the locations where the controller does the corrections. The controller obtains more information near the perigee point than at the apogee point. This integration process is used to express the motion of the maneuvering and reference satellite in the discrete domain for the trajectory and navigation systems.

9.7.4 Trajectory system of the pair of satellites

In Figure 9.4, the trajectory of the satellites is used to determine the desired conditions of the satellites. Equation (9.56) is used to supply the orbit propagator with the sampling time to determine the time in equation (9.57). Without perturbations, the satellites satisfy the separation distance and velocity conditions along the orbit and/or at certain points. The translational motion of the reference satellite is expressed with equation (9.58). For the reference satellite, the positions and velocities are defined as,

$$\vec{R} = [X_1^R \quad X_2^R \quad X_3^R]^T \quad \text{(km)}$$

Figure 9.6 **Numerical solution showing the locations for the sampling of the controller in the XY plane**

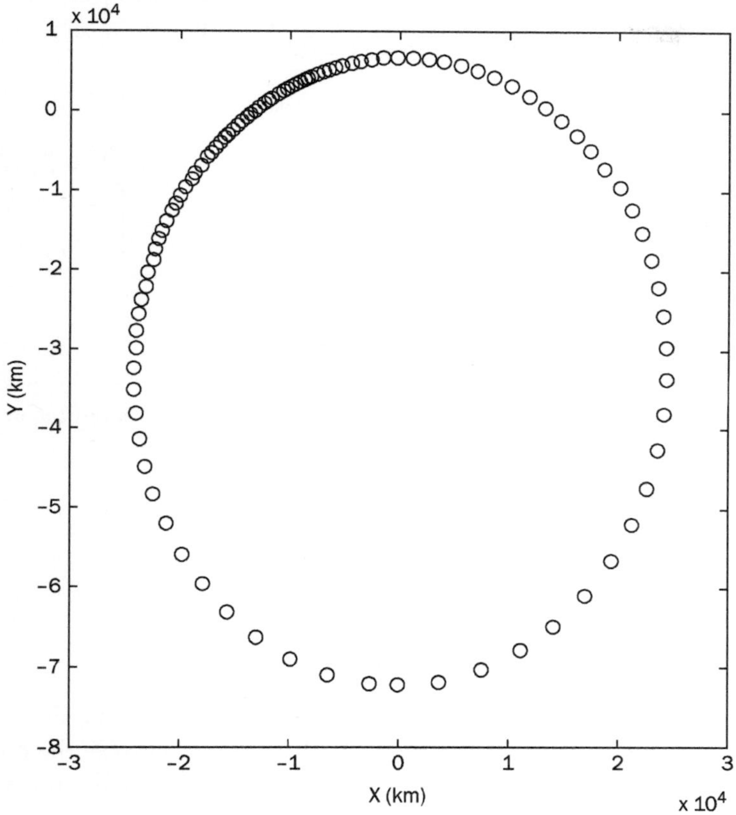

$$\dot{\vec{R}} = [\dot{X}_1^R \quad \dot{X}_2^R \quad \dot{X}_3^R]^T \quad \text{(km/sec)}$$

The maneuvering satellite is expressed as follows,

$$\ddot{\vec{r}}(t) + \mu \frac{\vec{r}}{|\vec{r}|^3} = 0 \tag{9.64}$$

where,

$$\vec{r} = [X_1^M \quad X_2^M \quad X_3^M]^T \quad \text{(km)}$$

$$\dot{\vec{r}} = [\dot{X}_1^M \quad \dot{X}_2^M \quad \dot{X}_3^M]^T \quad \text{(km/sec)}$$

Using equations (9.61) to integrate equations (9.58) and (9.64), the positions and velocities for the reference and maneuvering satellites are obtained at a sample k. This sample k is used to determine the separation

Figure 9.7 Block diagram for the trajectory

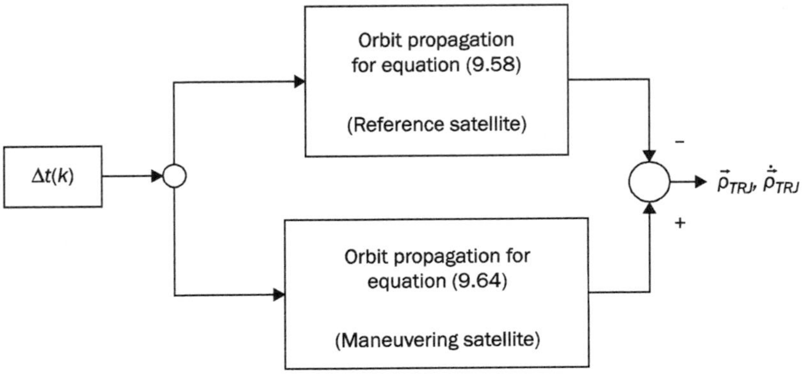

distances and the difference in velocities for the pair of satellites at one location in the HEO. These differences can be written as,

$$\vec{\rho}_{TRJ} = \vec{r}(k) - \vec{R}(k) \quad \dot{\vec{\rho}}_{TRJ} = \dot{\vec{r}}(k) - \dot{\vec{R}}(k) \tag{9.65}$$

where,

$$\vec{\rho}_{Traj} = \left[x_1^{TRJ} \quad x_2^{TRJ} \quad x_3^{TRJ}\right]^T \quad \text{(km)}$$

$$\dot{\vec{\rho}}_{TRJ} = \left[\dot{x}_1^{TRJ} \quad \dot{x}_2^{TRJ} \quad \dot{x}_3^{TRJ}\right]^T \quad \text{(km/sec)}$$

The superscript TRJ represents the trajectory subsystem. Equations (9.65) already contain the sample and holder which provide the output as a digital value. Figure 9.7 shows the process flow for the trajectory system.

9.7.5 Navigation system of the pair of satellites

The navigation system provides information about the location and velocity of the satellites along the HEO when the perturbations are included in the orbit propagator. The navigation system performs the same integration process as in the trajectory system but uses the output of the controller as an input for the propagation of the maneuvering satellite. This output of the controller is used to know the relative motion of the maneuvering satellite with respect to the reference satellite along the orbit.

The reference satellite in the navigation system can be written as,

$$\ddot{\vec{R}}_p(t) = -\mu\frac{\vec{R}_p}{\left|\vec{R}_p\right|^3} + \frac{3\mu\vec{R}_p}{\left|\vec{R}_p\right|^5}\left[\frac{\vec{R}_p}{2} - \frac{5}{2}\frac{\left(X_3^{R_C}\right)^2\vec{R}_p}{\left|\vec{R}_p\right|^2} + X_3^{R_C}\hat{k}\right] + \frac{\vec{f}_{R_C}}{m} \qquad (9.66)$$

where,

$$\vec{R}_p = [X_1^{Rc} \quad X_2^{Rc} \quad X_3^{Rc}]^T \quad (\text{km})$$

$$\dot{\vec{R}}_p = [\dot{X}_1^{Rc} \quad \dot{X}_2^{Rc} \quad \dot{X}_3^{Rc}]^T \quad (\text{km/sec})$$

m is the mass of the satellite. The perturbation is due to the oblateness of the Earth (or J2 perturbation) expressed with Brouwer's gravitational potential function [17]. $\vec{f}_{Rc}$ is the solar pressure force for the reference satellite defined by Karymov's theory [32]. Equation (9.66) is integrated with equations (9.61) to obtain the positions and velocities of the reference satellite at a sample k.

A similar expression can be written for the maneuvering satellite as follows,

$$\ddot{\vec{r}}_p(t) = -\mu\frac{\vec{r}_p}{\left|\vec{r}_p\right|^3} + \frac{3\mu\vec{r}_p}{\left|\vec{r}_p\right|^5}\left[\frac{\vec{r}_p}{2} - \frac{5}{2}\frac{\left(X_3^{M_C}\right)^2\vec{r}_p}{\left|\vec{r}_p\right|^2} + X_3^{M_C}\hat{k}\right] + \frac{\vec{f}_{M_C}}{m} \qquad (9.67)$$

where,

$$\vec{r}_p = \left[X_1^{Mc} \quad X_2^{Mc} \quad X_3^{Mc}\right]^T \quad (\text{km})$$

$$\dot{\vec{r}}_p = \left[\dot{X}_1^{Mc} \quad \dot{X}_2^{Mc} \quad \dot{X}_3^{Mc}\right]^T \quad (\text{km/sec})$$

$\vec{f}_{Mc}$ is the solar pressure for the maneuvering satellite. The integration process of equation (9.67) is the same as for the trajectory system, but the initial condition is defined differently for the maneuvering satellite in the navigational system than in the trajectory system. In this case, the initial condition for the maneuvering spacecraft contains the output of the controller as follows: (1) at $k = 0$, $\vec{r}_p(t_0) = \vec{R}_0 + \vec{\rho}_0$ and $\dot{\vec{r}}_p(t_0) = \dot{\vec{R}}_0 + \dot{\vec{\rho}}_0$, and (2) at $1 \le k \le N_f - 1$, $\vec{r}_p(t(k)) = \vec{r}_k + \vec{\rho}(k)$ and $\vec{r}_p\,(t(k)) = \dot{\vec{r}}_k + \dot{\vec{\rho}}(k)$. To use the separation distance and velocity drifts from the vehicle dynamics, a transformation from the time to the true anomaly angle domain and from the $\vec{y}$ to the $\vec{x}$ system is performed with equations (9.13) and (9.19a), respectively.

The separation distances and difference in the velocities obtained from the maneuvering and reference satellite in the navigation system can be described as,

$$\vec{\rho}_{NAV} = \vec{r}_p\,(k) - \vec{R}_p\,(k) \quad \dot{\vec{\rho}}_{NAV} = \dot{\vec{r}}_p\,(k) - \dot{\vec{R}}_p\,(k) \qquad (9.68)$$

Figure 9.8 **Block diagram for the navigation system**

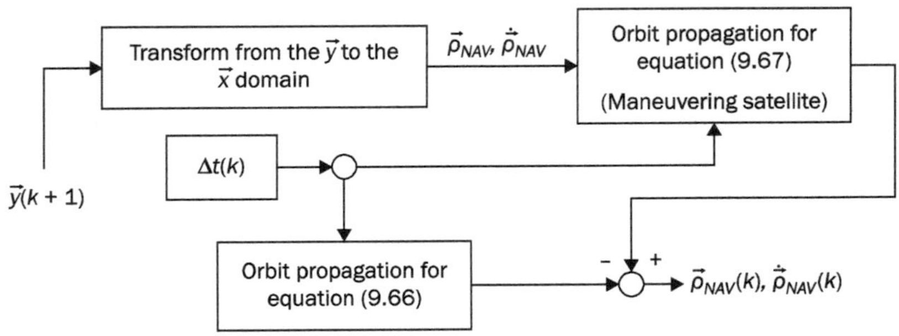

where,

$$\vec{\rho}_{NAV} = \begin{bmatrix} x_1^{NAV} & x_2^{NAV} & x_3^{NAV} \end{bmatrix}^T \quad \text{(km)}$$

$$\dot{\vec{\rho}}_{NAV} = \begin{bmatrix} \dot{x}_1^{NAV} & \dot{x}_2^{NAV} & \dot{x}_3^{NAV} \end{bmatrix}^T \quad \text{(km/sec)}$$

The superscript NAV refers to the navigation system. Figure 9.8 shows the block diagram for the navigation system.

9.7.6 Guidance and control interface

The Guidance and Control Interface (GCI) is used to provide the control input for the vehicle dynamics. The GCI contains two other subsystems which are used to supply the necessary information to determine the control inputs of the controller.

The discrete control scheme is explained to identify the different inputs and outputs of the GCI. The control scheme is based on a discretized version of the LQR which uses the cost function in equation (9.49) because of the highly elliptical orbit.

Equation (9.49) is used to express the hierarchical control scheme. The hierarchical control scheme takes care of the nonlinear terms by taking advantage of the linear form of the state equations, and, then, adds the effects of the nonlinear terms in a piecewise manner to obtain the control solution. This control scheme is a suboptimal controller because the control scheme does not solve the cost function and the state equations but takes advantage of the linear form of the equations.

The hierarchical control scheme is used to compensate for the drift correction but is re-expressed here as a steady-state system. The steady-state hierarchical control scheme reduces the computational time because the states in the control procedure are known and uses a 'look-up' table to obtain the control effort [103]. A 'look-up' table is a table that contains all the solutions for some of the equations in the control problem. The computer onboard the satellite accesses this 'look-up' table to obtain the solution for the control equation.

A discrete control system in steady-state conditions happens when the values for the co-state variables and other Lagrange multipliers are the same through the interval in the true anomaly angle. Section 7.4.4.2.2 shows the details for the development of the steady-state hierarchical control scheme. The control effort for the steady-state hierarchical control scheme can be expressed as,

$$\vec{u}(k) = -\boldsymbol{K}\vec{x}(k) - \boldsymbol{K}_{\psi}\left(\vec{\psi}[\vec{x}(k)] + \vec{\Gamma}(\mathrm{k})\right) + \boldsymbol{K}_D\vec{x}_D \tag{9.69a}$$

where,

$$\boldsymbol{K}(k) = (\Delta f\boldsymbol{R})^{-1}\,\hat{\boldsymbol{B}}^T\hat{\boldsymbol{P}}_{\infty}(k) + \boldsymbol{R}^{-1}\hat{\boldsymbol{B}}^T\,(\boldsymbol{M}(k))^{-1}\,\hat{\boldsymbol{J}}^T\hat{\boldsymbol{P}}_{\infty}(k) \tag{9.69b}$$

$$\boldsymbol{K}_{\psi}(k) = \boldsymbol{R}^{-1}\hat{\boldsymbol{B}}^T(\boldsymbol{M}(k))^{-1}\,\hat{\boldsymbol{A}}^T\hat{\boldsymbol{P}}_{\infty}(k)\,[\boldsymbol{I} + \boldsymbol{S}\hat{\boldsymbol{P}}_{\infty}(k)]^{-1} \tag{9.69c}$$

$$\boldsymbol{K}_D = \boldsymbol{R}^{-1}\hat{\boldsymbol{B}}^T\,(\boldsymbol{M}(k))^{-1}\,\boldsymbol{Q} \tag{9.69d}$$

$$\boldsymbol{M}(k) = \left[\boldsymbol{I} - \hat{\boldsymbol{A}}^T - \Delta f\hat{\boldsymbol{J}}^T + \hat{\boldsymbol{A}}^T\hat{\boldsymbol{P}}_{\infty}(k)\left[\boldsymbol{I} + \boldsymbol{S}\hat{\boldsymbol{P}}_{\infty}(k)\right]^{-1}\right] \tag{9.69e}$$

$$\hat{\boldsymbol{P}}_{\infty}(k) = \Delta f\boldsymbol{Q} + \hat{\boldsymbol{A}}^T\hat{\boldsymbol{P}}_{\infty}(k)[\boldsymbol{I} + \boldsymbol{S}\hat{\boldsymbol{P}}_{\infty}(k)]^{-1}\,\hat{\boldsymbol{A}} \tag{9.69f}$$

$$\hat{\boldsymbol{J}} = \frac{\partial\vec{\psi}\left[\vec{y}^L(k)\right]}{\partial\vec{y}^L(k)} \tag{9.69g}$$

$\boldsymbol{K}(k)$ is the control gain, $\boldsymbol{K}_{\psi}(k)$ is the gain due to the nonlinear terms in the perturbation vector and the solar pressure force, and $\boldsymbol{K}_D$ is the gain for the desired state. $\hat{\boldsymbol{P}}_{\infty}(k)$ is the steady-state Riccati equation and is obtained by solving a system of linear equations. In equations (9.69b) and (9.69f), $\boldsymbol{K}(k)$ and $\hat{\boldsymbol{P}}_{\infty}(k)$ are used to obtain the 'look-up' table. These values for $\boldsymbol{K}(k)$ and $\hat{\boldsymbol{P}}_{\infty}(k)$ are calculated in a different computer for different eccentricities and one complete orbit $0 \leq f \leq 2\pi$ and, then, stored in the computer onboard the satellite. After these values have been stored in the computer onboard the satellite, the gains in equations (9.69c–d) are obtained depending on the actual state, $\vec{y}(k)$, and the position of the satellite, $f(k)$. Substituting equation (9.69a) into equation (9.50a), the state equation is expressed as,

$$\vec{y}(k+1) = (\hat{A} - \hat{B}K(k))\vec{y}(k) + [\Delta f I - \hat{B}K_{\psi}(k)]\,(\vec{\psi}[\vec{y}(k)] + \vec{\Gamma}(k)) + \hat{B}K_D\vec{y}_D \tag{9.70}$$

All the states in equation (9.70) are known because they are defined in the present value, $\vec{y}(k)$.

The separation distance and velocity drifts shown in Figure 9.4 are defined as,

$$\vec{\rho}(k) = \vec{\rho}_{TRJ}(k) - \vec{\rho}_{NAV}(k) \quad \dot{\vec{\rho}}(k) = \dot{\vec{\rho}}_{TRJ}(k) - \dot{\vec{\rho}}_{NAV}(k) \tag{9.71}$$

where,

$$\vec{\rho} = [x_1 \quad x_2 \quad x_3]^T \quad \text{(km)}$$
$$\dot{\vec{\rho}} = [\dot{x}_1 \quad \dot{x}_2 \quad \dot{x}_3]^T \quad \text{(km/sec)}$$

Figure 9.9 shows the GCI. In Figure 9.9, the first subsystem transforms equations (9.71) from the $\vec{x}$ to the $\vec{y}$ system. The second subsystem outputs the values of $K(k)$ and $\hat{P}_{\infty}(k)$ from the look-up tables to accelerate the calculation process in equation (9.69). This information is sent to the vehicle dynamics system to determine the next separation distance and velocity drifts. These drifts are used by the navigation system to propagate the following separation distance and difference in velocity for the pair of satellites.

Figure 9.9 **Guidance and control interface**

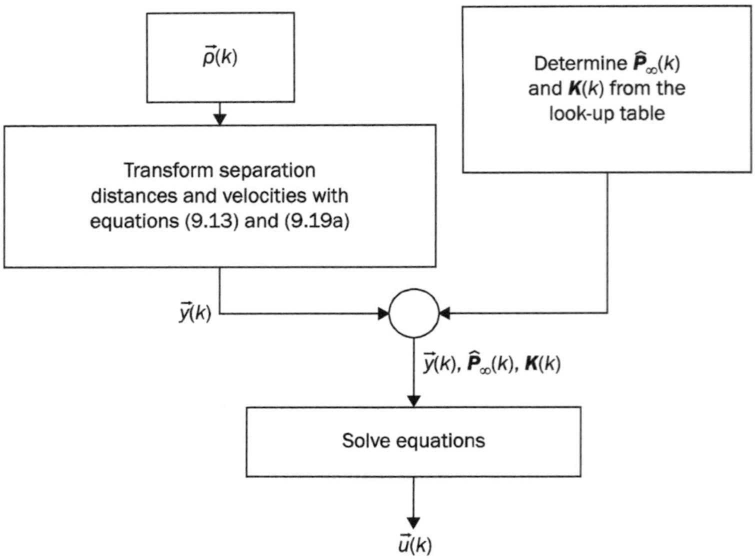

9.7.7 Example of controller for formation flying

The same specific size shown in Table 9.1 for the proposed tetrahedron constellation is used. For this HEO, Table 9.2 shows the initial conditions for the reference and maneuvering satellite. The initial conditions shown in Table 9.2 are assumed to be the same for the navigation and trajectory systems. In reality, the satellites are required to be near the perigee point to communicate their actual locations to the ground stations. Then, the computer onboard the satellites propagates the translational motion of the satellites to determine their positions and velocities at the apogee point. These values obtained at the apogee point for the positions and velocities of the satellites is used in the navigation system, but the trajectory system is specified by the values shown in Table 9.2 which are the desired locations of the satellites.

For the navigation system, the initial drifts are obtained from previous simulations with an orbit propagator [**148**]. The initial separation distance and velocity drifts in the system are obtained when the separation distance conditions between a pair of satellites in the proposed problem are violated for the first time [**148**]:

$$\vec{x}(f_0) = [-1.2\text{ km} \quad 0.3\text{ km} \quad 0.01\text{ km} \quad 2.2\times10^{-4}\text{ km/sec} \quad -1.4\times10^{-5}\text{ km/sec} \quad -9\times10^{-6}\text{ km/sec}]^T \tag{9.55}$$

The drift correction is performed faster when the separation distances for a pair of satellites are weighted more in the Q matrix than the constant coefficients of the separation velocity components [**100**]. This weighting for the Q matrix is used because the coefficients expressing the separation distance between a pair of satellites within a constellation

Table 9.2 Initial conditions for the pair of satellites

	Reference satellite	Maneuvering satellite
X_1 (km)	−8.66025403	0
X_2 (km)	−72,582.4525	−72,587.1941
X_3 (km)	−24,285.7489	−24,287.3354
$\dot{X}_1$ (km/sec)	0.973083288	0.972733623
$\dot{X}_2$ (km/sec)	0	0
$\dot{X}_3$ (km/sec)	0	0

are multiplied by the varying term, k. The weights used for the Q and R matrix are the following: $Q = diag([20, 20, 20, 1, 1, 1])$ and $R = diag([1, 1, 1])$. The same assumptions made in Section 9.6 about the spacecraft about size, mass, reflective properties and thruster capabilities are used in this solution.

For this simulation, the sampling in the true anomaly angle is set to 0.05 radians to obtain a better approximation of the system of equations. All the simulations begin at the apogee point and finish at the following apogee point. In addition, the initial time at the apogee point is zero.

Figure 9.10 shows the solution for the correction of the separation distance and velocity drifts for a pair of satellites. In Figure 9.10, the solution is shown in the $\vec{x}$ system. The separation distance and velocity drifts are corrected before the pair of satellites reaches the perigee point. Also, the control effort does not show a steady state solution because of the definition used for the cost function. Due to the true anomaly and eccentricity terms in the cost function, the control effort in the x_1 direction shows a periodic pattern. Also, the control effort is higher in the x_1 direction than in the other two directions.

Figure 9.10 Drift correction for the pair of satellites

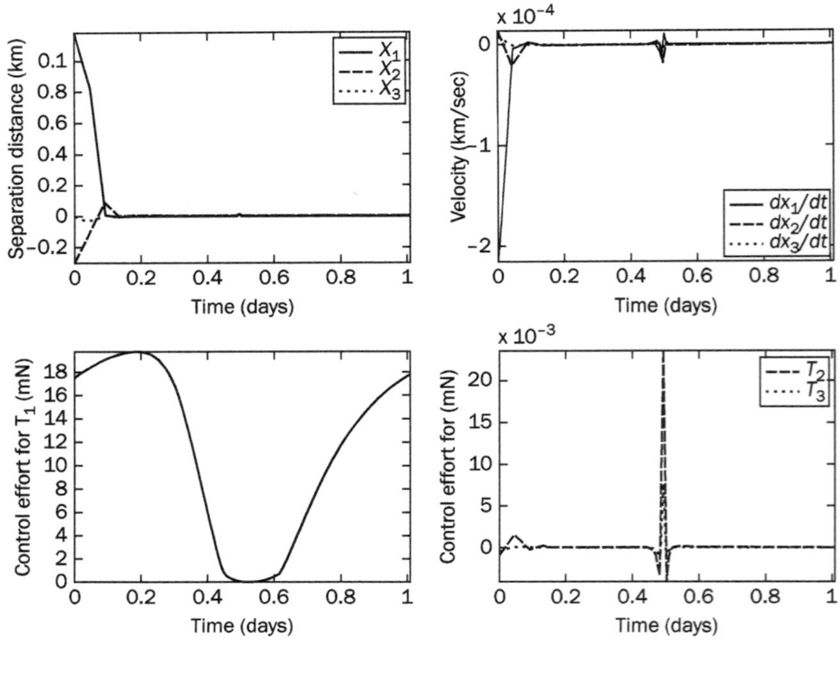

Figure 9.11 shows the x_1x_2 and x_1x_3 planes. In this figure, the solution of the trajectory and navigation systems is shown by circle and cross markers, respectively. The purpose of this graph is to illustrate that the separation distance of the pair of satellites has achieved the desired trajectory when the cross is inside of the circle. Therefore, the separation distance and velocity drifts are corrected, and the constellation satisfies the desired separation distance and velocity conditions at the following apogee point. The HEO for the maneuvering satellite from the navigation system does not show a relatively large variation in comparison to the HEO from the trajectory system; for this reason, the controller corrects the drifts without affecting the orbital motion of the satellites.

In the simulations, the relative position of the satellite is determined by the propagation of the satellite orbits with a high precision integrator. This integrator is used due to the highly elliptical orbit. There are other forms that can be used to measure the relative distance between spacecraft. Eishima [160] presents two options to determine the relative distance: 1) Radio Frequency Waves (RFW), and 2) Image Processing. With RFW, the reference satellite sends radio frequency signals to the maneuvering satellite. The time at which the signal arrives at the satellite can be used to determine the separation distance between spacecraft. This signal transmission may cause time delays in the information that

Figure 9.11 **Navigation and trajectory solution in the x_1x_2 and x_1x_3 planes**

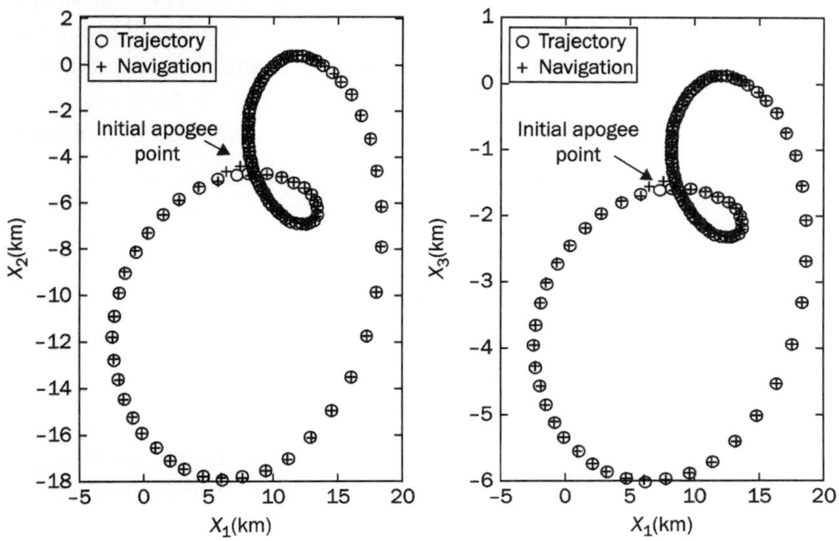

has to be added into the controller. With image processing, a camera has to be placed on the body of the maneuvering satellite. By relating the field of view of the camera and the pixel location of the reference satellite [161], the relative position of the satellite can be determined. The main advantage of this system is that there is no time delay due to the transmission of information. On the other hand, the maneuvering satellite requires a faster computational scheme to determine the relative position of the reference satellite. These are the two common methods that can be used to determine the relative position of the reference satellite with respect to the maneuvering satellite. By knowing this information, the controller shown in this section can be implemented on the satellite.

9.8 Suggested problems

Problem 9.1. Obtain the closed form solution for the CW equations (9.25) when all the disturbance and control forces are assumed to be absent. For the circular orbit recall that the mean motion, n, is a constant.

Problem 9.2. Obtain the closed form solution for the CW equations (9.25) when all the disturbance forces are assumed to be absent, and the control force is constant, Δ_T, in all the directions. For the circular orbit recall that the mean motion, n, is a constant.

Problem 9.3 The Tschauner–Hempel equations can be represented in a different frame. The frame can be defined as: 1. X_1 is positive along the radial position of the satellite; 2. X_2 is positive along the instantaneous positive velocity; 3. X_3 is positive when the right handed system is completed. Show that the Tschauner–Hempel equations can be written as [162],

$$y_1'' - 2y_2' - 3ky_1 = \frac{h^6}{\mu^4}k^3 T_1(f)$$

$$y_2'' + 2y_1' = \frac{h^6}{\mu^4}k^3 T_2(f)$$

$$y_3'' + y_3 = \frac{h^6}{\mu^4}k^3 T_3(f)$$

Also, show that if the satellite is in a circular orbit, the Tschauner–Hempel equations reduce to the well known Hill–Closhessy–Wiltshire equations as [163],

$$\ddot{x}_1 - 2n\dot{x}_2 - 3n^2 x_1 = \frac{T_1}{m}$$

$$\ddot{x}_2 + 2n\dot{x}_1 = \frac{T_2}{m}$$

$$\ddot{x}_3 + nx_3 = \frac{T_3}{m}$$

Problem 9.4. The Clohessy–Wiltshire (CW) equations can be developed as shown in this chapter, but there are cases in which the motion of a pair of satellites can be easily developed with Lagrange's equations.

a. Show that the kinetic energy equation for the CW equations can be written as,

$$T_K = \frac{m}{2}\left\{\dot{x}_1^2 + \dot{x}_2^2 + \dot{x}_3^2 + 2n\left[x_1\dot{x}_2 - \dot{x}_1\left(x_2 + R\right)\right] + n^2\left[\left(x_2 + R\right)^2 + x_1^2\right]\right\}$$

b. Show that the potential energy equation for the CW equations is defined as,

$$U = -\frac{\mu m}{R}\left[1 - \frac{x_2}{R} + \frac{2x_2^2 - x_1^2 - x_3^2}{2R^2}\right]$$

c. Solve Lagrange's equation to obtain the CW equations.

Note: R is the distance from the center of the Earth to the reference satellite in the circular orbit.

Problem 9.5. Using the orbital elements for Phase I of the proposed tetrahedron constellation,

a. Discretized the Tschauner-Hempel equations in **Problem 9.3.**

b. Using $\Delta f = 0.1$ rad, $\Delta f = 0.01$ rad, and $\Delta f = 0.5$ rad, solve the discrete linear quadratic regulator for the drift correction when the initial conditions are described by equation (9.55). Assume $T_m = 0.5$ N, $\boldsymbol{Q} = diag([20, 20, 20, 1, 1, 1])$, $\boldsymbol{R} = diag([1, 1, 1])$, and $m = 90$ kg.

c. Using $\Delta f = 0.1$ rad, solve the discrete linear quadratic regulator for the drift correction when the initial conditions are described by equation (9.55). Assume $T_m = 0.5$ N, and $m = 90$ kg. Also solve for the following weighting matrices: 1. $\boldsymbol{Q} = diag([1, 1, 1, 20, 20, 20])$ and $\boldsymbol{R} = diag([1, 1, 1])$; 2. $\boldsymbol{Q} = diag([20, 20, 20, 1, 1, 1])$ and $\boldsymbol{R} = diag([5, 5, 5])$; 3. $\boldsymbol{Q} = diag([20, 20, 20, 20, 20, 20])$ and $\boldsymbol{R} = diag([1, 1, 1])$.

10

Deployment procedure for a constellation

Abstract: The deployment procedure of a constellation is complex and depends on the separation distance between any pair of satellites within the constellation. The deployment procedure can be divided into two stages: the deployment from a circular parking orbit to an elliptical orbit, and the correction of the separation distance between pairs of satellites within the constellation. The solution of this problem is implemented with a combination of Hohmann transfer maneuvers and the digital linear quadratic regulator control scheme showing a minimum consumption of fuel. In summary, the combination of these two techniques provides a different approach to the deployment procedure of a constellation.

Key words: deployment procedure, tetrahedron constellation, digital linear quadratic regulator control scheme, Hohmann transfer maneuvers.

10.1 Introductory comments

One concern in a formation flying problem is the deployment and reconfiguration procedures. Some papers solved these procedures using different numerical schemes based on the pseudospectral methods [164] [165] [166]. This numerical method may take a longer period of time before the optimal control problem is solved. In addition, the pseudospectral methods involve a complex mathematical development to include different characteristics of the tetrahedron constellation. For these reasons, the objective of this chapter is to present a different solution to the deployment procedure of a constellation without the use of complex mathematical models.

The satellites are transferred from a circular orbit to an elliptical orbit with a Hohmann transfer maneuver [10]. This transfer maneuver represents the most fuel efficient procedure to obtain the desired elliptical orbit for the four satellites. The Hohmann transfer orbit has been used to deploy a different tetrahedron constellation as shown in reference [167]. Also, Bainum et al. [168] show that, by using a modified Hohmann transfer, an along-track constellation can be launched from a circular orbit to an elliptical orbit. With these maneuvers, the satellites in the along-track constellation reach the required configuration at the final apogee point. A similar procedure can be used here to achieve the final formation for the proposed tetrahedron constellation.

After the Hohmann transfer maneuver is used, the digital linear quadratic regulator (DLQR) is used to correct the separation distance and the velocities between any pair of satellites within the constellation. This DLQR control scheme can be used to provide a faster solution to the correction of the separation distances and velocities between any pair of satellites within the proposed constellation.

The purpose of this chapter is to present a combination of two techniques to finally deploy the proposed tetrahedron constellation [146] from an along-track circular orbit to a highly elliptical orbit. With this scheme, a different solution is provided to obtain the deployment procedure without the use of complex mathematical models and methods.

10.2 Desired conditions of the satellites in the proposed tetrahedron constellation

Figure G.2 in Appendix G shows a representation of the tetrahedron constellation at the apogee point. SB and SC are assumed to be located along the semimajor axis with a separation distance of 10 km. SA forms the equilateral triangle and is orbiting around the centroid of the equilateral triangle. SH is the fourth satellite located above the centroid of the equilateral triangle which forms the tetrahedron constellation.

For phase I, the initial positions and velocities are expressed in Table G.1 and G.3, respectively. These initial coordinates and velocities are the required conditions such that the final tetrahedron constellation can be obtained at the apogee point. Without perturbations [148], the

Table 10.1 **Orbital elements for the satellites in the tetrahedron constellation (Phase I)**

	SA	SB	SC	SH
a (km)	42,095.7	42,095.7	42,095.7	42,095.7
e	0.818182	0.818301	0.818064	0.818182
i (degrees)	18.5	18.5	18.5	18.494
Ω (degrees)	0	0	0	0
ω (degrees)	89.9921	90	90	89.9974

satellites in the constellation satisfies the separation distance constraints for a long period of time, and, with perturbations, the constellation maintains the separation distance conditions for a limited number of complete orbits. For phase I, Tables G.1 and G.3 can be used to define the orbital elements for every satellite. Table 10.1 shows the desired orbital elements at the final apogee point.

10.3 Transfer from a circular orbit to the elliptical orbit (stage 1)

The transfer procedure from a circular orbit to an elliptical orbit is complex and may take a period of time before it is achieved. Dow et al. [**167**] used a modified Hohmann transfer maneuver to transfer four satellites from a circular orbit to a final elliptical orbit. In Reference [**167**], a small consumption of fuel is obtained because the tetrahedron constellation is deployed using intermediate elliptical orbits. If the satellites are in a circular orbit and are transferred to an elliptical orbit with an eccentricity of 0.8, as an example, the intermediate elliptical orbit is defined as the chosen intermediate values of eccentricity (between 0 and 0.8) used to perform the Hohmann transfer maneuvers.

On the contrary, Bainum et al. [**168**] show that a modified Hohmann transfer orbit can be used to deploy an along-track constellation from a circular parking orbit to an elliptical orbit. In this technique, the satellites are deployed with restrictions on the period of the transfer orbit; in this way, the satellites can reach the apogee point at the same time in the along-track constellation. The required difference in velocity (ΔV) [**168**] to transfer the satellites from the circular to the elliptical transfer orbit is very similar for all of them which are a characteristic of the modified Hohmann transfer maneuvers.

This section uses similar modified Hohmann transfer maneuvers [**168**] to transfer the four satellites from a circular orbit into their respective elliptical orbits. After the satellites are released from a rocket, the four satellites are assumed to be in a circular orbit forming an along-track configuration; also, the separation distance between any pair of satellites within the constellation is assumed to be constant. There are techniques that can be used to deploy the satellite from the rocket as shown by Boutonnet et al. [**169**] to guarantee a desired constant separation distance between the satellites. It is also assumed that the circular orbit has an inclination angle of 18.5 degrees and a radius equal to $1.2R_E$. As shown in Table 10.2, this is the radius of perigee for phase I. Before the difference in velocity for the Hohmann transfer maneuvers is calculated, the period of the transfer orbit for every satellite is studied to determine the order in which the satellites are deployed. The period of a satellite is defined as,

$$T = 2\pi\sqrt{\frac{a^3}{\mu}} \tag{10.1}$$

To calculate the semimajor axis (a), the radius of perigee (r_p) for every satellite is set equal to $1.2R_E$, and the radius of apogee (r_a) is defined for every satellite depending on the desired eccentricity and semimajor axis as shown in Table 10.1. Table 10.2 illustrates the r_a, the semimajor axis for the transfer orbit (a_t), and the transfer period for every satellite (T_t).

In Table 10.2, SB and SC, respectively, has the highest and smallest period in comparison with the satellites SA and SH. The period of the four satellites provides the order in which the satellites are departing from the circular orbit. The first satellite to depart is SB because it has the highest period. The second satellite is SA because it must be ahead of SH to form the equilateral triangle. The third satellite to depart from the circular orbit is SH. This satellite is in the same plane as the other three satellites, but, after the transfer maneuvers, it is corrected with the DLQR to exhibit the out-of-plane motion. The last satellite to depart in the circular orbit is SC which has the smallest period. The separation distance

Table 10.2 **Radius of apogee, semimajor axis of the transfer orbit, and the period for every satellite**

	SA	SB	SC	SH
r_a (km)	76,537.64	76,545.338	76,532.638	76,537.64
a_t (km)	42,095.70	42,099.58	42,093.20	42,095.70
T_t (seconds)	85,910.72	85,922.59	85,903.07	85,910.72

between the satellites in the circular orbit is considered because, at the final apogee point, it makes a difference in the separation distance between any pair of satellites within the constellation in the final elliptical orbit.

The deployment procedure from the circular orbit to the elliptical orbit is defined as follows,

1. After the four satellites are released from the rocket, these satellites are assumed to be traveling in a circular orbit forming an along-track constellation. The along-track constellation can be obtained with the techniques shown in References [**168**] and [**169**]. The circular orbit has a radius equal to $1.2R_E$, and it is assumed that the orbit has an inclination angle equal to 18.5 degrees. Initially, the satellites are assumed to be separated by 0.5 degrees in the true anomaly angle. This difference in the true anomaly angle (δf) between any pair of satellites can be changed to higher angles, but, at the final apogee point of the transfer ellipse, the separation distance between any pair of satellites can be higher. In addition, the satellites may not reach the apogee point at the same time; for this reason, the separation in the true anomaly angle between some of the pairs of satellites in the along track constellation is constrained to angles between 0 and 1 degrees. Figure 10.1 shows the difference in the true anomaly angle between the satellites in the along-track constellation in the circular orbit. In Figure 10.1, δf is the difference in the true anomaly angle, f is the true anomaly angle, and the center of the Earth

Figure 10.1 **Location and separation of the four satellites in the circular orbit**

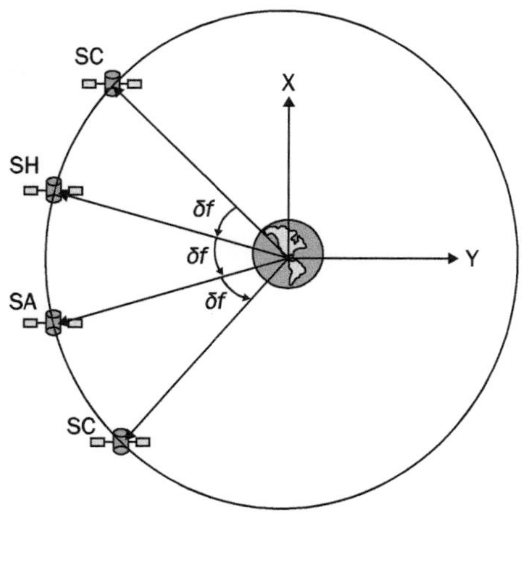

is denoted by the center of the Cartesian system X and Y. If the difference in the true anomaly angle is assumed equal to 0.5 degrees, the separation distance is approximately 66.79 km between the pairs SB-SA, SA-SH, and SH-SC in the along-track constellation. The four satellites have a velocity in the circular orbit equal to 7.2166 km/sec. As said earlier, the first satellite to be deployed is SB. The semimajor axis of the transfer orbit for SB is defined in Table 10.2, and the velocity at the perigee point in the elliptical transfer orbit is equal to,

$$V_{p,SB} = \sqrt{\mu\left(\frac{2}{r_p} - \frac{1}{a_t}\right)} = 9.73088\,(\text{km/sec}) \qquad (10.2\text{a})$$

The necessary ΔV to transfer SB from the circular orbit to the elliptical transfer orbit is equal to,

$$\Delta V_{p,SB} = 9.73088 - 7.2166 = 2.5148\ (\text{km/sec}) \qquad (10.2\text{b})$$

This ΔV maneuver is applied in the direction of the motion of the satellite to increase its velocity such that the satellite is transferred into the elliptical orbit. This ΔV procedure is performed when the true anomaly angle is equal to 90 degrees because the angle at which the satellite departs from the circular orbit is its argument of perigee (ω).

2. 9.25 seconds after SB has departed, SA is at the transfer point in the circular orbit ($f = 90°$). From Table 10.2, the difference in the period of the transfer orbit between SB and SA is 11.87 seconds. A correction to the period of the transfer orbit is not necessary because SA is 2.62 seconds ahead of SB. This difference in time causes SA to reach the position of SB in a short period of time. For this reason, a correction in the period of the elliptical transfer orbit for SA is not necessary. The velocity and ΔV to change SA from a circular orbit to the elliptical transfer orbit is,

$$V_{p,SA} = \sqrt{\mu\left(\frac{2}{r_p} - \frac{1}{a_{t,SA}}\right)} = 9.73083\,(\text{km/sec}) \qquad (10.3\text{a})$$

$$\Delta V_{p,SA} = 9.73083 - 7.2166 = 2.5142\ (\text{km/sec}) \qquad (10.3\text{b})$$

3. 9.25 seconds after SA has departed, SH has reached the transfer point in the circular orbit. The difference in the transfer period between SA and SH is zero (Table 10.2), but the time that SH takes to reach the transfer point provides the required condition to avoid a

collision between these two satellites. For this reason, the transfer period for SH is not altered. The velocity and ΔV to maneuver SH into the elliptical transfer orbit are the following:

$$V_{p,SH} = \sqrt{\mu\left(\frac{2}{r_p} - \frac{1}{a_{t,SH}}\right)} = 9.7307\,(\text{km/sec}) \tag{10.4a}$$

$$\Delta V_{p,SH} = 9.73083 - 7.2166 = 2.5141\ (\text{km/sec}) \tag{10.4b}$$

4. 9.25 seconds after SH has departed, SC reaches the transfer point ($f = 90°$). The difference in the transfer period between SB and SC is 19.52 seconds. Once more, a correction to the period of the elliptical transfer orbit is not necessary because the time that SC takes to reach the transfer point provides enough distance between the other three satellites to avoid a collision. The velocity and the ΔV at the transfer point required to maneuver SC into the elliptical transfer orbit can be defined as,

$$V_{p,SC} = \sqrt{\mu\left(\frac{2}{r_p} - \frac{1}{a_{t,SC}}\right)} = 9.7308\ (\text{km/sec}) \tag{10.5a}$$

$$\Delta V_{p,SC} = 9.7308 - 7.2166 = 2.5142\ (\text{km/sec}) \tag{10.5b}$$

5. Once the satellites have reached the apogee point, a second ΔV maneuver is performed to correct the semimajor axis and the eccentricity of the final elliptical orbit. To perform this maneuver, the velocity of the satellite at the apogee point in the elliptical transfer orbit is calculated and, then, is subtracted from the velocity at the apogee point defined in Table G.3 in Appendix G. Table 10.3 shows the velocity and ΔV required for the four satellites in the constellation at the apogee point.

   At the apogee point, this ΔV maneuver is also applied along the positive or negative tangential direction of the satellite. Figure 10.2 shows the ΔV maneuvers and the different orbits that are obtained with

Table 10.3 **Velocity at the apogee point in the transfer orbit and the ΔV required to correct the in-plane conditions of the final orbit**

	SB	SA	SH	SC
V_a (km/sec)	0.97299	0.97308	0.97308	0.97343
ΔV_a (km/sec)	-0.2888×10^{-3}	0	0	0.2828×10^{-3}

Figure 10.2 **Diagram of the deployment procedure of the constellation**

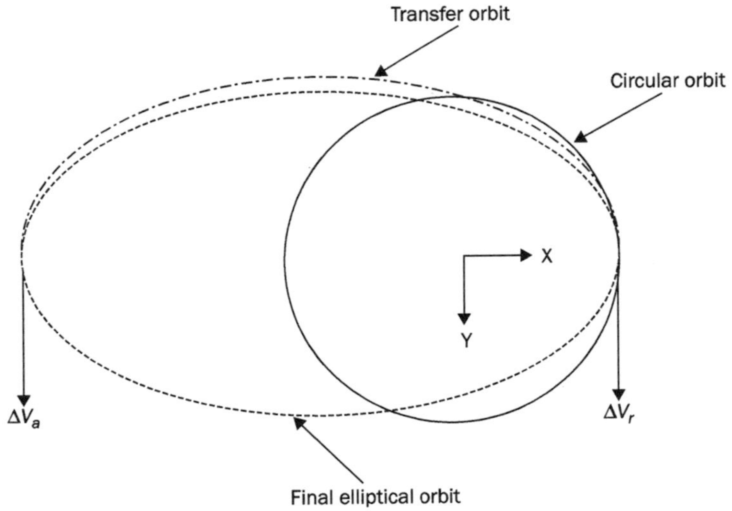

these modified Hohmann transfer maneuvers. The Cartesian axis at the center of the Earth is rotated because the elliptical orbit is created over the X axis in which the true anomaly angle is equal to 90 degrees. This angle defines the argument of perigee of the orbit of the four satellites.

10.4 Station-keeping procedure (stage 2)

These five steps define the elliptical orbit of the four satellites using Hohmann transfer maneuvers. Observing the transfer period of the four satellites, these satellites reach the apogee point at the same time. The tetrahedron formation is not obtained at the final apogee point; for this reason, a final correction of the separation distance between the satellites is required. This correction can be done in two ways: (1) solving a two point boundary value problem (TPBVP) with the linearized Tschauner–Hempel equations, or (2) correcting the drifts in the separation distance between any pair of satellites within the constellation with the DLQR controller defined in the previous chapter.

The solution of a TPBVP is complicated because it involves the theory of primer vector defined by Lawden [147]. The linearized Tschauner–Hempel (TH) equations defined by Carter–Humi (CH) [5] satisfy the primer vector relations, and the solution depends on the type of thrust arc. The solution of the TPBVP can be preferred for the formation of the tetrahedron constellation at the apogee point, but there are some other considerations related to the solution of this problem. For these reasons, the correction of the separation distance constraints for the proposed tetrahedron constellation is based on the DLQR control scheme. With this DLQR control scheme, the tetrahedron constellation can be formed with the required separation distance at the following apogee point.

10.5 Deployment procedure for the tetrahedron constellation

The complete deployment and station-keeping procedure are defined in the following steps: (1) two ΔV procedures are used to transfer the satellites from a circular orbit to their respective elliptical orbits; (2) then, the DLQR control scheme is used to correct the drifts in the separation distances and velocities between any pair of satellites such that the proposed tetrahedron constellation is obtained at the following apogee point.

10.5.1 Stage 1

Stage 1 of the deployment procedure can be performed with available commercial software. These software are called orbit propagators used to simulate a satellite in an orbit and are applied for the deployment procedure of the tetrahedron constellation. Reference [170] uses an orbit propagator to show the solution for the tetrahedron constellation. Using the commercial orbit propagator, reference [170] shows that the satellites arrive at the apogee point at the same time, but there are drifts in the separation distance between any pair of satellites within the constellation. In this case, the separation distance between any pair of satellites within the constellation ranges between 10 and 28 kilometers (for $\delta f = 0.5°$). For this reason, the constellation is corrected with the DLQR to obtain the proposed tetrahedron constellation at the following apogee point; in addition, SH must be located in its out-of-plane position to obtain the final formation.

The difference in the true anomaly angle (δf) can be analyzed with the orbit propagator [170] to determine how δf affects the separation distances between any pair of the satellites at the final apogee point. The difference in the true anomaly angle used in the previous simulations is 0.5 degrees, but this δf is increased to 1 degree and decreased to 0.3 degrees to determine the effects in the separation distance at the final apogee point. Table 10.4 shows the separation distance between any pair of satellites within the constellation for the chosen values of δf. In Table 10.4, δf must be constrained between 0 and 1 degree such that a minimum separation distance can be obtained at the apogee point. For the pair SA-SB, the separation distance does not significantly diminish with a decrease in the δf, but, for the other pairs of satellites, the separation distance is decreased as δf decreases. When $\delta f = 0.3$, the separation distance between any pair of satellites is near the 10 km range; in this way, the DLQR can be efficiently used to correct the drifts between any pair of satellites to obtain the desired configuration at the final apogee point.

Table 10.5 shows the orbital elements and the position of the four satellites at the final apogee point. In Table 10.5, the four satellites arrive at the apogee point at the same time. Comparing Table 10.1 with

Table 10.4 Different separation distances between any pair of satellites for different δf

	δf		
Pair of satellites	0.3 (degrees)	0.5 (degrees)	1 (degrees)
SA-SB	15.276 km	14.080 km	15.196 km
SA-SC	10.696 km	17.960 km	35.958 km
SA-SH	5.406 km	9.008 km	18.018 km
SB-SC	14.459 km	20.689 km	44.465 km
SB-SH	13.875 km	15.258 km	27.970 km
SC-SH	5.290 km	8.952 km	17.940 km

Table 10.5 Orbital elements for the four satellites within the constellation (Phase I) at the end of stage 1

	SA	SB	SC	SH
a (km)	42,091.02	42,095.44	42,093.37	42,091.02
e	0.818162	0.818359	0.818270	0.818124
i (degrees)	18.5	18.5	18.5	18.5
Ω (degrees)	0	0	0	0
ω (degrees)	90	90	90	90
f (degrees)	180.056	180.055	180.043	180.059

Table 10.5, SB can be used as the reference satellite because it has similar orbital elements as in Table 10.1. For this reason, SB is used as the target spacecraft to correct the other three satellites such that, at the following apogee point, the proposed tetrahedron constellation can be obtained.

10.5.2 Stage 2

The initial positions and velocities at the apogee point for the four satellites are obtained from the orbit propagator [**170**] or from the orbital elements in Table 10.5 [**9**] to determine the initial conditions for the DLQR active control scheme. Table 10.6 shows the initial coordinates and velocities of the four satellites at the final apogee point after the final transfer maneuver. The values shown in Table 10.6 are obtained for a difference in the true anomaly angle of 0.3 degrees and are used to calculate the difference between the reference and the maneuvering satellites because the separation distances are near the 10 km in comparison to the other values for the difference in the true anomaly angle shown in Table 10.4.

The reference satellite is SB, and the maneuvering satellites are SA, SC, and SH. The difference between the reference satellite and the maneuvering satellites for the nominal coordinates are defined in Table 10.7. This difference is obtained from Tables G.3 and G.4 in Appendix G that defines the desired coordinates and velocities for the four satellites. In Table 10.7, X_N, Y_N, and Z_N are the nominal separation distance between the reference and the maneuvering spacecraft for the X, Y, and Z direction, respectively. The same formulation is also used for the velocities. $\bar{\lambda}$ equals to B, C or H and represents SB, SC or SH, and N means 'nominal'. The same calculation is performed for the initial coordinates and velocities shown in Table 10.6.

Table 10.6 **Initial coordinates and velocities for the four satellites at the apogee point after the transfer maneuver**

	SA	SB	SC	SH
X (km)	410.539171	404.465887	399.847473	405.135249
Y (km)	−72,567.835478	−72,581.128031	−72,568.133913	−72,567.985691
Z (km)	−24,280.858097	−24,285.305723	−24,280.957952	−24,280.908358
V_x (km/sec)	0.973120	0.972665	0.973407	0.973122
V_y (km/sec)	0.027229	0.026829	0.026512	0.026871
V_z (km/sec)	0.009111	0.008977	0.008871	0.008991

Table 10.7 **Difference between the reference and maneuvering spacecraft for the nominal coordinates**

	SB-SA	SB-SC	SB-SH
$X_N = X_B - X_{\bar{\lambda}}$ (km)	8.66	0	2.89
$X_N = Y_B - Y_{\bar{\lambda}}$ (km)	–4.70	–10.23	–2.90
$Z_N = Z_B - Z_{\bar{\lambda}}$ (km)	–1.71	–3.17	–9.33
$V_{XN} = V_{XB} - V_{X\bar{\lambda}}$ (km/sec)	-1.36×10^{-4}	-6.99×10^{-4}	-3.5×10^{-4}
$V_{YN} = V_{YB} - V_{Y\bar{\lambda}}$ (km/sec)	0	0	0
$V_{ZN} = V_{ZB} - V_{Z\bar{\lambda}}$ (km/sec)	0	0	0

Table 10.8 **Difference between the reference and maneuvering spacecraft for the initial coordinates**

	SB-SA	SB-SC	SB-SH
$X_S = X_B - X_{\bar{\lambda}}$ (km)	–6.07	4.62	–0.67
$Y_S = Y_B - Y_{\bar{\lambda}}$ (km)	–13.29	–12.99	–13.14
$Z_S = Z_B - Z_{\bar{\lambda}}$ (km)	–4.45	–4.35	–4.40
$V_{XS} = V_{XB} - V_{X\bar{\lambda}}$ (km/sec)	-4.55×10^{-4}	-7.42×10^{-4}	-4.57×10^{-4}
$V_{YS} = V_{YB} - V_{Y\bar{\lambda}}$ (km/sec)	-400×10^{-4}	-3.17×10^{-4}	-4.20×10^{-5}
$V_{ZS} = V_{ZB} - V_{Z\bar{\lambda}}$ (km/sec)	-1.34×10^{-4}	1.06×10^{-4}	-1.40×10^{-5}

Table 10.8 shows the difference between the reference and maneuvering spacecraft for the initial coordinates and velocities shown in Table 10.6. The subscript S defines the difference in the positions and velocities for the initial coordinates.

From the schemes shown in Chapter 9, the stage 2 of the deployment procedure can be performed by either solving the two point boundary value problem or the drift correction. The problem is solved with the worst case which is correcting the relative separation distance between SB and SH. For SB and SH, SH has to be taken into the out of plane motion to form the proposed tetrahedron constellation. The same assumptions as in Chapter 9 are used here and are explained once more. The maximum thrust [10] (T_M) is assumed to be 0.5 N corresponding to the ion thruster. The $\boldsymbol{Q}$ and $\boldsymbol{R}$ matrix, respectively, is a 6×6 and 3×3 diagonal matrix. The weight in the $\boldsymbol{Q}$ matrix is *diag*([20,20,20,1,1,1]), and the weight of the $\boldsymbol{R}$ matrix is *diag*([1,1,1]). The mass of the satellite at the apogee point is 90 kg (assumed constant), and the inclination angle of the satellites is defined in Table 10.5. The sampling in the true anomaly angle is set to 0.05 radians to obtain a better approximation because of the nonlinearities in the TH equations for a perturbed motion. The maximum number of iterations in the second level is 60 (L = 60), and the tolerance value is equal to 1e-10. In addition, the satellite is assumed to have the same dimensions with a difference in the reflectivity coefficient equal to 0.01, and the satellite is assumed to be a right circular cylinder.

Figure 10.3 shows the solution of the TPBVP. Figure 10.3 shows once more the intermediate thrust explained by Lawden [**147**]. For this problem, the end conditions are satisfied such that SH is on the desired location to form the proposed tetrahedron constellation. In addition, the thrust arcs show a small consumption of fuel to correct the motion of the spacecraft.

Another simulation is performed with the drift correction. The drift correction is defined as the difference between the initial and final coordinates. The purpose of the controller is to correct that error as fast as possible. Table 10.9 shows the drifts for the separation distance and velocities for these pairs of satellites.

Figure 10.3 **Two point boundary value problem for the deployment procedure**

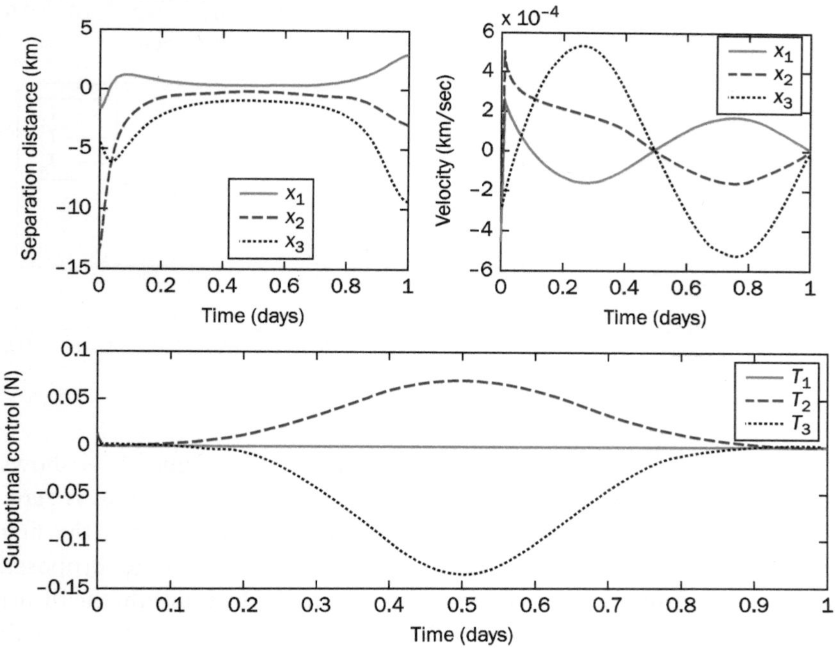

Table 10.9 Initial conditions for the drift correction in stage 2

	SB-SA	SB-SC	SB-SH
$x_1(f_0) = X_S - X_N$ (km)	–14.73	4.62	–3.56
$x_2(f_0) = Y_S - Y_N$ (km)	–8.59	–2.76	–10.24
$x_3(f_0) = Z_S - Z_N$ (km)	–2.74	–1.17	4.93
$x_1'(f_0) = V_{XS} - V_{XN}$ (km/sec)	-3.19×10^{-4}	-4.27×10^{-5}	-1.07×10^{-4}
$x_2'(f_0) = V_{YS} - V_{YN}$ (km/sec)	-4.00×10^{-4}	3.17×10^{-4}	-4.20×10^{-5}
$x_3'(f_0) = V_{ZS} - V_{ZN}$ (km/sec)	-1.34×10^{-4}	1.06×10^{-4}	-1.40×10^{-5}

Figure 10.4 Drift correction for the deployment procedure (stage 2)

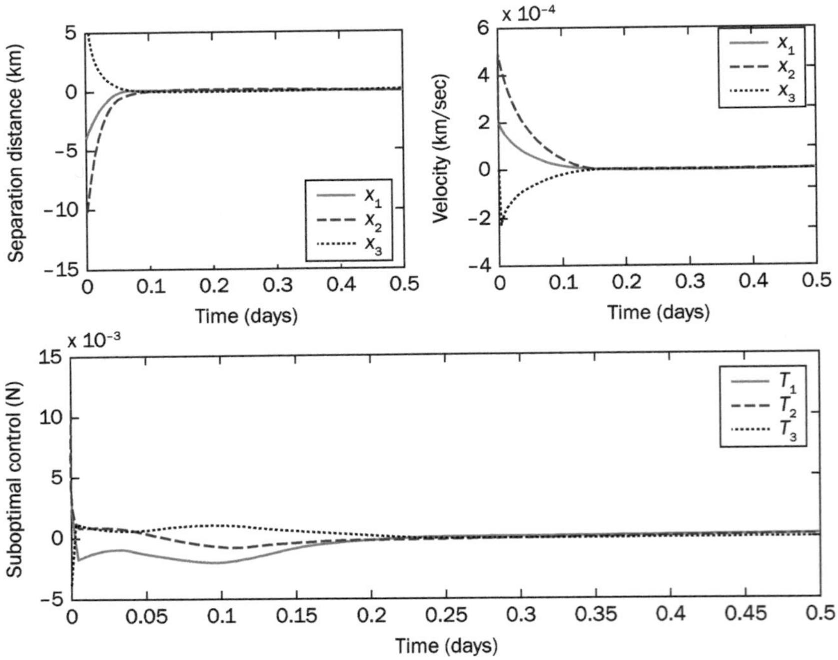

Using the drift for the pair of satellite SB and SH, Figure 10.4 shows the correction with the LQR controller when the desired state is zero. The drift correction is performed before the satellite reaches the first perigee point. Once the satellites reach the perigee point, the proposed tetrahedron constellation is configured once again. In addition, the thrust consumptions is smaller in the drift correction than in the TPBVP. In all the simulations, the correction for the separation distances and the velocities is performed such the constellation is configured at the apogee point. In addition, the optimal control shows a small consumption of fuel for the pair of satellites.

10.6 Remarks

In this chapter, a different solution for the deployment of a proposed tetrahedron constellation is provided. The deployment procedures shown here are much simpler than in references [**164**] [**165**] [**166**] that use the pseudospectral methods. All the techniques used to obtain the proposed tetrahedron constellation [**146**] are known and can be easily implemented for this problem.

In the deployment from the circular orbit to the elliptical orbit, the difference in the true anomaly angle between the satellites in the circular orbit must be less than 0.5 degrees to obtain a minimum value in the separation distance between any pair of satellites at the final apogee point of the transfer maneuver. After the final Hohmann maneuver is performed, the satellites reach the apogee point at the same time.

After the transfer maneuver is finished, the digital linear quadratic regulator is used to finally correct the drifts between the three pairs of satellites. It is shown that the satellites have a very small consumption of energy in which the mass inside of the satellites does not change rapidly. The satellite in the out-of-plane motion requires a higher thrust consumption to obtain the final tetrahedron constellation. In conclusion, a different solution for the deployment and station-keeping of the proposed constellation is defined when Hohmann transfer maneuvers and a DLQR are combined to obtain the desired formation. Using both techniques, a small consumption of fuel is obtained for the deployment procedure of the proposed tetrahedron constellation.

10.7 Suggested problems

Problem 10.1. In Section 10.3, the Stage 1 of the problem is explained such that a difference in velocity is used to take the satellite from a circular orbit to the elliptical orbit. Using a commercially available orbital propagator or writing your own program, show that Stage 1 can take the satellite to the desired elliptical orbit; also show that the difference in the true anomaly angle, δf, shows a similar solution as in Table 10.4.

Problem 10.2. Using the same satellite assumptions and satellite initial conditions, solve the two level hierarchical control scheme for the two point boundary value problem when the difference in the reflectivity coefficient is 0.001 and 0.1 for the pair SB and SA. Compare the magnitude in thrust for both reflectivity coefficients.

Problem 10.3. Using the same satellite assumptions and satellite initial conditions, solve the two level hierarchical control scheme for the drift correction when the difference in the reflectivity coefficient is 0.001 and 0.1 for the pair SB and SC. Compare the magnitude in thrust for both reflectivity coefficients. Also, compare the thrust magnitude to Problem 10.2.

11

Reconfiguration procedure for a constellation

Abstract: The purpose of this chapter is to demonstrate the use of an intelligent controller to perform reconfiguration procedures for a constellation. The intelligent controller can reconfigure all the satellites at the same time such that at the end of the maneuver the group of satellites is in formation. In addition, the intelligent controller is augmented with an adaptive control scheme to correct the orbital elements when additional perturbations, due to the Sun and Earth, are included to the process. In summary, this chapter demonstrates for the first time the use of an intelligent and adaptive control scheme to perform the reconfiguration for a single satellite such that a group of satellites can be in formation flying to achieve the mission goals.

Key words: constellation reconfiguration procedure, fuzzy logic controller, adaptive control scheme, Lagrange planetary equations, data mining process, discrete Lyapunov controller, impulsive maneuvers.

11.1 Introduction to the reconfiguration process of a constellation

The reconfiguration procedure of a constellation [146] is complex and may require many mathematical computations to determine an optimal solution for the problem. Reference [166] shows a numerical approach to solve the deployment and reconfiguration procedure of a tetrahedron constellation, but this mathematical approach is complicated for the solution of the problem. Reference [165] shows an optimal reconfiguration of the tetrahedron constellation using the Gauss Pseudospectral Method, but the mathematical approach may take a long computational time before an actual solution for the system is available. Reference [171] has tabulated for a simple case the computational time by using the Jacobi Pseudospectral Method. These numerical procedures may take a longer period of time before an actual solution for the reconfiguration procedure is obtained.

The objective of this chapter is to present a control strategy that does not depend on a computational approach of the type mentioned above, but is based on the knowledge extracted from the differential equations. This control scheme can be designed with the use of intelligent systems technology. As mentioned in Chapter 7, these intelligent systems adapt, learn, and take different decisions based on the particular problem [107]. Among these intelligent systems are the fuzzy logic (FL), neural networks (NN), and genetic algorithms (GA).

This chapter concentrates on the development of a FL controller because the FL is an intelligent system that does not require previous knowledge of the data and depends on the conditions and conclusions imposed on the system. To design a FL system, a data mining process is implemented. This process is performed to extract data and knowledge from equations, pictures, or a data base [107]. Since no complex mathematical approach is needed to develop the fuzzy control system, FL can be used as a nonlinear control for the reconfiguration procedure of a constellation. In this chapter, it is used for the first time for a formation flying reconfiguration procedure.

The FL controller is mainly used to control the in-plane motion of the satellite because, for the out-of-plane motion, the FL controller is constrained by the orbital dimensions. To avoid this constraint, a nonlinear controller based on the Lyapunov control theory is developed for a discretized system [118]. In addition, a control scheme based on impulsive maneuvers is defined to be used as a station-keeping procedure. The impulsive maneuvers happen at certain locations in the orbit such that a single impulse is applied to cause a difference in certain orbital elements defining the out-of-plane motion. The discrete nonlinear Lyapunov controller and the impulsive maneuvers are developed to correct the out-of-plane motion of a constellation. In addition, this chapter augments the FL controller with an adaptive control scheme to demonstrate the capability of both controllers to perform a reconfiguration procedure. The advantage of using an adaptive control scheme is that the system has the ability to correct for unknown or undefined forces.

To reach the next specific size, the dynamics for a single satellite is given by the Lagrange planetary equations [15]. Therefore, the chapter is divided into seven different sections that shows the data mining process [107], the development of the FL controller for in-plane motion, the development of the discrete Lyapunov controller and the impulsive maneuvers for the correction of the out-of-plane motion, the results of the reconfiguration procedure, and the adaptive control scheme. This nonlinear controller is applied for the reconfiguration procedure of the proposed tetrahedron constellation shown in Appendix G.

11.2 Data mining process of the Lagrange planetary equations

The Lagrange planetary (LP) equations [15] are a set of nonlinear equations that can be used to change the orbital elements of a single satellite using propulsive requirements. This set of equations use impulses to cause changes in the orbital elements. These equations are applied to the reconfiguration process of the proposed tetrahedron constellation [146] in Appendix G. In Section 3.2, equations (3.2) can be represented as:

$$\dot{\vec{X}} = \boldsymbol{B}(\vec{X})\vec{u} + \vec{\psi}[\vec{X}] \tag{11.1}$$

where,

$$\vec{X} = [a \quad e \quad i \quad \Omega \quad \omega]^T \qquad \vec{u} = [\bar{R} \quad \bar{B} \quad \bar{N}] \tag{11.2a}$$

$$\boldsymbol{B}(\vec{X}) = \begin{bmatrix} \dfrac{2na^3e}{\mu\sqrt{1-e^2}}\sin f & \dfrac{2na^3\sqrt{1-e^2}}{\mu R} & 0 \\ \dfrac{na^2\sqrt{1-e^2}}{\mu}\sin f & \dfrac{na^2\sqrt{1-e^2}}{\mu}(\cos f + \cos E) & 0 \\ 0 & 0 & \dfrac{naR}{\mu\sqrt{1-e^2}}\cos w \\ 0 & 0 & \dfrac{naR}{\mu\sqrt{1-e^2}}\sin w \operatorname{cosec} i \\ -\dfrac{na^2\sqrt{1-e^2}}{\mu e}\cos f & \dfrac{na^2\sqrt{1-e^2}}{\mu e}\left(1+\dfrac{R}{p}\right)\sin f & \dfrac{naR}{\mu\sqrt{1-e^2}}\sin w \operatorname{cosec} i \cos i \end{bmatrix} \tag{11.2b}$$

$$\vec{\psi}[\vec{X}] = \begin{bmatrix} 0 \\ 0 \\ 0 \\ -\frac{3}{2} J_2 \left(\frac{R_E}{p}\right)^2 n \cos i \\ \frac{3}{4} J_2 \left(\frac{R_E}{p}\right)^2 n\left(5\cos^2 i - 1\right) \end{bmatrix} \qquad R = \frac{p}{1+e\cos f} \tag{11.2c}$$

a is the semimajor axis, e is the eccentricity, i is the inclination angle, Ω is the right ascension of the ascending node (RAAN), and ω is the argument of perigee. The directions of the force impulse per unit mass ($\vec{u}$) are explained as: $\bar{R}$ represents the radial direction component, $\bar{B}$ represents the tangential component, and $\bar{N}$ represents the component normal to the orbit plane. w is the sum of f and ω, and E is the eccentric anomaly. $\vec{\psi}$ $(\vec{X})$ contains the perturbation due to the oblateness of the Earth [16] in which J_2 is approximated to 1.08263×10^{-3}.

As shown in Chapter 9, the control of satellites in an elliptical orbit should be approached in the true anomaly angle domain rather than in the time domain because the position and the eccentricity of the orbit are more important than the time that the correction may require. Using equations (9.13) and (9.17c), equation (11.1) is transformed into the true anomaly angle domain as,

$$\vec{X}' = \boldsymbol{B}\,[\vec{X}(f)]\,\vec{u} + \vec{\psi}\,[\vec{X}(f)] \tag{11.3a}$$

where,

$$\boldsymbol{B}[\vec{X}] = \begin{bmatrix} \frac{2\beta ae}{1-e^2}\sin f & \frac{2\beta a^2\sqrt{1-e^2}}{r} & 0 \\ \beta\sqrt{1-e^2}\sin f & \beta\sqrt{1-e^2}\left(\cos f + \cos E\right) & 0 \\ 0 & 0 & \frac{v}{\sqrt{1-e^2}}\cos w \\ 0 & 0 & v \sin w \operatorname{cosec} i \\ -\beta\sqrt{1-e^2}\cos f & \beta\sqrt{1-e^2}\left(1+\frac{r}{P}\right)\sin f & -v\sin w \operatorname{cosec} i \cos i \end{bmatrix} \tag{11.3b}$$

$$\vec{\psi}[\vec{X}(f)] = \begin{bmatrix} 0 \\ 0 \\ 0 \\ -\frac{3}{2}J_2\left(\frac{R_E}{p}\right)^2 \frac{h^3 n \cos i}{\mu^2\left(1+e\cos f\right)^2} \\ \frac{3}{4}J_2\left(\frac{R_E}{p}\right)^2 \frac{h^3 n\left(5\cos^2 i-1\right)}{\mu^2\left(1+e\cos f\right)^2} \end{bmatrix} \tag{11.3c}$$

$$\beta = \frac{h^3 n a^2}{\mu^3\left(1+e\cos f\right)^2} \qquad v = \frac{h^3 n a R}{\mu^3\left(1+e\cos f\right)^2} \tag{11.3d}$$

The fuzzy control system is designed in the true anomaly angle, and the data mining process is used to obtain the required knowledge from the equations.

Data mining [107] is defined as the process to extract data and knowledge of certain sources of information. For this application, the data mining is applied to the LP equations. For certain terms in equation (11.3b), the correction of the satellite depends on its location along the orbit defined by the f. Some conclusions can be formulated from equation (11.3b) to correct the satellite to its desired orbital elements. The correction of the orbit can be divided into the in-plane and out-of-plane motion corrections because of the direction of the applied thrust. If the in-plane motion is first corrected, $\bar{N} = 0$ while $\bar{R} \neq \bar{B} \neq 0$ until the system reaches the desired in-plane orbital elements. The out-of-plane motion can be corrected with $\bar{N}$ when $\bar{R} = \bar{B} = 0$, but $\bar{N} \neq 0$ until the desired orbital elements are obtained.

From equation (11.3b), the a and e can increase or decrease faster with a propulsive impulse along $\bar{B}$, but the e depends on the position of the satellite because of the function $\cos f + \cos E$, while a increases without restrictions on the position of the satellite. This correction in the a and e is performed when $\bar{R}$ is set to zero or to small values of thrust, but $\bar{R}$ can be used to maintain the desired orbital elements or to correct the e if it is a necessary maneuver. Any value of thrust applied in the $\bar{R}$ direction (when $\bar{B} = 0$) does not change the a, e, and ω faster because the terms are multiplied by the position of the satellite ($\cos f$ for ω, and $\sin f$ for a and e). Any thrust applied in the $\bar{B}$ and $\bar{R}$ direction causes the ω to increase or decrease depending on the position of the satellite. To correct the ω to the desired angle, a small thrust [172] along the $\bar{R}$ direction (while $\bar{B} = 0$) can be applied when the satellite is at the perigee or apogee point.

When cos $w = 0$ (in equation 11.3b), and the sin $w = 1$, the Ω can be easily corrected. The inclination angle is corrected using the same argument as in the correction of the Ω, and both angles can be adjusted at the same time. Any corrections made for the RAAN or i causes the ω to change, and the previously explained approach for the in-plane motion for correcting the ω can be used, once more, to correct this angle. From equation (11.3b), the correction for the in-plane motion does not change the out-of-plane motion with the thrust along the $\bar{B}$ and $\bar{R}$ direction, and the out-of-plane motion does not affect the in-plane motion with the propulsion along the $\bar{N}$ direction. The only orbital element affected by this correction with $\bar{R}$, $\bar{B}$, and $\bar{N}$ is the argument of perigee. This fact suggests that the in-plane and out-of-plane motion can be corrected at the same time. The out-of-plane motion is constrained by the dimension of the semimajor axis and the eccentricity, but an adaptive control scheme such as the use of nonlinear Lyapunov controllers can be implemented to improve the correction of these orbital elements.

The main problem is how much thrust is applied into the system such that the equations do not overflow. An overflow happens when the eccentricity reaches a value of 1. When $e = 1$, the other orbital elements result in imaginary values that do not correspond to the desired orbital elements, and a maximum thrust should be established to avoid this overflow condition in the differential equations. The maximum thrust acceleration (u_m) applied for an in-plane correction must be less than and/or equal to 10^{-3} (km/sec^2) such that the eccentricity does not increase faster and reach a value of one [173]. This value for the u_m is used for the reconfiguration procedure from phase I to II. For the reconfiguration procedure from phase II to III, the maximum thrust acceleration must be less than and/or equal to 10^{-5} (km/sec^2) because the semimajor axis increases while the eccentricity decreases. This u_m only applies to the in-plane motion correction because the out-of-plane motion is highly affected by the semimajor axis and the eccentricity; for this reason, a different nonlinear controller is proposed.

11.3 Fuzzy logic controller

The FL controller is designed depending on the knowledge extracted from the data mining process. The FL system [107] is considered as a

Figure 11.1 **Schematic of the fuzzy logic system**

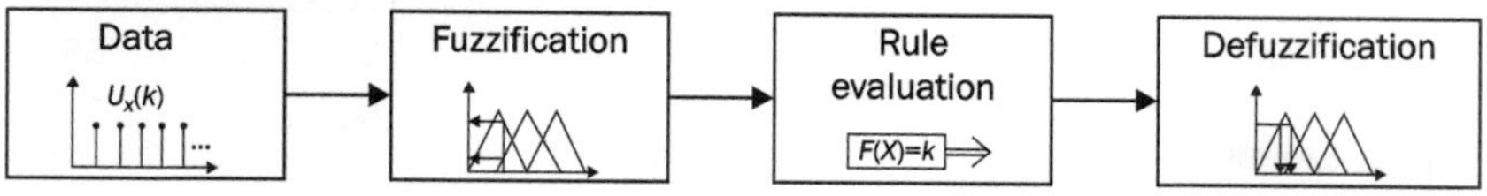

'black box' because the main problem resides in how many rules the system requires to obtain the desired state, but the extraction of the data given in the last section provides the necessary knowledge to formulate the correct decisions on the control effort for the reconfiguration procedure.

The FL follows a simple procedure shown in Figure 11.1. The FL system is based on the Mandami [107] style fuzzy inference method. This method is composed of the input data, the fuzzification process, the rule evaluation and aggregation, and finally the defuzzification process. The first step is to input the data as numerical values limited to a unit value [115] between –1 and 1. For this reason, the data entered into the system is scaled down with the following equation,

$$U_X(k) = \frac{X_D - X(k)}{X_D} \tag{11.4}$$

where X represents the orbital element to be corrected, k is the sample taken at a certain true anomaly angle, X_D is the desired orbital element for the next phase, and $U_X(k)$ is the unit value for a given orbital element. X_D must be different from zero, and only the RAAN is defined as the difference between the X_D and $X(k)$ (see Appendix G).

The unit values are evaluated through different membership functions. A membership function [107] is a mathematical function that defines the input data but depends on the conditions to correct the orbital elements of the orbit. Membership functions are defined in different ways, but the most common are the triangles and trapezoids. For the fuzzification process, the triangle membership functions in Figure 11.2 are used. In Figure 11.2, every membership function has a linguistic value. A linguistic value [107] is defined as a variable that contains the fuzzy set. These linguistic variables refer to the type of error encountered when $U_X(k)$ is input into the membership functions. For this system, the fuzzy sets are classified into five different errors. The errors are defined as follows: NBE is a negative bad error, NSE is a negative satisfactory error, ZE is a zero error, PSE is a positive satisfactory error, and PBE is a positive bad error. The membership function has a fifty percent of overlap such that a 'good'

Figure 11.2 **Membership functions**

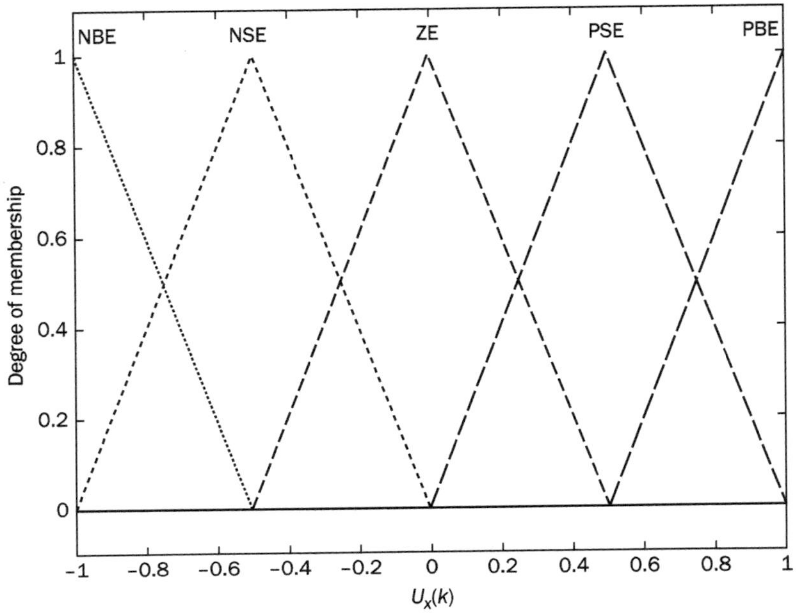

decision can be made by the system. The recommended overlap between membership functions is between 25% and 50%. The y-axis of the membership function gives the degree of the membership function represented by $U_X(k)$. The output of the membership function is known as a fuzzy set.

The third step is the rule evaluation and aggregation of the fuzzy sets. A rule in the fuzzy system is a statement presented in the form: IF (antecedent) THEN (consequent). The antecedent rules [107] have the relations for the change in the orbital dimensions as a unit value, and the consequence rules [107] have the decision of how much force per unit mass is applied. As an example, IF $U_a(k)$ is NBE AND $U_e(k)$ is NBE THEN $\bar{B}$ is NFT. In this case, the rule says that IF the unit error in a is NBE and the unit error in e is NBE THEN a negative full thrust (NFT) must be applied into the $\bar{B}$ direction to correct the satellite. The number of rules depends on the number of inputs and membership functions that are used to define the fuzzy system, and a specific number of these rules may be needed. If two inputs and five membership functions per input are used, the system involves 25 rules but not all the rules may be used if the data mining process is performed.

In FL, the 'AND' statement [107] is defined as the minimum value between two input values. This process of deciding the minimum value between two fuzzy variables is completed after the degree of the membership function is determined. Then, this value is sent to the defuzzification process. In the defuzzification process [107], the minimum value is imposed into a different set of membership functions representing the level of thrust applied into any direction of $\vec{u}(f)$. Using a center of gravity equation, a centralized decision is obtained about the amount of applied thrust. This centralized decision is the output value which is represented in a range between –1 and 1. The membership functions in a defuzzification process are shown in Figure 11.3. The linguistic variables in Figure 11.3 are defined as: NFT is a negative full thrust, NST is negative satisfactory thrust, ZT is a zero thrust, PST is a positive satisfactory thrust, and PFT is a positive full thrust. These membership functions use the Gaussian membership functions because the decision of the thrust magnitude is high near the center of the membership function and less in its corners, but the Gaussian membership function is also referred to as 'indeed'. This is a different terminology used in fuzzy logic systems and is explained in Reference [107].

The output value obtained from the defuzzification process is a unit value which is the degree of the applied thrust. This output value is

Figure 11.3 Membership functions for the defuzzification process

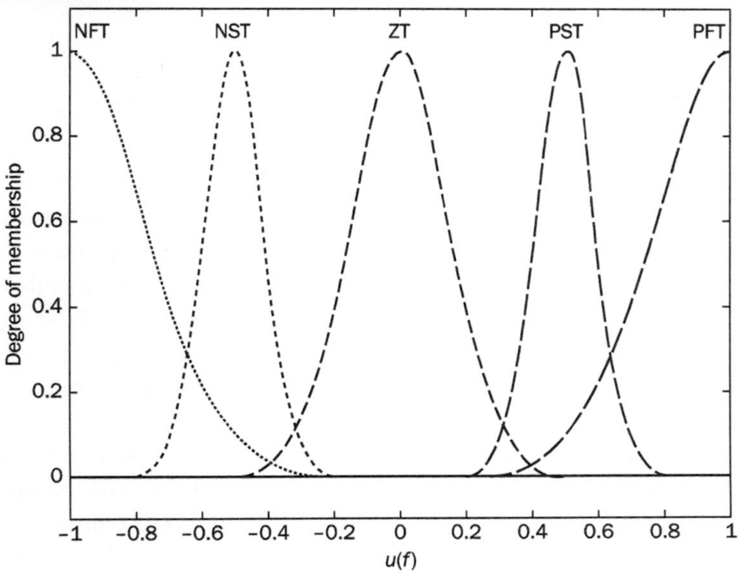

Figure 11.4 Fuzzy logic controller schematic

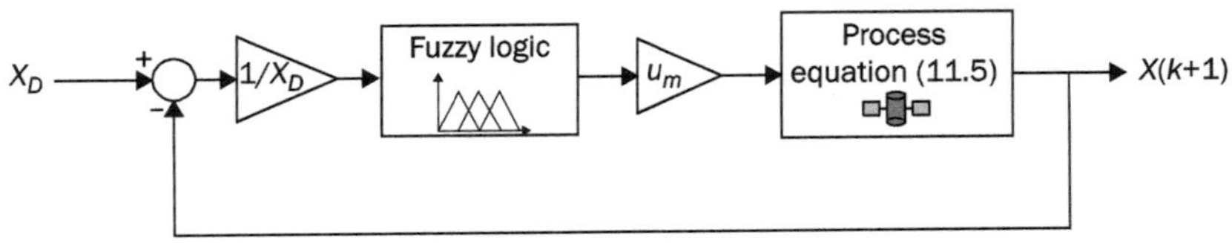

multiplied by the maximum applied thrust (u_m) because the output value is between 1 and –1. The maximum thrust is chosen depending on the correction made to the in-plane motion.

The thrust calculated from the FL control system is substituted in the LP equations. An integration procedure is performed to determine the next value for the orbital elements of the orbit. The integration procedure is executed in the discrete domain because, in equation (11.4), the unit value is expressed for every sample taken through the orbit. For this reason, the system of equations is represented in discrete form as:

$$\vec{X}(k+1) = \vec{X}(k) + \Delta f \mathbf{B}[\vec{X}(k)]\vec{u}(k) + \Delta f \vec{\psi}[\vec{X}(k)] \tag{11.5}$$

where Δf is the sampling interval in the true anomaly angle. The true anomaly angle is represented as: $f(k) = f_0 + k\Delta f$ where f_0 is the initial true anomaly angle. With this definition, the schematic for the FL control system is shown in Figure 11.4. The FL control system for the in-plane motion is explained in the following sections.

11.4 Phase I to II in-plane motion fuzzy logic control system

The FL controller for the in-plane motion in Figure 11.5 involves three inputs, 25 rules, and 2 outputs. The unit values of the a and e (using equation (11.4)) are input into the FL system, and the membership functions have the same definitions as shown in Figure 11.2. The third input includes the position of the satellite along the orbit. As specified in the data mining process, the e increases or decreases depending on the position of the satellite. To describe the position of the satellite, Figure 11.6 shows the trapezoidal membership functions. The linguistic values for the membership functions defining the position of the satellite are defined as: NR is the negative region, ZR is the zero region, and PR is the positive region. Since it is the sum of the cos $f(k)$ and cos $E(k)$, the

Figure 11.5 **Fuzzy logic control system for the in-plane motion correction from phase I to II**

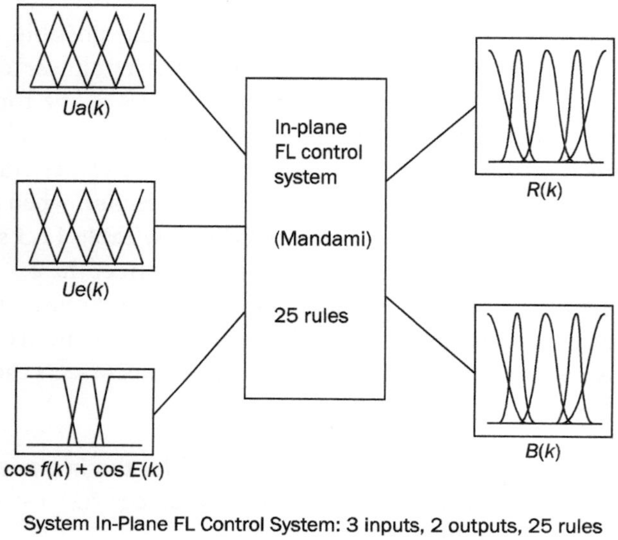

Figure 11.6 **Membership function for the position of the satellite**

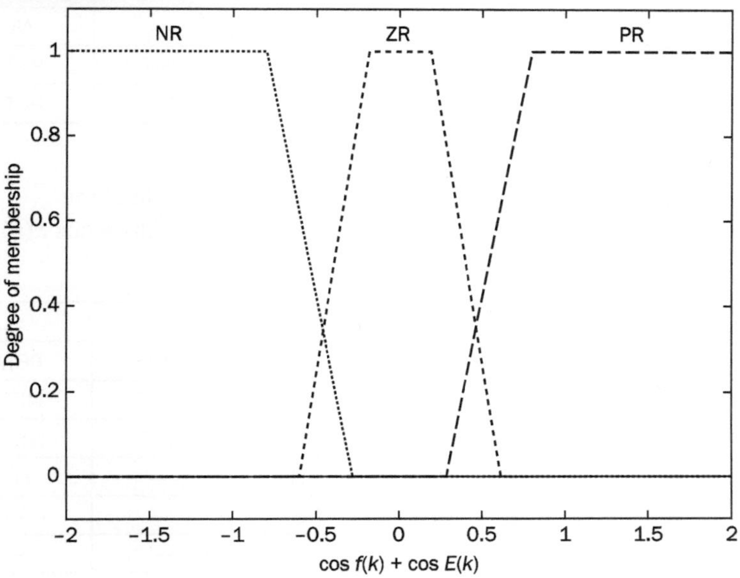

range of the membership function is between –2 and 2. The output uses the Gaussian membership function for the control in the $\bar{R}$ and $\bar{B}$ direction as shown in Figure 11.3.

The FL controller involves 150 rules if all the possible rules are used, but only 25 rules are used to develop the nonlinear controller. Tables 11.1 through 11.3 show the linguistic values at the top and left side. At the center of the tables, the consequence rules (applied thrust along a direction) are shown in terms of its linguistic values. Tables 11.1 and 11.2 show the thrust levels applied in the $\bar{B}$ direction when $\cos f(k) + \cos E(k)$ is in the *PR* and *NR* regions, respectively. Table 11.3 shows the thrust levels for the $\bar{R}$ direction which is used to correct the *e*.

With these conditions, the FL control system corrects the *a* and *e* to its desired value, but the ω increases. As pointed out in the data mining process, a small thrust applied along the $\bar{R}$ direction corrects the ω to the

Table 11.1 **Antecedent and consequence for the in-plane motion for the $\bar{B}$ direction when $\cos f(k) + \cos E(k)$ is in the *PR***

$U_a(k)$	$U_e(k)$				
	NBE	NSE	ZE	PSE	PBE
NBE	NFT	NFT	NST	—	—
NSE	NFT	NST	NST	—	—
ZE	—	—	ZT	—	N
PSE	—	—	PST	PST	PFT
PBE	—	—	PST	PFT	PFT

Table 11.2 **Antecedent and consequence for the in-plane motion for the $\bar{B}$ direction when $\cos f(k) + \cos E(k)$ is in the *NR***

$U_a(k)$	$U_e(k)$				
	NBE	NSE	ZE	PSE	PBE
NBE	—	—	NST	NST	NFT
NSE	—	—	NST	NST	NST
ZE	—	—	ZT	—	N
PSE	PST	PST	PST	—	—
PBE	PFT	PST	PST	—	—

Table 11.3 Antecedent and consequence for the in-plane motion for the $\bar{R}$ direction

$U_a(k)$	$U_e(k)$				
	NBE	NSE	ZE	PSE	PBE
NBE	—	—	—	—	—
NSE	—	—	—	—	—
ZE	NST	NST	ZT	PST	PST
PSE	—	—	—	—	—
PBE	—	—	—	—	—

desired angle at the perigee point and $\bar{B} = 0$. Setting $\bar{B} = 0$ and setting the $f = 0$, the thrust applied [97] in the $\bar{R}$ direction is given as,

$$\bar{R} = \frac{\omega(k) - \omega_D}{\sqrt{1-e^2}\,\beta\Delta f} \tag{11.6}$$

where ω_D is the desired argument of perigee, $\omega(k)$ is the actual argument of perigee at sample k, and β is evaluated at the perigee or apogee point. Then, the value obtained from equation (11.6) is substituted into the LP equation to correct the ω.

11.5 Phase II to III in-plane motion fuzzy logic controller

This FL controller does not use the same rules as the previous intelligent controller. Figure 11.7 shows the FL controller for the reconfiguration procedure from phase II to III. This FL controller is composed of two inputs, 17 rules, and 2 outputs. The two inputs are the unit values of the a and e which are defined by the membership functions in Figure 11.5. The outputs are the unit values representing the degree of thrust applied along the $\bar{R}$ and $\bar{B}$ direction, and the membership functions defining the outputs are defined by Figure 11.3. With two inputs and two outputs, the FL system requires 75 rules per output to cover all the possibilities, but only 17 rules are required to satisfy the solution for this reconfiguration problem. Tables 11.4 and 11.5 show the antecedent rules at the left and top side and the consequent rules.

Figure 11.7 **Fuzzy logic control system for the in-plane motion correction from phase II to III**

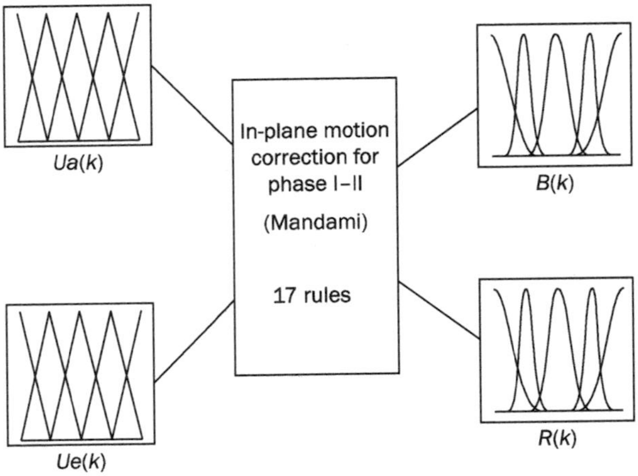

System dadep2: 2 inputs, 2 outputs, 17 rules

Table 11.4 **Antecedent and consequence rules for the in-plane motion for the $\bar{B}$ direction**

	$U_e(k)$				
$U_a(k)$	$U_e(k)$	NSE	ZE	PSE	PBE
NBE	—	—	NST	NST	NFT
NSE	—	—	NST	NST	NFT
ZE	—	—	ZT	—	N
PSE	PST	PST	PST	—	—
PBE	PFT	PST	PST	—	—

Table 11.5 **Antecedent and consequence rules for the in-plane motion for the $\bar{R}$ direction**

	$U_e(k)$				
$U_a(k)$	NBE	NSE	ZE	PSE	PBE
NBE	—	—	—	—	—
NSE	—	—	—	—	—
ZE	NST	NST	ZT	PST	PST
PSE	—	—	—	—	—
PBE	—	—	—	—	—

The argument of perigee is corrected by using a propulsion requirement along the $\bar{R}$ direction defined by equation (11.6). This correction is performed when the fuzzy logic controller finishes the correction for the in-plane motion.

11.6 Out-of-plane motion correction

The out-of-plane motion correction is defined differently because the reconfiguration procedure using a fuzzy logic controller is constrained by the orbital dimensions. For this reason, the correction of the out-of-plane motion is performed with two different techniques: 1) Digital nonlinear Lyapunov controller, or 2) impulsive maneuvers. With the digital nonlinear Lyapunov controller (DNL), this discrete nonlinear controller adapts itself to every new value of the semimajor axis and eccentricity. The impulsive maneuvers is performed in similar manner as the correction of the argument of perigee, but this type of correction is performed as a station-keeping process because a small amount of thrust is necessary. The correction of the inclination angle and the RAAN is performed at certain locations defined by the angle w. The DNL and the impulsive maneuvers are used for the correction of the out-of-plane motion from phase II to III of the proposed tetrahedron constellation [**146**].

11.6.1 Digital nonlinear Lyapunov controller

Lyapunov [**118**] developed a criterion for the stability of certain nonlinear systems. This Lyapunov's direct method does not require a solution to the differential equations. If the following homogeneous quadratic function is positive definite,

$$V_L = \delta \vec{X}^T Q \delta \vec{X} \tag{11.7}$$

the corresponding Q matrix is also positive definite. $\delta \vec{X}$ is the difference between the actual ($\vec{X}$) and the desired ($\vec{X}_D$) orbital elements of the satellite. In Lyapunov theory, the system is asymptotically stable if,

$$\dot{V}_L < 0 \tag{11.8}$$

and Q is a positive definite matrix. The nonlinear Lyapunov controller is expressed in discrete format because the fuzzy logic is defined in the discrete domain. Using Euler's theorem [**14**], equation (11.8) is expanded in discrete format as,

$$V_L(k + 1) < V_L(k) \tag{11.9}$$

With this condition, the discrete system defined by equations (11.3) is asymptotically stable as expressed in Lyapunov's theory. References [**174**] and [**175**] used a discrete form of the Lyapunov controller for a linear state system; instead, reference [**175**] established that the optimal state feedback control law satisfying inequality (11.9) for the worst uncertainty case with respect to $\vec{u}$ is written as,

$$\vec{u}(k) = -(\boldsymbol{B}^T \boldsymbol{Q}\boldsymbol{B})^{-1}\, \boldsymbol{B}^T\, \delta\vec{X}(k) \tag{11.10}$$

This Lyapunov control equation is described in the present value. References [97] and [3] showed that equation (11.10) can be used to obtain an asymptotically stable solution to a system of nonlinear equations, but the $\boldsymbol{Q}$ matrix was defined as an identity matrix. Equation (11.10) is substituted into equation (11.5) to define the control vector for the discrete system of differential equations defined by the LP equations. Equation (11.10) is only implemented for the out-of-plane motion. This controller is used at the same time as the fuzzy logic controller because the correction of the out-of-plane motion does not affect the semimajor axis and the eccentricity. This correction only affects the argument of perigee and is corrected with a single impulse at the perigee point with equation (11.6) when the FL controller and the DNL finish the correction. Equation (11.10) is written for the out-of-plane motion in the following format:

$$\mathrm{N}(k) = -(\boldsymbol{B}_{OM}^T \boldsymbol{B}_{OM})^{-1} \boldsymbol{B}_{OM}^T \delta\vec{X}(k) \tag{11.11}$$

where,

$$\boldsymbol{B}_{\mathrm{OM}}\left[\vec{X}\left(f\left(k\right)\right)\right] = \begin{bmatrix} \dfrac{v\cos w}{\sqrt{1-e^2}} \\ v\sin w \operatorname{cosec} i \end{bmatrix}$$

$$\delta\vec{X}\left(k\right) = \begin{bmatrix} \delta i\left(k\right) \\ \delta\Omega\left(k\right) \end{bmatrix} = \begin{bmatrix} i\left(k\right) - i_D \\ \Omega\left(k\right) - \Omega_D \end{bmatrix}$$

where i_D and Ω_D are the desired inclination angle and the desired RAAN, respectively. Equation (11.11) is used to correct the inclination and the RAAN for the reconfiguration procedure from phase II to III.

11.6.2 Impulsive maneuvers for the out-of-plane motion

From the data mining process, different characteristics are established to define a change to the inclination angle and the RAAN. When $\cos w(k) = 1$ *or* -1, the $\sin w(k) = 0$, and the inclination angle is corrected without changing the RAAN and the argument of perigee (equation 11.3b). The correction is not performed when $f = 0°$ *or* $180°$ because of the angle $w(k)$. The thrust applied along the $\bar{N}$ direction to change the inclination angle is,

$$\bar{N}(k) = \frac{i_D - i(k)}{\Delta f} \frac{\sqrt{1 - e^2(k)}}{v(k)\cos w(k)} \tag{11.12}$$

where i_D is the desired inclination angle, $i(k)$ is the inclination at the sample k, and $v(k)$ is evaluated at sample k. The $\cos w(k)$ is maintained to define the change in sign of the thrust $\bar{N}$ which depends on the location of the satellite.

When $\sin w(k) = 1$ *or* -1, then $\cos w(k) = 0$, and the RAAN is corrected without changing the inclination. This correction to the RAAN does not greatly affect the argument of perigee because the main perturbation of the RAAN is due to the oblateness of the Earth. In addition, the correction is not performed when $f = 90°$ *or* $270°$ because of the variations in the argument of perigee and the angle w. For this case, the thrust applied along the $\bar{N}$ direction is represented as,

$$\bar{N}(k) = \frac{\Omega_D - \Omega(k)}{\Delta f\, v(k) \sin w(k) \operatorname{cosec} i(k)} \tag{11.13}$$

where $\Omega(k)$ is the RAAN at sample k, and the Ω_D is the desired RAAN. This correction for the inclination angle and the RAAN can be performed at the same time as the in-plane correction. Once the in-plane and out-of-plane motion correction are finished, the argument of perigee is corrected to its desired orbital element with equation (11.6). These equations are mainly used as a station-keeping process to apply a small amount of thrust at certain places. The DNL controller is used to cause variations in the out-of-plane motion for the reconfiguration procedure from phase II to III.

11.7 Some solutions for the reconfiguration procedures

The FL controller for the in-plane motion is tested for the reconfiguration procedure of the proposed tetrahedron constellation [146]. To perform the reconfiguration procedure, every satellite in the tetrahedron constellation contains a FL and DNL controller or impulsive maneuvers. The reconfiguration procedure begins at the perigee point because the velocity of the satellite is high, and a small amount of energy can be used. Also, the satellite can be monitored for a period of time at the perigee point.

The reconfiguration procedure performed for the three phases of the proposed tetrahedron constellation [1] takes the following steps:

1. Correct the separation distance between any pair of satellites within the constellation to the nominal separation distance of 10 km before the reconfiguration procedure is performed.
2. Allow the satellites to orbit from the apogee point to the perigee point to obtain a similar tetrahedron constellation at the perigee point.
3. Use the FL controller implemented in every satellite within the constellation to correct its orbital dimensions to the desired orbital dimensions.

At the perigee point, the FL control system is used to correct the four satellites to the desired orbital elements. The satellites are assumed to have an orbit that does not collide with the other satellites during the reconfiguration procedure. This assumption is used to understand the response of the FL controller for the different reconfiguration procedures. For these simulations, the sampling in the true anomaly angle is equal to 0.1 radians.

11.7.1 Reconfiguration procedure from phase I to phase II

For this reconfiguration procedure, Tables G.9 and G.10 in Appendix G show the initial and desired conditions for the four satellites, respectively. Using the FL controller, Figure 11.8 shows the response of the four satellites when the FL control system for the in-plane motion is used. For

Figure 11.8 Reconfiguration procedure from phase I to II using fuzzy logic control

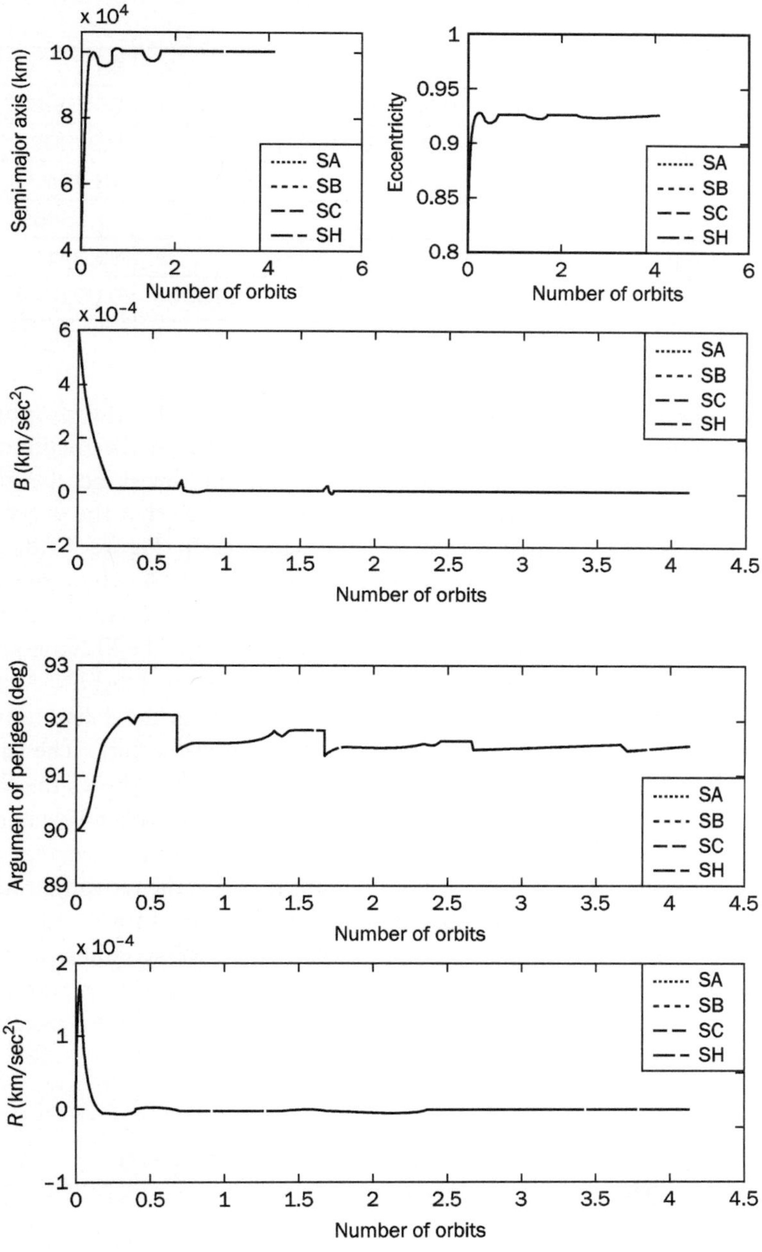

Table 11.6 **Final orbital elements and thrust along the $\bar{R}$ direction to correct the argument of perigee after the reconfiguration procedure**

	SA	SB	SC	SH
a (km)	99498.9077	99498.9077	99498.9077	99498.9077
e	0.92308	0.92313	0.92303	0.92308
i (degrees)	18.5	18.5	18.5	18.494
Ω (degrees)	–0.15661	–0.15671	–0.15652	–0.15659
ω (degrees)	89.997	90	90	89.991
$\bar{R}$ (km/sec^2)	0.0017901	0.0017992	0.0017992	0.0018095

a reconfiguration procedure from phase I to phase II, the proposed tetrahedron constellation does not require a change in the inclination angle between these two phases. The maximum thrust acceleration applied to the $\bar{R}$ and $\bar{B}$ directions is 10^{-3} (km/sec^2) such that the overflow condition is not reached in the control scheme. In Figure 11.8, the satellites begin to reach a steady state in approximate 4 orbits, but the simulation does not stop until the difference in the semimajor axis and eccentricity is less than 0.01 km and 10^{-5}, respectively. The FL controller finishes the correction at the perigee point in exactly 8 orbits. At the perigee point, the satellites applied a small thrust along the $\bar{R}$ direction to correct the argument of perigee back to their desired values. The same controller is used for the other satellites, and Table 11.6 shows the final orbital dimensions after the FL has corrected the desired orbital elements. This table shows the thrust along the $\bar{R}$ direction used to correct the argument of perigee for the four satellites at the end of the reconfiguration maneuver. One important observation from Figure 11.8 is that the satellites take similar decisions to reach their corresponding desired orbital elements, and this situation guarantees that the satellites reach the final perigee point at the same time with the required relative separation distance explained in the proposed problem.

11.7.2 Reconfiguration procedure form phase II to phase III

The reconfiguration procedure from phase II to phase III is performed in two steps. First, the fuzzy logic controller from phase II to III is used to correct the semimajor axis and the eccentricity of the orbit for the four satellites. At the same time, the discrete nonlinear Lyapunov controller is used to correct the inclination angle and the RAAN of the four satellites. In the data mining process, the semimajor axis and the eccentricity are not affected with the out-of-plane motion correction, and the only orbital element affected is the argument of perigee. When the fuzzy logic controller finishes the correction to the orbital elements of phase III, the fuzzy logic controller for the reconfiguration procedure for phase I to II is used to correct the semimajor axis and the eccentricity to their desired values. This fuzzy logic controller performs a station-keeping procedure but is used at the same time as the DNL to finally correct the orbital dimensions of the four satellites. For the fuzzy logic controller for the reconfiguration procedure from phase II to III, the maximum thrust acceleration is maintained to 10^{-5} (km/sec^2), and the simulation stop when the difference between the actual orbital element and the desired orbital element for the a is less than or equal to 0.1 km. For the fuzzy logic controller for the reconfiguration procedure from phase I to II, the maximum thrust acceleration is defined to be equal to 10^{-3} (km/sec^2) for both directions. The simulation stops when the difference between the actual and desired orbital element for the a and e is less than 0.01 (km) and 10^{-5}, respectively.

Figure 11.9 shows the correction of the semimajor axis, the eccentricity, and the argument of perigee for the reconfiguration procedure from phase II to III. In the first five orbits, Figure 11.9 shows a similar solution as in Figure 11.8. The correction of the semimajor axis and the eccentricity is performed in approximately 2 orbits, but the fuzzy logic controller does not finish the correction until 5 orbits. The maximum value of thrust for the $\bar{R}$ and $\bar{B}$ direction is 10^{-7} and 10^{-6} (km/sec^2), respectively. After 5 orbits, the fuzzy logic controller for the reconfiguration procedure from phase II to III stops the correction, and the fuzzy logic for the reconfiguration procedure from phase I to II performs the station-keeping process. This fuzzy logic controller for the reconfiguration procedure from phase I to II applies a second small impulse to obtain the desired values for the semimajor axis and the eccentricity. This correction of the in-plane motion begins after six orbits and comes to rest in approximately 16 complete orbits. With this correction, the argument of perigee decreases from 100 degrees to 99 degrees, but the impulsive maneuver to correct this orbital element is used at the perigee point when the different control schemes finish the correction from phase II to phase III. In both figures, the satellites take very similar 'decisions' in the amount of thrust applied to correct the orbital elements. In addition, the four satellites

Figure 11.9 **Reconfiguration procedure from phase II to III using fuzzy logic control**

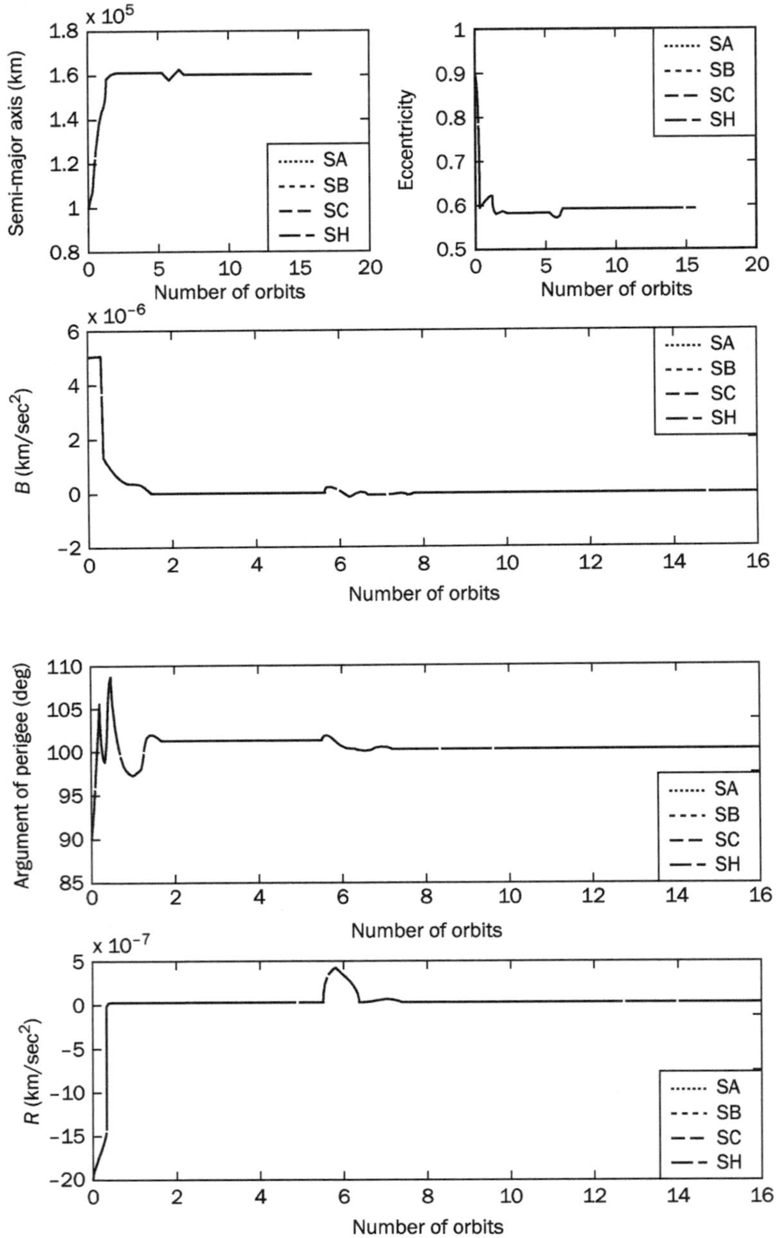

finish at the same time the correction of the orbital elements for the in-plane motion; in this way, the constellation is formed at the final perigee point.

Figure 11.10 shows the simulations for the out-of-plane motion correction. The inclination angle is corrected before 2 complete orbits, and the RAAN comes into a stable motion in approximately 2 complete orbits. In this case, the maximum value of thrust acceleration along the $\bar{N}$ direction is 10^{-3} (km/sec^2) before 5 orbits. After 5 orbits, the approximate maximum value of thrust acceleration is 10^{-6} (km/sec^2). The DNL is only used to correct the inclination angle and the RAAN in which an asymptotically stability condition is obtained. The impulsive maneuvers are not used for this reconfiguration procedure of the out-of-plane motion. In Figure 11.10, the discrete nonlinear Lyapunov control performs similar corrections to the inclination angle and the RAAN in which the satellites within the constellation finish at the same time at the final perigee point.

Another simulation is performed for the reconfiguration procedure from phase II to phase III, but, in this case, the impulsive maneuvers is used to correct the out-of-plane motion of the proposed tetrahedron

Figure 11.10 Out-of-plane motion reconfiguration procedure from phase II to III using the digital nonlinear Lyapunov controller

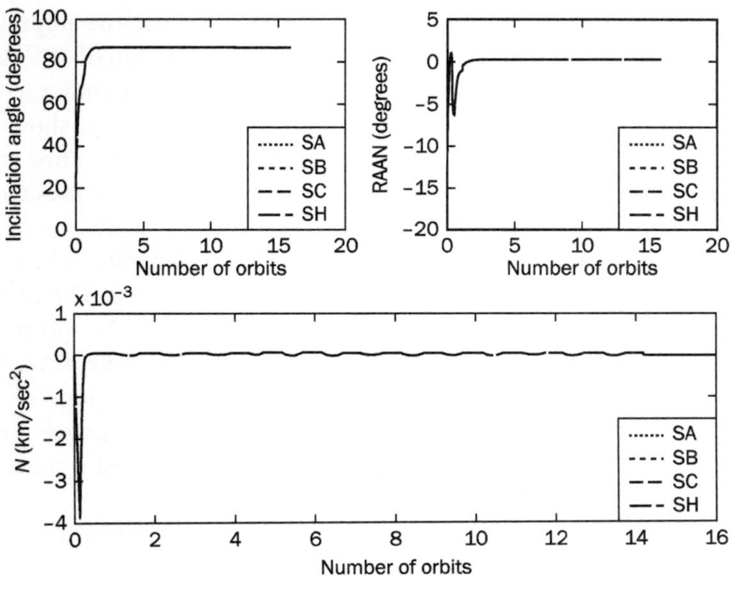

constellation. First, the in-plane motion is corrected with the fuzzy logic controller for the in-plane motion from phase II to phase III. Once this correction is performed, the fuzzy logic controller for the in-plane motion from phase I to phase II is used to correct the in-plane orbital elements of the four satellites. At the same time, the impulsive maneuvers are used to correct the out-of-plane position of the four satellites within the proposed tetrahedron constellation. In the data mining process, the semimajor axis and the eccentricity are not perturbed by the impulsive maneuvers for the correction of the inclination angle and the RAAN. The only perturbed orbital element is the argument or perigee. After the in-plane and out-of-plane orbital elements of the four satellites have been corrected, one impulsive maneuver is performed to correct the argument of perigee at the following perigee point.

Figure 11.11 shows the correction of the out-of-plane position for the four satellites in the proposed constellation using impulsive maneuvers. The in-plane motion correction is not shown because it shows the same results as in Figure 11.9. Figure 11.11 shows the correction of the out-of-plane position before (top) and after (bottom) six complete orbits when the impulsive maneuvers are used. The inclination angle is corrected after 5 complete orbits with two impulses. Before 6 complete orbits, the RAAN decreases as the true anomaly angle increases because of the J2 perturbation which only affects the RAAN and the argument of perigee. Before six complete orbits, the impulsive maneuver requires a maximum impulsive thrust acceleration of 10^{-4} (km/sec^2) and is used mainly to correct the inclination angle. After six complete orbits, the impulsive maneuvers require a maximum impulsive thrust acceleration of 10^{-7} (km/sec^2) to correct the out-of-plane motion. This impulse is mainly used to correct the RAAN rather than the inclination angle because, as shown in Figure 11.11, the inclination angle is corrected using the first two impulses. In addition, the RAAN is corrected with small impulses along the $\bar{N}$ direction when the cos w is equal to 1 or –1.

Table 11.6 shows the values for the orbital elements when the reconfiguration procedure from phase II to III has finished. The orbital elements satisfied the desired conditions to obtain the proposed tetrahedron constellation at the final perigee point. Also, the RAAN is corrected to a small value which does not affect the positions of the satellites. When the satellites orbit from the perigee to the apogee point, the tetrahedron constellation is formed back to the required separation distance constraints at the following apogee point.

Table 11.8 shows the final values for the orbital elements for the same reconfiguration process but for the impulsive maneuvers. In this case,

Figure 11.11 **Correction of the out-of-plane position with impulsive maneuvers**

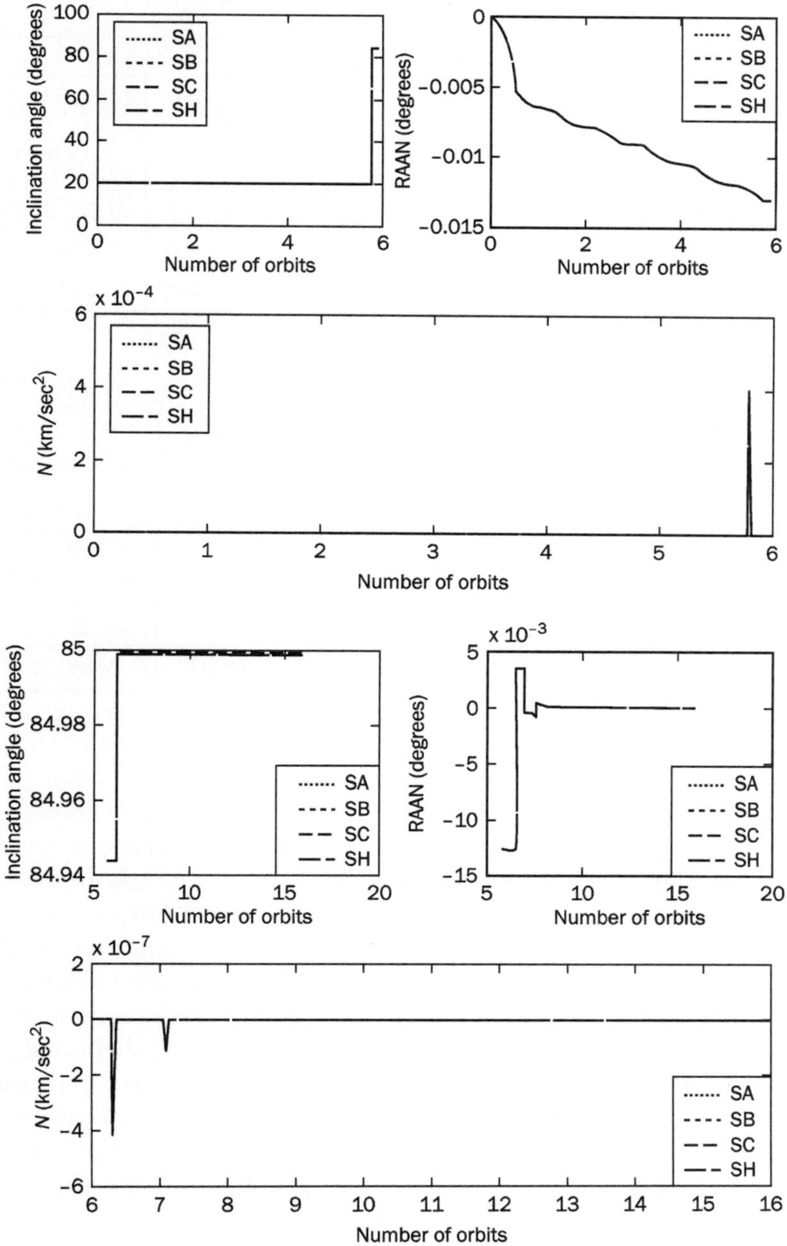

Table 11.7 **Final orbital elements and thrust after the reconfiguration procedure from phase II to III using the digital nonlinear Lyapunov controller**

	SA	SB	SC	SH
a (km)	159,453.01	159,453.01	159,453.01	159,453.01
e	0.6	0.60003	0.59997	0.6
i (degrees)	85	85	85	84.999
Ω (degrees)	−0.0002257	−0.0002257	−0.0002257	−0.0002257
ω (degrees)	89.997	90	90	89.999
$\bar{R}$ (km/sec^2)	1.0125×10^{-4}	1.0123×10^{-4}	1.0125×10^{-4}	1.0115×10^{-4}

Table 11.8 **Final orbital elements and thrust after the reconfiguration Procedure from phase II to III using the impulsive maneuvers**

i (degrees)	85	85	85	84.999
Ω (degrees)	−0.0001762	−0.00017621	−0.00017619	−0.00017624
ω (degrees)	89.997	90	90	89.999
$\bar{R}$ (km/sec^2)	2.2524×10^{-5}	2.2493×10^{-5}	2.2545×10^{-5}	2.2438×10^{-5}

only the orbital elements related to the out-of-plane position are shown because the values for the semimajor axis and the eccentricity are the same as in Table 11.7. Comparing Table 11.7 with Table 11.8, the RAAN shows smaller values using the impulsive maneuvers rather than using the DNL controller. In addition, the last impulse to correct the argument of perigee is smaller in the impulsive maneuvers than in the DNL controller. This difference in the RAAN and the impulse along the $\bar{R}$ direction is due to the definition of the control strategies. The DNL controller is used as an active control scheme correcting the out-of-plane orbital elements trough the entire reconfiguration procedure; the impulsive maneuvers use single impulses at certain locations to instantly correct the out-of-plane motion to the desired value. For this reason, the impulsive maneuvers for the correction of the out-of-plane motion can provide a better control strategy in comparison to the DNL controller, but both controllers can be used for a station-keeping process.

11.8 Implementation of the fuzzy logic controller

The fuzzy logic system controls the amount of energy expended while the thrust is maintained to a maximum value. This maximum value of thrust is in the same range as the electrostatic and electrothermal thrusters [10]. The propulsion systems mentioned can be easily connected to the output of these intelligent controllers to behave as the maximum value of thrust to avoid the singularity problems in the Lagrange planetary equations.

The implementation of the fuzzy logic system is very simple with electronic devices [**115**], but Negnevitsky [107] suggested that a C language program can be used to recreate the fuzzy logic system. With C language programming, a microcontroller processor can be programmed to have the fuzzy logic controller, the digital nonlinear Lyapunov controller, and the impulsive maneuvers for the reconfiguration procedure of a constellation. This microcontroller is the central processing unit regulating the propulsive conditions applied to the $\bar{R}$, $\bar{B}$, and $\bar{N}$ directions for the correction of the orbital elements.

11.9 Adaptive control scheme for reconfiguration procedure

There are disturbances that are unknown and can cause additional effects to the controller. During the data mining process, the LP equations are analyzed based on undisturbed motion for the in-plane motion. When the J2 perturbation is added, the fuzzy logic controller can take care of the perturbation. If there are other forces acting on the system, the fuzzy logic controller cannot compensate for these forces; for this reason, the adaptive control scheme is shown here to demonstrate its capabilities.

Before developing the adaptive control law, equation (11.3a) has to be rewritten to include additional disturbances. The input force for the $\bar{R}$, $\bar{B}$, and $\bar{N}$ directions can be defined as,

$$\vec{u} = \vec{u}_C + \vec{u}_A + \vec{u}_{SP} \tag{11.14}$$

where $\vec{u}_C$ is the input force for the controller, $\vec{u}_A$ is the input force due to the atmospheric density, and $\vec{u}_{SP}$ is the force due to the solar pressure. From Chapter 3, the atmospheric force can be written as,

$$\vec{F}(R) = \frac{1}{2}\rho(R)v^2 C_D S \hat{V}$$

and,

$$\rho(R) = \rho_0 e^{-\frac{R-r_p}{H}}$$

Assuming that the atmospheric force acts in direct opposition to the velocity vector $\hat{V} = -\hat{l}_{\bar{B}}$, $\vec{u}_A$ can be written as,

$$\vec{\psi}_A = \boldsymbol{B}(\vec{X})\vec{u}_A = \begin{bmatrix} \dfrac{\rho(R)v^2 C_D S\beta a^2\sqrt{1-e^2}}{R} \\ \dfrac{n\rho(R)v^2 C_D S a^2\sqrt{1-e^2}}{2\mu}(\cos f + \cos E) \\ 0 \\ 0 \\ \dfrac{\rho(R)v^2 C_D S\beta\sqrt{1-e^2}}{2}\left(1+\dfrac{R}{p}\right)\sin f \end{bmatrix} \quad (11.15)$$

From Chapter 9, the solar pressure force for a right circular cylinder is written as follows,

$$\vec{u}_{SP} = \frac{(1-\varepsilon)}{m}\vec{F}^{+} + \frac{\varepsilon}{m}\vec{F}^{-} \quad (11.16)$$

where,

$$\vec{F}^{+} = -R_C\bar{h}_0\,(4l_C b_0 + \pi R_C c_0)\hat{\sigma}$$

$$\vec{F}^{-} = R_C\bar{h}_0\begin{bmatrix} -\dfrac{16}{3}l_C a_0 b_0 \\ \dfrac{4l_C}{3}\left(2a_0^2 + 8b_0^3\right) \\ -\pi R_C c_0^2 \end{bmatrix}$$

Then,

$$\vec{\psi}_{SP} = \boldsymbol{B}(\vec{X})\vec{u}_{SP} \quad (11.17)$$

Equation (11.3) is written as,

$$\vec{X}' = \boldsymbol{B}\,(\vec{X}(f))\,\vec{u}_C + \vec{\psi}\,(\vec{X}(f)) + \vec{\psi}_A\,(\vec{X}(f)) + \vec{\psi}_{SP}\,(\vec{X}(f)) + \vec{\psi}_{Unknown} \quad (11.18)$$

$\vec{\Psi}_{Unknown}$ are unknown disturbances that can be described by random process or other functions.

The adaptive control system is developed from the adaptive reference model shown in Figure 11.12 and is only applied to the reconfiguration procedure from phase I to II. The process model is defined with equation (11.18), containing only the perturbations due to the Sun and Earth. As shown in Chapter 7, the control input function is defined in terms of the baseline controller ($\vec{u}_{Base}$) and the adaptive control system ($\bar{\vec{u}}$). The control input function is written as,

$$\vec{u}_C = \vec{u}_{Base} + \bar{\vec{u}} \tag{11.20}$$

The reference model is defined by a simple first order differential equation in the true anomaly angle domain as,

$$\vec{x}' = A_m\vec{x} + B_m\vec{r} \tag{11.21}$$

where,

$$A_m = \begin{bmatrix} -3 & 0 \\ 0 & -4 \end{bmatrix} \quad B_m = \begin{bmatrix} 3 & 0 \\ 0 & 4 \end{bmatrix} \quad \vec{x} = \begin{bmatrix} a & e \end{bmatrix}^T \quad \vec{r} = \begin{bmatrix} a_D & e_D \end{bmatrix}$$

Equation (11.21) can be easily transformed into the discrete domain by any of the techniques shown in Chapter 7. The reference model shown in equation (11.21) can vary depending on the vehicle dynamics but can be determined from the linear system of equations. The baseline control is the fuzzy logic controller for the in-plane motion from Phase I to II. The out-of-plane motion is corrected by using impulsive maneuvers if necessary.

Figure 11.12 **Adaptive reference model for the fuzzy logic controller**

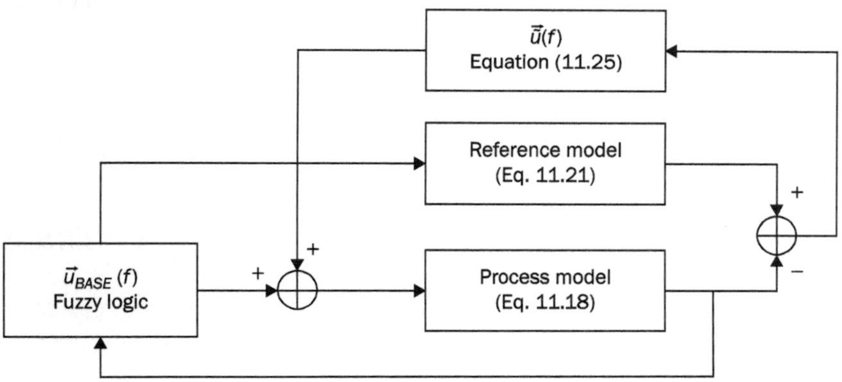

The adaptive control scheme is more involved, and Reference [**176**] shows the first steps to determine the adaptive control law for the system. The definition of the adaptive control law is left as an exercise to the reader. The adaptive control law is written as,

$$\vec{\tilde{u}}(f) = \tilde{K}_P e_x + \tilde{K}_D e'_x \tag{11.22}$$

where,

$$\tilde{K}'_P = -\gamma_L e_x^2 - \theta_A \tilde{K}_P \tag{11.23a}$$

$$\tilde{K}'_D = -\gamma_L (e'_x)^2 - \theta_A \tilde{K}_P \tag{11.23b}$$

$\tilde{K}_P$ and $\tilde{K}_D$ are, respectively, the proportional and derivative gain for the adaptive scheme. γ_L and θ_A is the learning and momentum parameter, respectively. γ_L has to be larger than zero. If γ_L is chosen close to zero, the adaptive system corrects the control at a slower pace while large values of γ_L makes the adaptive system to correct itself faster. Larger values of the learning parameter cause equations (11.23) to create errors in the control gains; and smaller values for the learning parameter slows down the process but provide more accurate solutions. The purpose of θ_A is to accelerate the convergence process. e_x is the error between the process and reference model, i.e. $e_x = \vec{X} - \vec{x}$; and e'_x is the derivative in the true anomaly angle of e_x· γ_L and θ_A can be different for both equations but are used here with the same values.

For this reconfiguration procedure, $\vec{\tilde{u}}(f) = [\tilde{R} \quad \tilde{B}]^T$ in which $\tilde{K}_P$, $\tilde{K}_D$, and e_x are 2×1 column vectors. In e_x, the error associated with the semimajor axis has larger values in comparison to the error in the eccentricity. For this reason, the error is expressed as a unit value as was done for the fuzzy logic. The unit value is expressed as,

$$U_e(k) = \frac{e_x}{X_D} \tag{11.24}$$

Then, the adaptive control is expressed as,

$$\vec{\tilde{u}}(t) = u_m(\tilde{K}_P U_e + \tilde{K}_D \dot{U}_e) \tag{11.25}$$

And,

$$\tilde{K}'_P = -\gamma_L U_e^2 - \theta_A \tilde{K}_P \tag{11.26a}$$

$$\tilde{K}'_D = -\gamma_L \dot{U}_e^2 - \theta_A \tilde{K}_P \tag{11.26b}$$

where u_m is the applied maximum acceleration.

11.9.1 Results for the reconfiguration procedure with adaptive scheme

The same initial conditions for SA in the reconfiguration procedure from Phase I to II shown in Section 11.7.1 are used here. The difference is the addition of the perturbations due to the Sun and the atmospheric effects of the Earth. The spacecraft is assumed to weight 90 kg, with a radius of 0.508 m and a length of 0.559 m. The reflectivity coefficient for the satellite is 0.8. The radius of perigee is 1.2 R_E which describes an atmospheric density of 2.6×10^{-15} kg/m^3 and a scale height of 417 km. For the adaptive scheme, $\gamma_L = 0.1$ and $\theta_A = 1$. The sampling in the true anomaly angle is 0.1 rad. $\tilde{K}_P$ and $\tilde{K}'_d$ are equal to zero at the initial true anomaly angle. The simulation begins at the perigee point. The reconfiguration process stops when the error in the semimajor axis and the eccentricity is less than 1 km and 0.01, respectively. These conditions are different than in Section 11.7 because it is desired to study the adaptive control scheme.

Figures 11.13 and 11.14 show a comparison of the reconfiguration procedure for the in-plane motion with and without the adaptive scheme.

Figure 11.13 In-plane motion reconfiguration for adaptive scheme

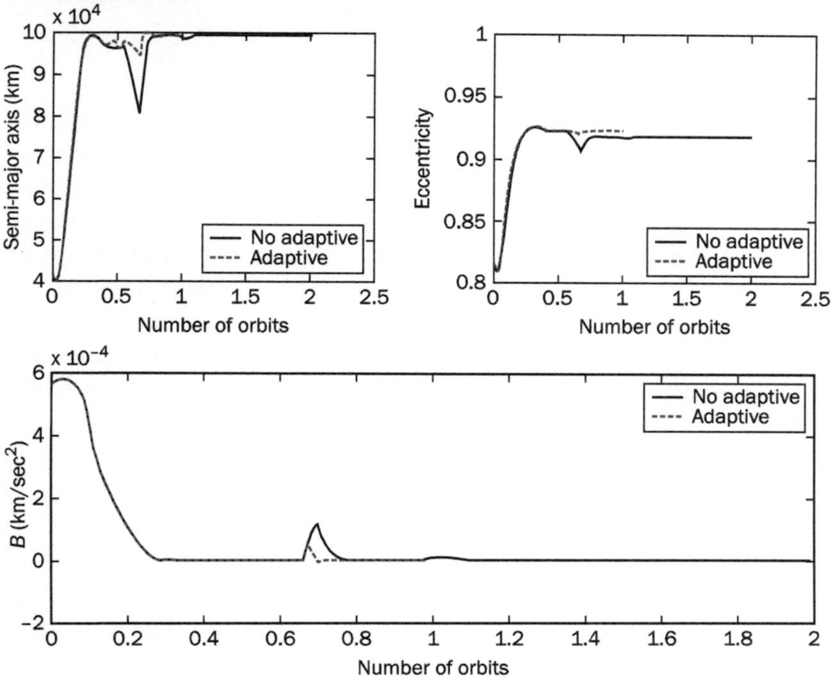

Figure 11.14 In-plane reconfiguration procedure with adaptive scheme for the argument of perigee

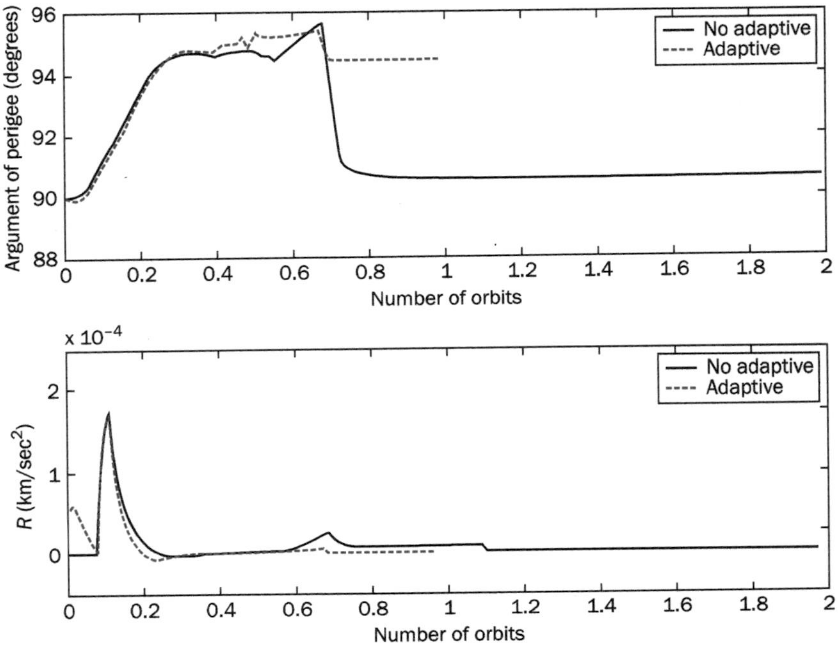

The reconfiguration with an adaptive scheme took 1 orbit to satisfy the criteria to stop the process while the reconfiguration without the adaptive scheme took two orbits. The impulse for $\bar{B}$ is very similar for both control cases, but at approximately 0.7 orbits, a smaller use of thrust is observed in the reconfiguration with the adaptive scheme than in the reconfiguration procedure without the adaptive scheme. This happens because the adaptive scheme is taking care of the disturbances such that the input force can be controlled. In Figure 11.14, $\bar{R}$ shows that at the initial true anomaly angle, a single impulse is applied to cause a variation to the semimajor axis and the eccentricity. Once the system is reaching steady state, $\bar{R}$ is less in the reconfiguration with the adaptive scheme than without the adaptive scheme. As mentioned before, the argument of perigee is corrected with a single impule once the satellite is either at the perigee or apogee point.

Figure 11.15 shows very similar patterns for both cases. In the reconfiguration procedure from Phase I to II, the control process takes more in consideration the in-plane motion than the out of plane motion.

Figure 11.15 Out-of-plane correction for the adaptive scheme

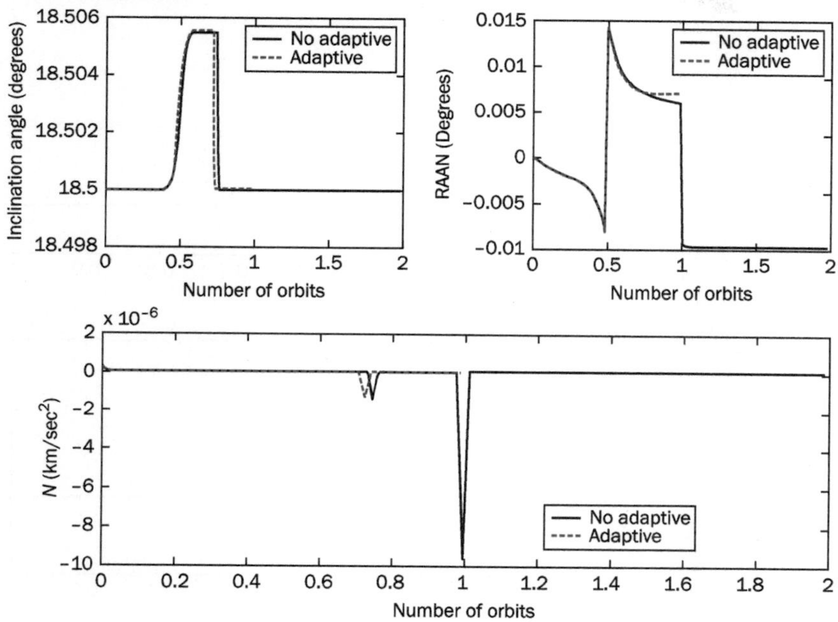

For this case, the out of plane motion is performed with the impulsive maneuvers. For this reason, certain peaks can be observed at some locations along the orbit in both cases.

Figure 11.16 shows how the control input function keeps varying with respect to the true anomaly angle. In Figure 11.16, the adaptive function for $\bar{R}$ shows a larger value than the adaptive function for $\bar{B}$ because $\bar{R}$ is mainly related to the semimajor axis while $\bar{B}$ is related to the eccentricity. In Figure 11.16, the applied impulses highly affect the fuzzy logic controller at the initial true anomaly angle. Once the system is reaching a steady state, the effects of the adaptive scheme are being reduced while the fuzzy logic controller takes care of the errors. The combination of the controllers provides a good control approach to perform the reconfiguration procedure for the other satellites in the constellation. Table 11.9 shows a comparison between the end conditions for both cases. The difference in the orbital elements between the cases is not relatively large, but the orbital elements are satisfied with the adaptive scheme; on the other hand, the adaptive control scheme demonstrates that the correction can be performed in a short number of orbits.

Figure 11.16 Control input function for the adaptive scheme

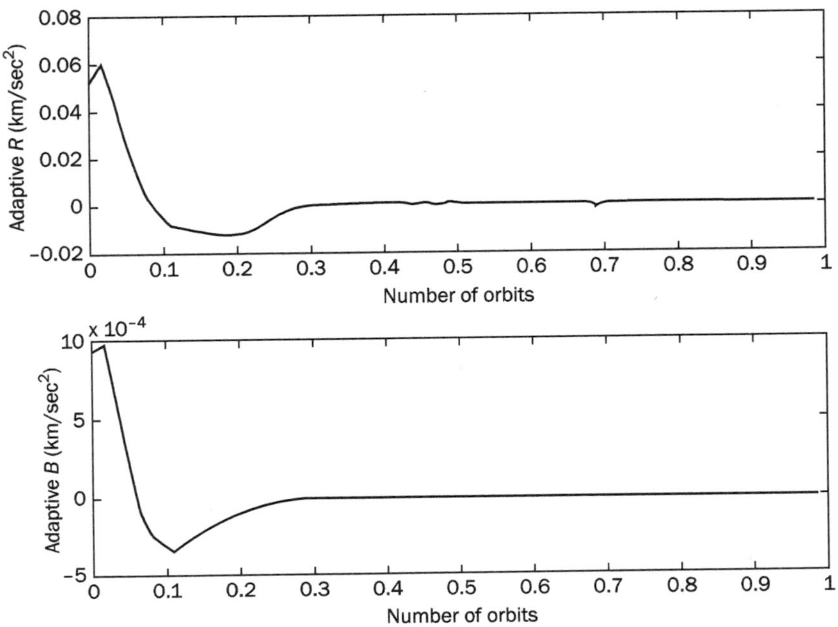

Table 11.9 Comparison of the end conditions for the reconfiguration procedure without and with adaptive control scheme

	Without adaptive scheme	With adaptive scheme
a (km)	99,498.3	99,498.1
e	0.918	0.923
i (degree)	18.5	18.5
Ω (degree)	–0.00958	0.00720
ω (degree)	89.997	90
$\bar{R}$(km/sec^2)	5.081×10^{-4}	4.875×10^{-3}
Number of orbits	2	1

11.10 Remarks

This chapter proposes the use of a fuzzy logic controller and an adaptive control scheme to reconfigure a constellation. By performing a data mining process, the FL controller is successfully used as a nonlinear controller to reach the desired orbital elements for the in-plane motion from phase I to II and phase II to III. The FL controller does not require a complex mathematical model because of the knowledge extracted from the Lagrange planetary equations. By including the adaptive scheme, both controllers take care of the effects of the perturbations due to the Earth and Sun.

In the reconfiguration procedures with only the FL controller, the satellites in the proposed tetrahedron constellation took similar decisions, and the satellites should reach the final perigee point at the same time guarantying that a similar tetrahedron constellation is formed back as shown in Appendix G. In all the simulations, the satellites used a small amount of thrust because of the degree of the decision in the defuzzification process. This degree of the decision is multiplied by the maximum thrust to avoid the overflow condition in which the overflow condition is never obtained. For the out-of-plane motion, the discrete nonlinear Lyapunov controller provides a faster response to the correction of the inclination angle and the RAAN, but, to maintain the desired values, the impulsive maneuvers can be used to finally correct the separation distance constraints.

It can be suggested that the fuzzy logic controller and the adaptive control scheme can be used as reconfiguration and station-keeping processes. These techniques require very small consumptions of energy, but the fuzzy logic controller provides the satellites with some 'intelligence' while the adaptive scheme provides a 'learning capability' to compensate for the outside perturbation. Also, the three techniques can be used in real time to correct the orbital elements of the four satellites because these techniques do not require previous knowledge of the data.

11.11 Suggested problems

Problem 11.1. Equations (11.1) are the Lagrange planetary equations in the time domain. Show that the Lagrange planetary equations in the true anomaly angle domain equations (11.3) are obtained with the following transformations,

$$\upsilon_S = k_h^2\left(1+e\cos f\right)^2 \frac{d\varsigma}{dt} = \upsilon_S \frac{d\varsigma}{df} = \upsilon_S \varsigma'$$

Problem 11.2. As mentioned before, the data mining process is performed to obtain information about the equations of motion. This is performed by setting certain conditions on the thrust requirement to understand the behavior of the orbital elements.

a. Develop an algorithm that integrates the Lagrange planetary equations in the true anomaly angle domain using the initial conditions for phase I of the proposed tetrahedron constellation for satellite SA (Appendix G, Table G.9). Also, assume that the initial true anomaly angle is at the perigee point.

b. What happened to the orbital elements applying the following constant thrust value

   1. $\bar{B} = 10^{-3}$ km/sec^2 while $\bar{R} = \bar{\mathcal{N}} = 0$?
   2. $\bar{B} = 10^{-6}$ km/sec^2 while $\bar{R} = \bar{\mathcal{N}} = 0$?
   3. $\bar{R} = 10^{-3}$ km/sec^2 while $\bar{B} = \bar{\mathcal{N}} = 0$?
   4. $\bar{R} = 10^{-6}$ km/sec^2 while $\bar{B} = \bar{\mathcal{N}} = 0$?
   5. $\bar{N} = 10^{-6}$ km/sec^2 while $\bar{B} = \bar{R} = 0$?

c. Which conditions can be used to reconfigure the tetrahedron constellation from phase I to II and from phase II to III?

d. How can the out of plane rotation be affected by the impulses in the out of plane motion?

e. What will happen if the initial true anomaly angle is at the apogee point?

f. Show that the in-plane motion is affected by the eccentric anomaly and true anomaly angle. Also, show that the out of plane motion is affected by the angle w.

Problem 11.3. Using the solution in Problem 11.2, show that the fuzzy logic rules shown in Table 11.1 through 11.5 satisfy the data mining process and the maximum thrust selected for the reconfiguration process.

Problem 11.4. Adaptive control schemes can be very complicated because it depends on the type of process model, reference model, and Lyapunov function to develop the adaptive control law [176]. In Section 11.9, the adaptive control law provides enough capability to take care of the errors associated with the process model. The adaptive control law can be developed for a single process system to understand its development. Consider the following unknown process model,

$$\dot{x} = ax + u$$

where a is an unknown constant, x is the state variable, and u is the control input function equal to $f(x,t)$. Also assume that the reference model can be represented as,

$$\dot{x} = -a_m x$$

a. Adding and subtracting k^*x to the process model, show that $a - k^* = -a_m$.

b. If $u = -kx$, show that $\hat{k} = k - k^*$.

c. Knowing that $e_x = x - \hat{x}$, show that

$$\dot{e}_x = -a_m e_x + u$$

where $u(t) = -\hat{k}x$, and $u(t)$ is the adaptive control law, and $\hat{k}$ is the adaptive gain.

d. Because $\hat{k}$ is the adaptive gain, it can be represented as $\dot{\hat{k}} = \dot{k} = f(x, t)$. Using the following Lyapunov function,

$$V_L\left(e_x, \hat{k}\right) = \frac{e_x^2}{2} + \frac{\hat{k}^2}{2\gamma_L}$$

Show that $\dot{k} = \gamma_L e_x^2$ by the Lyapunov function and satisfies the Lyapunov stability condition: $\dot{V}_L \leq -a_m e_x^2$. Also $k(0) = k_0$ in which k_0 is the initial condition for the gain.

e. Assume $a = 4$, $a_m = 2$, $k(0) = 0$, and $\gamma_L = 2$, develop and run an algorithm that uses the adaptive control scheme. Show the results.

f. Run another simulation using the following function:

$$\dot{x} = ax + u + \sin(2t)$$

Show the results of the adaptive control scheme.

g. Is it possible to extend the adaptive control law to a PID system? If so, please demonstrate the solution with the Lyapunov function.

Appendix A. Solar ephemeris model

The algorithm to determine the solar ephemeris can be obtained from Vallado [**36**]. The algorithm is shown here to let the reader know about its existence. The variable T_{JD} is the input in the algorithm and is expressed in Julian dates [**10**]. The following algorithm provides the solution for the direction of the incident solar rays on the surface of the satellite [**36**]:

1. Transform the time in Julian dates which is the input variable T_{JD}.
2. Transform the time from Julian dates to Julian centuries with,

$$T_{JC} = \frac{T_{JD} - 2451545}{36525}$$

3. Determine the mean anomaly of the Sun and express its value in a reference angle.

$$M_{sun} = 357.5277233 + 35999.05034 * T_{JC}$$

M_{sun} is expressed in degrees. If the remainder in the division of M_{sun} by 360 is less than zero, add 360 to M_{sun}.

4. Determine the mean longitude of the Sun and express its value in a reference angle.

$$\lambda_{sun} = 280.4606184 + 36000.77005361 * T_{JC}$$

λ_{sun} is expressed in degrees. If the remainder in the division of λ_{sun} by 360 is less than zero, add 360 to λ_{sun}.

5. Determine the longitude of the ecliptic plane.

$$\lambda_{ecliptic} = \lambda_{sun} + 1.914666471 * \sin\left(\frac{M_{sun}\pi}{180}\right) + 0.01994643 * \sin\left(\frac{2M_{sun}\pi}{180}\right)$$

$\lambda_{ecliptic}$ is expressed in radians.

6. Determine the inclination angle of the ecliptic plane.

$$i_{sun} = 23.439291 - 0.0130042 * T_{JC}$$

i_{sun} is expressed in degrees.

7. Determine the distance from the Earth to the Sun.

$$R_{sun} = 1.000140612 - 0.016708617 * \cos\left(\frac{M_{sun}\pi}{180}\right) - 0.000139589 * \cos\left(\frac{2M_{sun}\pi}{180}\right)$$

R_{sun} is expressed in astronomical units.

8. Determine the direction of the sun light with respect to the Earth Centered Inertial (ECI) frame.

$$\hat{\sigma} = \begin{bmatrix} a_0 \\ b_0 \\ c_0 \end{bmatrix} = \begin{bmatrix} \cos\left(\frac{\pi\lambda_{ecliptic}}{180}\right) \\ \cos\left(\frac{\pi i_{sun}}{180}\right)\sin\left(\frac{\pi\lambda_{ecliptic}}{180}\right) \\ \sin\left(\frac{\pi i_{sun}}{180}\right)\sin\left(\frac{\pi\lambda_{ecliptic}}{180}\right) \end{bmatrix}$$

If it is desired to determine the distance from the center of the Earth to the center of the Sun, multiply R_{sun} times $\hat{\sigma}$. Remember that $\hat{\sigma}$ resembles a unit vector. The output of step 8 is in the J2000 frame [36].

Appendix B. Quaternion algebra and operations

This appendix presents the quaternion algebra and operations. Reference [42] provides more operations associated with quaternions. This information is important for the development of the rotational sequences. Express the following two quaternions,

$$q = q_4 + \vec{q} = q_4 + q_1\hat{\imath} + q_2\hat{\jmath} + q_3\hat{k}$$

$$p = p_4 + \vec{p} = p_4 + p_1\hat{\imath} + p_2\hat{\jmath} + p_3\hat{k}$$

The addition of two quaternions follows very similar rules as the addition of two vectors. The addition can be written as,

$$p + q = (p_4 + q_4) + (\vec{p} + \vec{q})$$

The multiplication of a constant and a quaternion is equal to,

$$cq = cq_4 + c\vec{q} = cq_4 + cq_1\hat{\imath} + cq_2\hat{\jmath} + cq_3\hat{k}$$

Two quaternions can be multiplied as follows,

$$pq = p_4q_4 - \vec{p} \cdot \vec{q} + p_4\vec{q} + q_4\vec{p} + \vec{p} \times \vec{q}$$

The complex conjugate of a quaternion is denoted as,

$$q^* = q_4 - \vec{q}$$

For any two quaternions, the complex conjugate is written as,

$$(pq)^* = q^*p^*$$

The sum of a quaternion and its conjugate is equal to twice its magnitude i.e. $q + q^* = 2q_4$.

The norm of a quaternion is explained in the following manner,

$$|q| = \sqrt{qq^*} = \sqrt{q^*q}$$

The unit or normalized quaternions have a norm equal to one and satisfies the following relation,

$$|q|^2 = q_1^2 + q_2^2 + q_3^2 + q_4^2 = 1$$

The norm of two quaternions is defined as follows,

$$|pq|^2 = |p|^2|q|^2$$

The inverse of a quaternion is equal to,

$$q^{-1} = \frac{q^*}{|q|^2}$$

If the quaternion is normalized, $q^{-1} = q^*$. Basically, this property corresponds to the orthogonality condition for the rotational matrices.

Appendix C. Magnetic unit conversion

The following are conversions from the centimeter-gram-second (cgs) unit system to the metric system:

$1\ pole = 125.6637 \times 10^{-9}\ Wb$

$1\ gamma = 10^{-9}\ ^{Wb}/_{m^2}$

$1\ oersted = 1\ ^{A}/_{m}$

$1\ gauss = 0.1\ \mu J$

$1\ cycle = 2\pi\ rad$

$1\ Wb = 1\ ^{A}/_{s}$

Appendix D. Common Laplace and Z-transforms

The following table shows the common Laplace transforms [13]:

$f(t)$	$F(s)$
$\delta(t)$	1
$u(t)$	$\frac{1}{s}$
$tu(t)$	$\frac{1}{s^2}$
$e^{-at}u(t)$	$\frac{1}{s+a}$
$\sin(\omega t)u(t)$	$\frac{\omega}{s^2+\omega^2}$
$\cos(\omega t)u(t)$	$\frac{s}{s^2+\omega^2}$
$\mathcal{L}\{kf(t)\}$	$kF(s)$
$\mathcal{L}\{e^{-at} f(t)\}$	$F(s+a)$
$\mathcal{L}\{f(t-T)\}$	$e^{-sT} F(s)$
$\mathcal{L}\{f(at)\}$	$\frac{1}{a}F\left(\frac{s}{a}\right)$
$\mathcal{L}\left\{\frac{d^n f}{d^n t}\right\}$	$s^n F(s) - \sum_{k=1}^{n} s^{n-k} f^{k-1}(0)$
$\mathcal{L}\left\{\int_0^t f(\tau)d\tau\right\}$	$\frac{F(s)}{s}$

The inverse Laplace transform can be written as [76],

$$\mathcal{L}^{-1}\{F(s)\} = \frac{1}{2\pi j}\int_{\sigma-j\infty}^{\sigma+j\infty} F(s)e^{st}dt$$

The following table shows the common Z-transforms:

F(s)	F(z)
1	z^{-n} where $\delta(k-n)$
$\frac{1}{s}$	$\frac{z}{z-1}$
$\frac{1}{s^2}$	$\frac{\Delta t z}{(z-1)^2}$
$\frac{1}{s+a}$	$\frac{z}{z-e^{-a\Delta t}}$
$\frac{\omega}{s^2+\omega^2}$	$\frac{z\sin(\omega\Delta t)}{z-2z\cos(\omega\Delta t)+1}$
$\frac{s}{s^2+\omega^2}$	$\frac{z(z-\cos(\omega\Delta t))}{z-2z\cos(\omega\Delta t)+1}$

The Z-transform can be written as:

$$E(z) = Z[\{e(k)\}] = \sum_{k=0}^{\infty} e(k) z^{-k}$$

The inverse integral is defined as,

$$e(k) = \frac{1}{2\pi j} \oint_{\Gamma} E(z) z^{k-1}\, dz$$

where $j = \sqrt{-1}$. The translation property is described as,

$$Z\{e(k+n)u(k)\} = z^n \left[E(z) - \sum_{k=0}^{n-1} e(k) z^{-k} \right]$$

Appendix E. Algorithm to determine the eigenvalues of a matrix

The algorithm to determine the eigenvalues of a matrix uses a technique for matrix decomposition. First, the matrix decomposition technique is explained, and then the algorithm to determine the eigenvalues for a matrix is discussed.

There exist different methods to decompose a matrix such as singular value decomposition, polar decomposition, QR method, and LU methods. All these methods are used to decompose a large matrix as simple manageable matrices that can be used to obtain the inverse of a matrix or take advantage of their properties. To determine the eigenvalues of a matrix, the QR method is used in the algorithm. Given a real matrix, the matrix can be decomposed as follows [14],

$$A = QR \tag{e.1}$$

where Q is an orthogonal matrix written as,

$$Q = [\vec{q}_1 \ \vec{q}_2 \ldots \quad \ldots \vec{q}_{n-1} \ \vec{q}_n] \tag{e.2}$$

$\vec{q}_n$ is the column vector of the Q matrix. Also, the columns of the A must be linearly independent. The column vectors, $\vec{q}_n$, can be obtained with the Gram–Schmidt process or the Householder matrices [14]. The Gram–Schmidt process can be written as follows,

$$\vec{v}_n = \vec{a}_n - \sum_{j=1}^{n-1} \frac{\vec{a}_n^T \vec{v}_j}{\vec{v}_j^T \vec{v}_j} \vec{v}_j \tag{e.3}$$

where $\vec{a}_n$ are the column vectors of A, and n is the number of column vectors in A. The column vector of the orthogonal matrix in equation (e.2) is determined from equation (e.3) as,

$$\vec{q}_n = \frac{\vec{v}_n}{|\vec{v}_n|} \tag{e.4}$$

The $\boldsymbol{R}$ matrix is an upper triangular matrix that is written as,

$$\boldsymbol{R} = \begin{bmatrix} \vec{q}_1^T\vec{a}_1 & \vec{q}_1^T\vec{a}_2 & \vec{q}_1^T\vec{a}_3 \\ 0 & \vec{q}_2^T\vec{a}_2 & \vec{q}_2^T\vec{a}_3 \\ 0 & 0 & \vec{q}_3^T\vec{a}_3 \end{bmatrix} \tag{e.5}$$

Equation (e.5) shows the $\boldsymbol{R}$ matrix for the $\boldsymbol{A}$ matrix of 3 columns. Equation (e.5) can be expanded for larger $\boldsymbol{A}$ matrices. If equation (e.2) and (e.5) are substituted into equation (e.1), the $\boldsymbol{A}$ matrix can be obtained.

The objective of using the QR method is to determine the eigenvalues of the A matrix. Reference [14] provides a sequence that can be used to determine the eigenvalues with the QR method. Call $\boldsymbol{A}_1 = \boldsymbol{A}$, and define the sequential matrices as $\boldsymbol{A}_m$, $\boldsymbol{Q}_m$, and $\boldsymbol{R}_m$; then, the process can be written as,

$$\boldsymbol{A}_m = \boldsymbol{Q}_m\boldsymbol{R}_m \tag{e.6}$$

And,

$$\boldsymbol{A}_{m+1} = \boldsymbol{R}_m\boldsymbol{Q}_m \tag{e.7}$$

If the algorithm keeps iterating between equations (e.6) and (e.7), the final $\boldsymbol{A}_{m+1}$ contains in the principal diagonal the eigenvalues for the $\boldsymbol{A}$ matrix. Equation (e.7) provides a slow convergence in the algorithm. To accelerate the process, equation (e.7) can be rewritten as,

$$\boldsymbol{A}_{m+1} = \boldsymbol{C}_m + \boldsymbol{R}_m\boldsymbol{Q}_m \tag{e.8}$$

where $\boldsymbol{C}_m$ is a matrix containing the principal diagonal of the $\boldsymbol{A}_m$ matrix. This technique is known as the shift method for the QR method [14]. This method accelerates the process to determine eigenvalues of the $\boldsymbol{A}$ matrix.

The algorithm to determine the eigenvalue of a matrix can be described as follows:

1. Set $m = 1$, $\Gamma_1 = 0$ and $\boldsymbol{A}_1 = \boldsymbol{A}$.
2. Use the QR method explained with equations (e.1) through (e.5) to decompose the matrix $\boldsymbol{A}_m$.
3. Use equation (e.7) or (e.8) to obtain $\boldsymbol{A}_{m+1}$.
4. If equation (e.8) is used, set $\boldsymbol{C}_1$ as a zero matrix, identity matrix, or random numbers.
5. Obtain the principal diagonal of $\boldsymbol{A}_{m+1}$ and save it in Γ_{m+1}.

6. Compare the values in Γ_{m+1} and Γ_m with respect to a tolerance using the mean square error.
7. If equation (e.8) is used, set $\boldsymbol{C}_{m+1} = \Gamma_{m+1}\boldsymbol{I}$.
8. Stop the algorithm if the error is less than the tolerance value; otherwise, go to step 2 and $m = m + 1$.

Γ_{m+1} is a column vector containing the principal diagonal of the A_{m+1} matrix. By implementing this algorithm, the eigenvalues of a matrix can be determined. In addition, the poles of the state matrix can be obtained such that the stability of the system can be understood. There is another technique that can be used to determine the eigenvector but appears in References [**14**] and [**84**].

Appendix F. Determination of the co-state equation for the linear quadratic regulator

The co-state equation for the linear quadratic regulator (LQR) is important for the development of the control procedure. Using equation (7.63d) and equation (7.87), the end condition for the co-state variables can be written as,

$$\vec{\lambda}(t_1) = \frac{1}{2}\frac{\partial}{\partial \vec{x}(t_1)}\left(\vec{x}(t_1) - \vec{x}_D\right)^T \boldsymbol{F}\left(\vec{x}(t_1) - \vec{x}_D\right) = \boldsymbol{F}\vec{x}(t_1) - \boldsymbol{F}\vec{x}_D \tag{F.1}$$

where $\boldsymbol{F}$ is a constant $n \times n$ positive definite matrix which is used as a weighting matrix for the end condition. Letting $\Phi\,(t, t_1)$ be a fundamental $2n \times 2n$ matrix for the system in equation (7.93) and $\vec{\lambda}(t_0)$ be the initial (unknown) co-state variable, the solution of the differential equation in equation (7.93) equals to,

$$\begin{bmatrix} \vec{x}(t) \\ \vec{\lambda}(t) \end{bmatrix} = \boldsymbol{\Phi}(t, t_1)\begin{bmatrix} \vec{x}(t_0) \\ \vec{\lambda}(t_0) \end{bmatrix} + \begin{bmatrix} \vec{\psi}(t_0) \\ \boldsymbol{Q}\vec{x}_D \end{bmatrix} \tag{F.2}$$

At $t = t_1$,

$$\begin{bmatrix} \vec{x}(t_1) \\ \vec{\lambda}(t_1) \end{bmatrix} = \boldsymbol{\Phi}\left(t_1, t\right)\begin{bmatrix} \vec{x}(t) \\ \vec{\lambda}(t) \end{bmatrix} + \begin{bmatrix} \vec{\psi}(t) \\ \boldsymbol{Q}\vec{x}_D \end{bmatrix} \tag{F.3}$$

Partitioning $\Phi(t_1, t)$ into four $n \times n$ matrices,

$$\boldsymbol{\Phi}(t_1, t) = \begin{bmatrix} \boldsymbol{\Phi}_{11}(t_1, t) & \boldsymbol{\Phi}_{12}(t_1, t) \\ \boldsymbol{\Phi}_{21}(t_1, t) & \boldsymbol{\Phi}_{22}(t_1, t) \end{bmatrix} \tag{F.4}$$

With the use of equation (F.1) and (F.4), equation (7.93) can be written in the following form as:

$$\vec{x}(t_1, t) = \Phi_{11}(t_1, t)\vec{x}(t) + \Phi_{12}(t_1, t)\vec{\lambda}(t) + \vec{\psi}(t) \quad \text{(F.5a)}$$

$$\vec{\lambda}(t_1, t) = \Phi_{21}(t_1, t)\vec{x}(t) + \Phi_{22}(t_1, t)\vec{\lambda}(t) + \boldsymbol{Q}\vec{x}_D = \boldsymbol{F}\vec{x}(t) - \boldsymbol{F}\vec{x}_D \quad \text{(F.5b)}$$

Substituting equation (F.5a) into equation (F.5b), the resulting equation can be solved for the co-state vector $\vec{\lambda}(t)$ as:

$$\vec{\lambda}(t) = \boldsymbol{P}(t)\vec{x}(t) + \vec{G}(t) \quad \text{(F.6)}$$

where,

$$\boldsymbol{P}(t) = (\Phi_{22}(t_1, t) - \boldsymbol{F}\Phi_{12}(t_1, t))^{-1}\,(\boldsymbol{F}\Phi_{11}(t_1, t) - \Phi_{12}(t_1, t)) \quad \text{(F.7a)}$$

$$\vec{G}(t) = (\Phi_{22}(t_1, t) - \boldsymbol{F}\Phi_{12}(t_1, t))^{-1}\,(\boldsymbol{F}\vec{x}_D + \boldsymbol{F}\vec{\psi}(t) - \boldsymbol{Q}\vec{x}_D) \quad \text{(F.7b)}$$

The $\boldsymbol{P}(t)$ and $\vec{G}(t)$ matrices, respectively, are $n \times n$ matrix and $n \times 1$ matrix which depend on the final time t_1 and the matrix $\boldsymbol{F}$, but they do not depend on the initial state. The solutions of the state and co-state vectors are related by equation (F.6).

Appendix G. NASA Benchmark Tetrahedron constellation phases

The NASA Benchmark Tetrahedron configuration [**146**] is similar to a pyramid, but the base of this configuration is an equilateral triangle with an apex point above the centroid of the triangle in a different orbital plane. Figure G.1 shows the tetrahedron constellation. The points SA, SB, SC, and SH are the nominal position of the satellites in the constellation. For convenience, points SB and SC are assumed to be nominally situated along the line of apsides, and points SH and SA are assumed to be the satellites nominally orbiting around the centroid in a different orbital plane, and, in the equilateral triangle, respectively (Figure 2.2). The nominal separation distance between any two sub-satellites at apogee is 10 km, and the separation error at subsequent apogees should be within 10%, giving a range between 9 and 11 kilometers [**146**]. At other points in the orbit, the minimum separation distance between any pairs of sub-satellites should be 1 km [**146**].

The positions of the satellites are determined from the reference point with respect to the Earth Coordinate Inertial (ECI) frame. In this problem, the configuration is assumed such that the satellites arrive at the initial apogee point by some predetermined launch sequence. Figure G.2 shows the tetrahedron configuration in the XY plane where r_a is the radius of apogee. SB and SC are situated along the line of apsides (Y direction), and the remaining two satellites are located with their corresponding separation distances. Table G.1 shows the initial positions of the constellation at the apogee point with respect to the ECI frame [2].

The benchmark problem has four phases with a mission period of two years, but the present work is only concerned with the three phases that contain the restrictions to maintain the separation distance constraints. The three phases of the proposed problem are detailed in Table G.2 in terms of the mean orbital elements. The fourth phase for the NASA benchmark problem is a lunar swing-by which is not considered here.

Figure G.1 Tetrahedron configuration

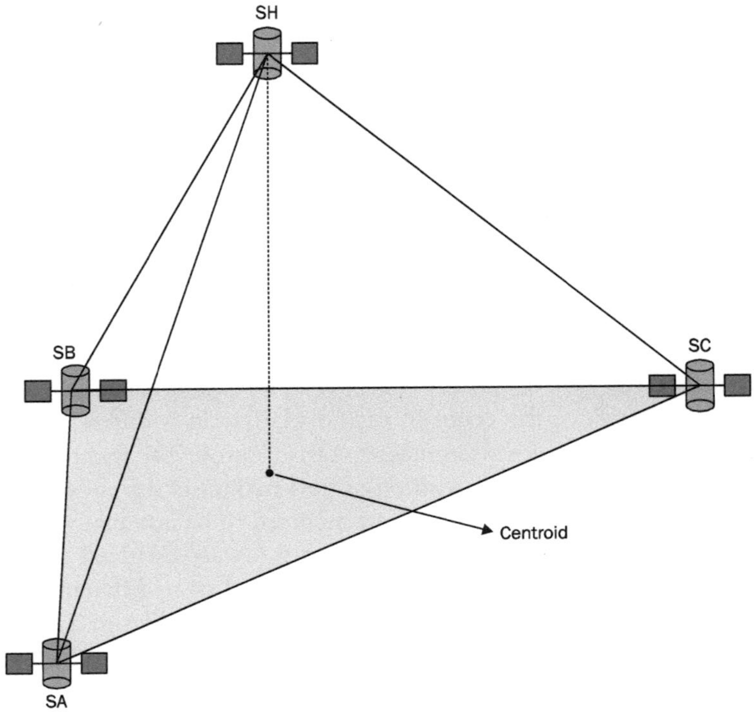

Figure G.2 Two dimensional view of the configuration at apogee point

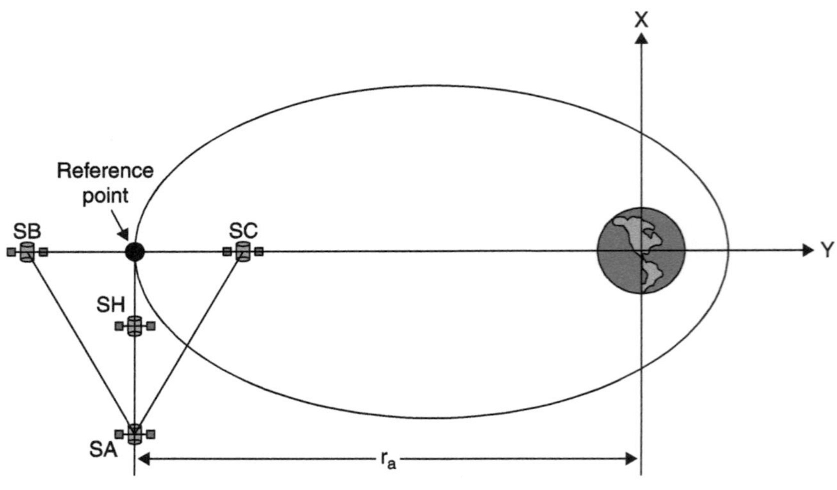

Table G.1 **Initial position at apogee for the satellites with respect to the ECI frame**

Axis (ECI frame)	Reference point (E) (km)	SA (km)	SB (km)	SC (km)	Centroid (km)	SH (km)
X	0	$-5\sqrt{3}$	0	0	$-5\sqrt{3}$	$-5\sqrt{3}$
Y	$-r_a$	$-r_a$	$-(r_a + 5)$	$-(r_a - 5)$	$-r_a$	$-r_a$
Z	0	0	0	0	0	$10\sqrt{3}$

Table G.2 **Dimensions for the three phases expressed as mean orbital elements**

Dimensions	First phase	Second phase	Third phase
Radius of perigee *(r_p)*	1.2 ER	1.2 ER	10 ER
Radius of apogee *(r_a)*	12 ER	30 ER	40 ER
Semimajor axis *(a)*	42,095.7 km	99,498.92 km	159,453.4 km
Eccentricity *(e)*	0.818	0.923	0.6
Inclination angle *(i)*	18.5°	18.5°	85.0°
Period (*days*)	1	3.6	7

Table G.2 shows the dimensions for the three phases that are considered here. The inclination angle in the third phase is not specified in the benchmark problem [**146**] because the constellation must be in a near-polar orbit, so the orbital inclination angle is chosen here to be 85 degrees. From Table G.2, the last phase has the larger orbit, but the eccentricity for the orbit in the third phase is small as compared with the other two orbits; in addition, the period of this orbit is the highest one in comparison with the other two phases.

Substituting the values of Table G.2 into Table G.1 and performing an axis rotation along the X direction, Tables G.3–G.5 detail the initial positions at the apogee point with respect to the ECI frame for the three phases.

Tables G.3–G.5 show the initial velocities, per phase, of the constellation using techniques based on the satellite period and centroid location [**149**]: Using Tables G.3 to G.8, Tables G.9 to G.11 are the orbital elements of the satellites per phase.

Table G.3 Satellite initial positions for phase I

Axis	SA (km)	SB (km)	SC (km)	SH (km)
X	−8.66025403	0	0	−2.88675134
Y	−72,582.4525	−72,587.1941	−72,577.7109	−72,585.0433
Z	−24,285.7489	−24,287.3354	−24,284.1624	−24,278.0058

Table G.4 Satellite initial positions for phase II

Axis	SA (km)	SB (km)	SC (km)	SH (km)
X	−8.66025403	0	0	−2.88675134
Y	−179,772.5402	−179,777.2379	−179,767.8426	−179,775.3366
Z	−65,531.6439	−65,533.3563	−65,529.9315	−65,523.9727

Table G.5 Satellite initial positions for phase III

Axis	SA (km)	SB (km)	SC (km)	SH (km)
X	−8.66025403	0	0	−2.88675134
Y	−22,235.6489	−22,236.0847	−22,235.2131	−22,243.7828
Z	−25,4154.6302	−25,4159.6112	−25,4149.6492	−25,4153.9186

Table G.6 Semimajor axis, radius and initial velocities for the satellites (phase I)

Satellite	Semimajor axis (km)	Radius (r_a) (km)	Velocity (km/sec)
SA	42,095.700834	76,537.63837	0.973083288
SB	42,095.700834	76,542.63788	0.972733623
SC	42,095.700834	76,532.63788	0.973432881
SH	42,095.7010934	76,537.63837	0.973083324

Table G.7 Semimajor axis, radius and initial velocities for the satellites (phase II)

Satellite	Semimajor axis (km)	Radius (r_a) (km)	Velocity (km/sec)
SA	99,498.929244	191,344.094896	0.400303769
SB	99,498.929244	191,349.0947	0.400167766
SC	99,498.929244	191,339.0947	0.400439734
SH	99,498.929527	191,344.094896	0.400303783

Table G.8 Semimajor axis, radius and initial velocities for the satellites (phase III)

Satellite	Semimajor axis (km)	Radius (r_a) (km)	Velocity (km/sec)
SA	159,453.41225	255,125.459616	0.790536603383
SB	159,453.41225	255,130.4596	0.790497870578
SC	159,453.41225	255,120.4596	0.790575336054
SH	159,453.41229	255,125.459616	0.790536603504

Table G.9 Orbital elements for the four satellites within the constellation (phase I)

	SA	SB	SC	SH
a (km)	42,095.7	42,095.7	42,095.7	42,095.7
ϵ	0.818182	0.818301	0.818064	0.818182
i (degrees)	18.5	18.5	18.5	18.494
Ω (degrees)	0	0	0	0
ω (degrees)	89.9921	90	90	89.9974

Table G.10 Orbital elements for the four satellites within the constellation (phase II)

	SA	SB	SC	SH
a (km)	99,498.9	99,498.9	99,498.9	99,498.9
ϵ	0.923077	0.923127	0.923027	0.923077
i (degrees)	18.5	18.5	18.5	18.494
Ω (degrees)	0	0	0	0
ω (degrees)	89.9972	90	90	89.991

Table G.11 **Orbital elements for the four satellites within the constellation (phase III)**

	SA	SB	SC	SH
a (km)	159,453	159,453	159,453	159,453
ϵ	0.6	0.600031	0.599969	0.6
i (degrees)	85	85	85	84.9982
Ω (degrees)	0	0	0	0
ω (degrees)	89.9968	90	90	89.9989

References

[1] Alfriend, Kyle T., Vadali, Srinivas R., Gurfil, Pini, How, Johnathan P., and Breger, Louis S., *Spacecraft Formation Flying: Dynamics, Control, and Navigation*, 1st ed., Cambridge, UK: Elsevier Ltd., 2010.

[2] Schaub, Hanspeter and Junkins, John L., *Analytical Mechanics of Space Systems*, Reston, Virginia, USA: AIAA Education Series, 2003.

[3] Schaub, Hanspeter, Vadali, Srinivas R., Junkins, John L., and Alfriend, Kyle T., 'Spacecraft Formation Flying Control Using Mean Orbital Elements', *The Journal of the Astronautical Sciences*, Vol. 48, Issue no. 1, pp. 69–87, January–March 2000.

[4] Carpenter, Russell J. and Alfriend, Kyle T., 'Navigation Accuracy Guidelines for Orbital Formation Flying', *The Journal of the Astronautical Sciences*, Vol. 53, Issue no. 2, pp. 207–219, April–June 2005.

[5] Carter, T. and Humi, M., 'Fuel-Optimal Rendezvous Near a Point in General Keplerian Orbit', *Journal of Guidance, Control and Dynamics*, Vol. 10, Issue no. 6, pp. 567–573, November–December 1987.

[6] Vadali, S. R., Vaddi, S. S. and Alfriend, K. T., 'A New Concept for Controlling Formation Flying Satellite Constellations', in *AAS/AIAA Space Flight Mechanics Meeting*, Santa Barbara, California, February 11–14, 2001, AAS 01–218.

[7] Thomson, William T., *Introduction to Space Dynamics*, New York, USA: Dover Publications, Inc., 1986.

[8] Bate, Roger R., Mueller, Donald D. and White, Jerry E., *Fundamental of Astrodynamics*, New York, USA: Dover Publications, Inc., 1971.

[9] Chobotov, Vladimir A., *Orbital Mechanics*, 3rd ed., Reston, Virginia, USA: AIAA Education Series, 2002.

[10] Wertz, James R. and Larson, Wiley J., *Space Mission Analysis and Design*, 3rd ed., California, USA: Microcosm Press and Kluwer Academic Publishers, 1999.

[11] Goldstein, Herbert, Poole, Charles, and Safko, John, *Classical Mechanics*, 3rd ed., San Fransisco, USA: Addison Wesley, 2002.

[12] O'Neil, Peter V., *Advanced Engineering Mathematics*, 4th ed., Boston, USA: PWS-Kent Publishing Company, 1995.

[13] Fogiel, M., *Handbook of Mathematical, Scientific, and Engineering Formulas, Tables, Functions, Graphs, Transforms*, New Jersey, USA: Research & Education Association, 2001.

[14] Atkinson, Kendall E., *An Introduction to Numerical Analysis*, 2nd ed., New York, USA: John Wiley & Sons, 1989.

[15] Danby, J. M. A., *Fundametals of Celestial Mechanics*, New York, USA: The MacMillan Company, 1964.

[16] Battin, Richard H., *An Introduction to the Mathematics and Methods of Astrodynamics*, Revised ed., Virginia, USA: AIAA Education Series, 1999.
[17] Brouwer, D. and Clemence, G. M., *Methods of Celestial Mechanics*, New York, USA: Academic Press, 1961.
[18] Sidi, Marcel J., *Spacecraft Dynamics and Control, A Practical Engineering Approach*, New York, USA: Cambridge University Press, 1997.
[19] Harris, I. and Spencer, N. W., *Introduction to Space Science*, New York, USA: Gordon and Breach Science Publishers, 1965.
[20] Wolverton, R. W., *Flight Performance Handbook for Orbital Operations*, New York, USA: John Wiley, 1963.
[21] Anderson, John D., *Fundamentals of Aerodynamics*, 4th ed., New York, USA: McGraw Hill, 2007.
[22] De Lafontaine, J., 'Orbital Dynamics in a Stochastic Atmosphere', *Journal of Guidance, Control, and Dynamics*, Vol. 13, Issue no. 3, pp. 483–490, 1990.
[23] Meyer, Rudolf X., *Elements of Space Technology for Aerospace Engineers*, San Diego, California, USA: Academic Press, 1999.
[24] Mishne, David, 'Relative Formation Keeping of LEO Satellites Subject to Small Drag Differences', in *AAS/AIAA Astrodynamics Specialist Conference*, San Diego, California, July 30-August 2, 2001, AAS 01–455.
[25] Schulz, Walkiria, Bertachini de Almeida Prado, Antonio Fernando and Vilhena de Moraes, Rodolpho, 'An Analytical and Numerical Study of Plane Change Maneuvers Using Aerodynamic Force', *The Journal of Astronautical Sciences*, Vol. 50, Issue no. 3, pp. 289–303, July–September 2002.
[26] Frik, Martin A., 'Attitude Stability of Satellites Subjected to Gravity Gradient and Aerodynamic Torques', George C. Marshall Space Flight Center, Huntsville, Alabama, USA: NASA Technical Memorandum X-53796, October 14, 1968.
[27] Polyaka, Ye N., 'Solar Radiation Pressure and the Motion of Earth Satellites', *AIAA Journal*, Vol. I, Issue no. 12, pp. 2893–2909, December 1963.
[28] Agrawal, Brij N., *Design of Geosynchronous Spacecraft*, New Jersey, USA: Prentice-Hall, Inc., 1986.
[29] Stuch, B. W., 'Solar Pressure Three-Axis Attitude Control', *Journal of Guidance, Control and Dynamics*, Vol. 3, Issue no. 2, pp. 132–139, 1980.
[30] Williams, Trevor and Wang, Zhong-Sheng, 'Potential Uses of Solar Radiation Pressure in Satellite Formation Flying', in *AAS/AIAA Spaceflight Mechanics Meeting*, Clearwater, Florida, 2000, AAS 00–204, pp. 1599–1611.
[31] Dev Kumar, Krishna, Bang, Hy-Choong, and Tahk, Min-Jea, 'Formation Flying Using Solar Radiation Pressure', in *24th International Symposium on Space Technology and Science*, Miyazaki, Japan, May 30-June 6, 2004, ISTS 2004-d-25.
[32] Karymov, A. A., 'Determination of Forces and Moments due to Light Pressure Acting on a Body in Motion in Cosmic Space', *Prikl. Mat. Mekh. (Journal of Applied Mathematics and Mechanics)*, Vol. 26, Issue no. 5, pp. 867–876, 1962.
[33] Fuechsel, P. G., Bainum, P. M. and Grunberger, P. J., 'Attitude Motion of a Nutationally Damped Dual-Spin Spacecraft in the Presence of Near-Earth

Environment', *Journal of Spacecraft and Rockets*, Vol. 8, Issue no. 9, pp. 913–914, September 1971.

[34] Burns, Rich, Gabor, Michael J., McLaughlin, Craig A., Luu, K. Kim, and Sabol, Chris, 'Solar Radiation Pressure Effects on Formation Flying of Satellites with Different Area to Mass Ratios', in *AIAA/AAS Astrodynamics Specialist Conference*, Denver, Colorado, USA: August 14–17, 2000, AIAA 2000-4132.

[35] Wong, Chang-Hee and Ahn, Hyo-Sung, 'Nonlinear Orbital Dynamic Equations and State-Dependent Riccati Equation Control of Formation Flying Satellites', *The Journal of the Astronautical Sciences*, Vol. 51, Issue no. 4, pp. 433–449, October–December 2003.

[36] Vallado, David A., *Fundamentals of Astrodynamics and Applications*, 2nd ed., El Segundo, California, USA: Kluwer Publications, 2004.

[37] Chamberlin, Alan B. (1996, October 29) Horizons System http://ssd.jpl.nasa.gov/?horizons

[38] Giorgini, J. D. et al., 'JPL's On-Line Solar System Data Service', *Bulletin of the American Astronomical Society*, Vol. 28, Issue no. 3, p. 1158, 1996.

[39] Vulpetti, Giovanni, Johnson, Les, and Matloff, Gregory L., *Solar Sails, A Novel Approach to Interplanetary Travel*, New York, USA: Praxis Publishing Ltd., 2008.

[40] Chulaki, Anna. Model Catalog and Archives. http://ccmc.gsfc.nasa.gov/modelweb/

[41] Griffin, Michael D. and French, James R., *Space Vehicle Design*, 2nd ed., Reston, Virginia, USA: AIAA Education Series, 2004.

[42] Kuipers, Jack B., *Quaternions and Rotation Sequences, A Primer with Applications to Orbits, Aerospace, and Virtual Reality*, New Jersey, USA: Princeton University Press, 1999.

[43] Wie, Bong, *Space Vehicle Dynamics and Control*, Reston, Virginia, USA: AIAA Education Series, 1998.

[44] Hughes, Peter C., *Spacecraft Attitude Dynamics*, New York, USA: John Wiley & Sons, 1986.

[45] Luna Imaging. (2009) NASA Images. http://www.nasaimages.org/index.html

[46] Shuster, Malcolm D., 'Constraint in Attitude Estimation Part 1: Constrained Estimation', *The Jornal of Astronautical Sciences*, Vol. 51, Issue no. 1, pp. 51–74, January–March 2003.

[47] Wie, B., Weiss, H. and Arapostathis, A., 'Quaternion Feedback Regulator for Spacecraft Eigenaxis Rotations', *Journal of the Guidance, Control, and Dynamics*, Vol. 12, Issue no. 3, pp. 375–381, May–June 1989.

[48] Atkinson, Kendall and Han, Weimin, *Theoretical Numerical Analysis: A Functional Analysis Framework*, 3rd ed., New York, USA: Springer, 2009.

[49] Edwards, T. L. and Kaplan, M. H., 'Automatic Spacecraft Detumbling by Internal Mass Motion', *AIAA Journal*, Vol. 12, Issue no. 4, pp. 496–502, 1974.

[50] Kaplan, Marshall H., *Modern Spacecraft Dynamics and Control*, New York, USA: John Wiley & Sons, 1976.

[51] Grubin, C., 'Dynamics of a Vehicle Containing Moving Parts', *Transactions of ASME, Journal of Applied Mechanics*, Vol. 29, pp. 486–488, September 1962.

[52] Malvern, Lawrence E., *Introduction to the Mechanics of a Continuous Medium*, 1st ed., New Jersey, USA: Prentice Hall, 1977.
[53] Bainum, Peter M. and Mackison, Donald, 'Gravity-Gradient Stabilization of Synchronous Orbiting Satellites', *Journal of the British Interplanetary Society*, Vol. 21, Issue no. 4, pp. 341–369, December 1968.
[54] Meirovitch, Leonard, *Methods of Analytical Dynamics*, New York, USA: McGraw-Hill, 1970.
[55] Fonseca, Ijar M. and Bainum, Peter M., 'Semi-analytical Approach for the Integrated Structural/Control Optimization for a Space Station Model', in *AAS/AIAA Astrodynamics Specialists Conference*, Big Sky, Montana, August 3–7, 2003, AAS 03–562.
[56] Fonseca, Ijar M., Bainum, Peter M. and Da Silva, Adenilson R., 'Structural Control Interaction for a LSS Attitude Control System Using Thrusters and Reaction Wheels', *Acta Astronautica*, Vol. 60, Issue no. 10–11, pp. 865–872, May–June 2007.
[57] Bainum, Peter M., Fuechsel, P. G. and Mackison, D. L., 'Motion and Stability of a Dual-Spin Satellite with Nutation Damping', *Journal of Spacecraft and Rockets*, Vol. 7, Issue no. 6, pp. 690–696, June 1970.
[58] Boen, Brooke. (2009, December) Overview: Ares V Cargo Launch Vehicle. http://www.nasa.gov/mission_pages/constellation/ares/aresV/index.html
[59] Seltzer, Sherman M., 'An Acceleration Vector Attitude Control System', *Journal of Spacecraft*, Vol. 5, Issue no. 6, pp. 666–671, June 1968.
[60] Seltzer, Sherman M., 'An Attitude Control System for Large Launch Vehicles', *Journal of Spacecraft*, Vol. 6, Issue no. 6, pp. 748–751, June 1969.
[61] Ashenberg, Joshua and Lorenzini, Enrico C., 'Active Gravity-Gradient Stabilization of a Satellite in Elliptical Orbits', *Acta Astronautica*, Vol. 45, Issue no. 10, pp. 619–627, March 1999.
[62] Shrivastava, S. K. and Modi, V. J., 'Satellite Attitude Dynamics and Control in the Presence of Environmental Torques – A Brief Survey', *Journal of Guidance*, Vol. 6, Issue no. 6, pp. 461–472, November–December 1983.
[63] Sadiku, Matthew N. O., *Elements of Electromagnetics*, 3rd ed., New York, USA: Oxford University Press, 2001.
[64] Maus, S. et al., 'The 10th Generation International Geomagnetic Reference Field', *Physics of the Earth and Planetary Interiors*, Vol. 151, pp. 320–322, March 2005.
[65] Nair, Manoj. (2010, January) International Geomagnetic Reference Field. http://www.ngdc.noaa.gov/IAGA/vmod/igrf.html
[66] Oldenburg, James A. and Tragesser, Steven G., 'Minimizing the Effects of Transverse Torques During Thrusting for Spin-Stabilized Spacecraft', *Journal of Guidance, Control, and Dynamics*, Vol. 25, Issue no. 3, pp. 591–595, May–June 2002.
[67] da Fonseca, Ijar M., Arantes, Gilberto, Carvalho, Himilcon C. and Bainum, Peter M., 'On the EQUARS Attitude Dynamics and Control', in *55th International Astronautical Congress*, Vancouver, Canada, October 4–8, 2004, IAC-04-A.4.06.
[68] Psiaki, Mark L., 'Magnetic Torquer Attitude Control via Asymptotic Periodic Linear Quadratic Regulation', in *AIAA Guidance, Navigation, and*

Control Conference and Exhibit, Denver, Colorado, August 14–17, 2000, AIAA-2000-4043.

[69] White, John E. and Robinet III, Rush D., 'Principal Axis Misalignment Control for Deconing Spinning Spacecraft', *Journal of Guidance Control, and Dynamics*, Vol. 17, Issue no. 4, pp. 823–830, July–August 1994.

[70] Harduvel, John T., 'Continuous Momentum Management of Earth-Oriented Spacecraft', *Journal of Guidance, Control, and Dynamics*, Vol. 15, Issue no. 6, pp. 1417–1427, November–December 1992.

[71] Comisel, H. et al., 'Attitude Estimation of a Near-Earth Satellite Using Magnetometer Data', *Acta Astronautica*, Vol. 40, Issue no. 11, pp. 781–788, 1997.

[72] Powell, Barry K., 'Gravity-Gradient Momentum Management', *Journal of Spacecraft*, Vol. 9, Issue no. 6, pp. 385–386, June 1972.

[73] Choi, Yoonhyuk, Leeghim, Henzeh, and Bang, Hyochoong, 'Efficient Control Torque Distribution Approach for Spacecraft Attitude Control', in *AIAA Guidance, Navigation and Control Conference and Exhibit*, Honolulu, Hawaii, August 18–21, 2008, AIAA 2008–7234.

[74] Auclair, Gary F., 'Advanced Reaction Wheel Controller for Spacecraft Attitude Control', in *AIAA Guidance, Control and Flight Mechanics Conference*, Princeton, New Jersey, August 18–20, 1969, AIAA 69–855.

[75] Guru, Bhag S. and Hiziroglu, Huseyin R., *Electric Machinery and Transformers*, 3rd ed., New York, USA: Oxford University Press, 2001.

[76] Nise, Norman S., *Control Systems Engineering*, 4th ed., Massachusetts, USA: John Wiley & Sons, 2004.

[77] Fischell, Robert E., 'Magnetic Damping of the Angular Motions of Earth Satellites', *ARS Journal*, Vol. 31, Issue no. 9, pp. 1210–1217, September 1961, also, The Johns Hopkins University, Applied Physics Laboratory, Report Section T, CM 983, November 1960.

[78] Phillips, Charles L. and Nagle, H. Troy, *Digital Control System Analysis and Design*, 3rd ed., New Jersey, USA: Prentice Hall, 1995.

[79] White, Frank M., *Viscous Fluid Flow*, 3rd ed., New York, USA: McGraw-Hill, 2005.

[80] Bejan, Adrian, *Advanced Engineering Thermodynamics*, 2nd ed., New York, USA: John Wiley & Sons, 1997.

[81] Strang, Gilbert, *Introduction to Applied Mathematics*, Massachusetts, USA: Wellesley-Cambridge Press, 1986.

[82] Snider, Arthur D., *Partial Differential Equations, Sources and Solutions*, New Jersey, USA: Prentice Hall, 1999.

[83] Routh, E. J., *A Treatise on the Stability of a Given State of Motion, Particularly Steady Motion*, Michigan, USA: University of Michigan Library, 2009.

[84] Press, William H., Teukolsky, Saul A., Vetterling, William T. and Flannery, Brian P., *Numerical Recipes in C, The Art of Scientific Computing*, 2nd ed., New York, USA: Cambridge University Press, 1992.

[85] Pinch, Enid R., *Optimal Control and the Calculus of Variations*, New York, USA: Oxford Science Publications, 2002.

[86] Roberts, Sanford M. and Shipman, Jerome S., *Two Point Boundary Value Problems: Shooting Methods*, New York, USA: American Elsevier Publishing Company, 1972.

[87] Athans, Michael and Falb, Peter L., *Optimal Control, An Introduction to the Theory and its Applications*, New York, USA: McGraw-Hill Book Company, 1966.
[88] Andry, A. N. and Shapiro, E. Y. and Chung, J.C., 'Eigenstructure Assignment for Linear Systems', *IEEE Transaction on Aerospace and Electronic Systems*, Vol. AES-19, Issue no. 5, pp. 711–729, September 1983.
[89] Vu, Hung V. and Esfandiari, Ramin S., *Dynamic Systems, Modeling and Analysis*, New York, USA: McGraw-Hill, 1997.
[90] Wie, Bong and Barba, Peter M., 'Quaternion Feedback for Spacecraft Large Angle Maneuvers', *Journal of Guidance, Control, and Dynamics*, Vol. 8, Issue no. 3, pp. 360–366, 1984.
[91] Li, Feiyue and Bainum, Peter M., 'Numerical Approach for Solving Rigid Spacecraft Minimum Time Attitude Maneuvers', *Journal of Guidance, Control, and Dynamics*, Vol. 13, Issue no. 1, pp. 38–45, January–February 1980.
[92] Pontryagin, L. S., Botyanskii, V. G., Gamkrelidze, R. V. and Mishchenko, E. F., *The Mathematical Theory of Optimal Process*, London, UK: Pergamon Press, 1964.
[93] Stengel, Robert F., *Optimal Control and Estimation*, New York, USA: Dover Publications Inc., 1984.
[94] Pierre, Donald A., *Optimization Theory with Application*, New York, USA: Dover Publications, 1986.
[95] Flugge-Lotz, I. and Marbach, H., 'The Optimal Control of Some Attitude Control Systems for Different Performance Criteria', *The Journal of Basic Energy*, Vol. 85, pp. 165–176, 1963.
[96] Minter, C. F. and Fuller-Rowell, T. J., 'A Robust Algorithm for Solving Unconstrained Two-Point Boundary Value Problems', in *AAS/AIAA Space Flight Mechanics Conference*, Vol. 120, Part 1, Copper Mountain, Colorado, January 23–27, 2005, AAS 05–129, pp. 409–444.
[97] Tan, Zhaozhi, Bainum, Peter M. and Strong, Avaine, 'The Implementation of Maintaining Constant Distance between Satellites in Coplanar Elliptic Orbits', *The Journal of Astronautical Sciences*, Vol. 50, Issue no. 1, pp. 53–69, January–March 2002.
[98] Campbell, Mark E., Zanon, Darren, and Kulkami, Jayant, 'Cluster Planning and Control of Spacecraft Formations', in *AAS/AIAA Space Flight Mechanics Conference*, Maui, Hawaii, February 8–12, 2004, AAS 04–254.
[99] Franco, Sergio, *Design with Operational Amplifiers and Analog Integrated Circuits*, 2nd ed., Boston, USA: McGraw-Hill, 1998.
[100] Capó-Lugo, Pedro A. and Bainum, Peter M., 'Digital LQR Control for Maintaining NASA Benchmark Tetrahedron Constellation Separation Distance Constraints', *Acta Astronautica*, Vol. 67, Issue no. 7–8, pp. 1058–1067, November 2009.
[101] Hassan, Mohamed F. and Singh, Madan, 'A Two Level Co-state Prediction Algorithm for Non-linear Systems', *Automatica*, Vol. 13, Issue no. 6, pp. 629–634, November 1977.
[102] Hassan, Mohammed F. and Singh, Madan, 'The Optimization of Non-linear Systems Using a New Two-Level Method', *Automatica*, Vol. 12, Issue no. 4, pp. 359–363, July 1976.

[103] Capó-Lugo, Pedro A. and Bainum, Peter M., 'Steady-state hierarchical control for the drift correction of a constellation in highlly elliptical orbits', *Proceedings of the Institution of Mechanical Engineers, Part G, Journal of Aerospace Engineering*, Vol. 223, Issue no. G8, pp. 1125–1139, 2009.

[104] Capó-Lugo, Pedro A. and Bainum, Peter M., 'Development and Control Solution for the Perturbed Motion of a Constellation in an Elliptical Orbit', in *AAS/AIAA Astrodynamics Specialist Conference and Exhibit*, Mackinac Island, Michigan, August 19–23, 2007, AAS 07–298.

[105] Donato, Peter, Abdallah, Chaoki and Cerone, Vito, *Linear-Quadratic Control, An Introduction*, New Jersey, USA: Prentice Hall, 1995.

[106] Antsaklis, Panos J. and Passino, Kevin M., *An Introduction to Intelligent and Autonomous Control*, Massachusetts, USA: Kluwer Academic Publishers, 1992.

[107] Negnevitsky, Michael, *Artificial Intelligence, A Guide to Intelligent Systems*, 2nd ed., Harlow, UK: Addison-Wesley, 2005.

[108] Passino, Kevin M. and Yurkovich, Stephen, *Fuzzy Control*, California, USA: Addison-Wesley, Inc., 1998.

[109] Rubaai, A. and Kotaru, R., 'Adaptation Learning Control Scheme for a High-Performance Permanent-Magnet Stepper Motor Using Online Random Training of Neural Networks', *IEEE Transactions in Industrial Applications*, Vol. 37, Issue no. 2, pp. 495–502, March–April 2001.

[110] Rubaai, A. and Kotaru, R., 'Online Identification and Control of a DC Motor Using Learning Adaptation of Neural Networks', *IEEE Transactions in Industrial Applications*, Vol. 36, Issue no. 3, pp. 935–942, May–June 2000.

[111] Kelly, Thomas J. and Nordwall, Jill A., 'Compression of Orbital Ephemerides Using Neural Networks', in *AAS/AIAA Spaceflight Mechanics Meeting*, Vol. 82, Part II, Pasadena California, February 22–24, 1993, AAS 93–108, pp. 727–764.

[112] Smith, James E., Proulx, Ronald J., Cefola, Paul J. and Draim, John E., 'Optimal Station-keeping Strategies via Parallel Genetic Algorithms', in *AAS/AIAA Space Flight Mechanics Meeting*, Vol. 102, Breckenridge, Colorado, February 7–10, 1999, AAS 99–123, pp. 303–322.

[113] Ely, Todd A., Bishop, Robert H. and Crain, Timothy P., 'Adaptive Interplanetary Navigation Using Genetic Algorithms', in *Richard H. Battin Astrodynamics Conference*, Vol. 106, College Station, Texas, March 20–21, 2000, AAS 00–271, pp. 147–160.

[114] Bennis, R. J. M., Ortega, G., Chu, Q. P. and Mulder, J. A., 'Feasibility Study of a Fuzzy Logic Based Controller for Rendezvous and Docking of the Automated Transfer Vehicle (ATV) to the International Space Station (ISS)', in *50th International Astronautical Congress*, Amsterdam, The Netherlands, October 4–8, 1999, IAF-99-A.1.05.

[115] Bennis, R. J. M., Chu, Q.P. and Mulder, J. A., 'Adaptive Fuzzy Control by Reinforcement Learning for Rendezvous and Docking', in *52nd International Astronautical Congress*, Toulouse, France, October 1–5, 2001, IAF-01-A.3.09.

[116] Rubaai, A., Ofoli, A. R., Burge III, L., and Garuba, M., 'Hardware Implementation of Adaptive Network-Based Fuzzy Controller for DC-DC

Converters', *IEEE Transaction in Industrial Applications*, Vol. 41, Issue no. 6, pp. 1557–1565, November–December 2005.

[117] Sastry, Shankar and Bodson, Marc, *Adaptive Control, Stability, Convergence and Robustness*, New Jersey, USA: Prentice Hall, 1989.

[118] La Salle, Joseph and Lefchetz, Solomon, *Stability by Lyapunov's Direct Method with Applications*, New York, USA: Academic Press, 1961.

[119] Johnson, Eric N. and Kannan, Suresh K., 'Adaptive Trajectory Control for Autonomous Helicopters', *Journal of Guidance, Control, and Dynamics*, Vol. 28, Issue no. 3, pp. 524–538, May–June 2005.

[120] Boskovic, Jovan D., Chen, Lingki and Mehra, Raman K., 'Adaptive Control Design for Nonaffine Models Arising in Flight Control', *Journal of Guidance, Control, and Dynamics*, Vol. 27, Issue no. 2, pp. 209–217, March–April 2004.

[121] Fiorentini, Lisa, Serrani, Andrea, Boleder, Michael A. and Doman, David A., 'Nonlinear Robust Adaptive Control of Flexible Air-Breathing Hypersonic Vehicles', *Journal of Guidance, Control, and Dynamics*, Vol. 32, Issue no. 2, pp. 402–417, March–April 2009.

[122] Suresh, S., Omkar, S. N., Mani, V. and Sunaraarajan, N., 'Nonlinear Adaptive Neural Controller for Unstable Aircraft', *Journal of Guidance, Control, and Dynamics*, Vol. 28, Issue no. 6, pp. 1103–1111, November–December 2005.

[123] Green, A. and Sasiadek, J. Z., 'Adaptive Control of a Flexible Robot Using Fuzzy Logic', *Journal of Guidance, Control, and Dynamics*, Vol. 28, Issue no. 1, pp. 36–42, January–February 2005.

[124] Wertz, James R., *Attitude Determination and Control*, Dordecht, Holland: Reidel Publishing Company, 2002.

[125] Pilasch, Daniel W., 'Data Analysis of a Spinning Payload Using Sun Sensors', in *AIAA Aerodynamic Decelerator Systems Technology Conference*, Clearwater Beach, Florada, May 15–18, 1995, AIAA-95-1543-CP, pp. 66–76.

[126] Psiaki, Mark L., 'Autonomous Low-Earth-Orbit Determination from Magnetometer and Sun Sensor Data', *Journal of Guidance, Control, and Dynamics*, Vol. 22, Issue no. 2, pp. 296–304, April 1999.

[127] Vega, Karla, Auslander, David and Pankow, David, 'Design and Modeling of an Active Attitude Control System for CubeSat Class Satellites', in *AIAA Modeling and Simulation Technologies Conference*, Chicago, Ilinois, August 10–13, 1009, AIAA-2009-5812.

[128] Psiaki, Mark L., 'Estimation of a Spacecraft's Attitude Dynamics Parameters by Using Flight Data', *Journal of Guidance, Control, and Dynamics*, Vol. 28, Issue no. 4, pp. 594–604, July–August 2005.

[129] Kitamura, Shoki, Miyazaki, Katsuhiro, Hayakawa, Yukio, Yoshida, Hideki and Akai, Kouseki, 'Performance Improvement of 150-mN Xenon Ion Thrusters', *Acta Astronautica*, Vol. 52, Issue no. 1, pp. 11–20, January 2003.

[130] Oland, E. and Schlanbusch, R., 'Reaction Wheel Design for Cubesats', in *4th International Conference on Recent Advances in Space Technologies*, Instanbul, Turkey, 2009, pp. 778–783.

[131] Lovera, Marco and Astolfi, Alessandro, 'Global Magnetic Attitude Control of Spacecraft in the Presence of Gravity Gradient', *IEEE Transaction on*

Aerospace and Electronic Systems, Vol. 42, Issue no. 3, pp. 796–806, July 2006.

[132] Kawasaki, Naoya and Shimemura, Etsujiro, 'Determining Quadratic Weighting Matrices to Locate Poles in a Specified Region', *Automatica*, Vol. 19, Issue no. 5, pp. 557–560, September 1983.

[133] Broussard, John R., 'A Quadratic Weight Selection Algorithm', *IEEE Transaction on Automatic Control*, Vol. AC-27, Issue no. 4, pp. 945–947, August 1982.

[134] Becus, Georges A. and Sonmez, Gokhan, 'Symbolic Manipulation and Numerical Optimization Techniques for Robust Eigenstructure Assignment in LQR Design of Multivariable Control Systems', in *IEEE Control Systems Society Workshop on Computer Aided Control System Design*, 1989, pp. 102–109.

[135] Choi, Jae Weon and Seo, Young Bong, 'LQR Design with Eigenstructure Assignment Capability', *IEEE Transactions on Aerospace and Electronic Systems*, Vol. 35, Issue no. 2, pp. 700–708, April 1999.

[136] Luo, Jia and Lan, C. Edward, 'Determination of Weighting Matrices of a Linear Quadratic Regulator', *Journal of Guidance, Control, and Dynamics*, Vol. 18, Issue no. 6, pp. 1462–1463, November–December 1995.

[137] Ohta, H., Kakinuma, M. and Nikiforuk, P. N., 'Use of Negative Weights in Linear Quadratic Regulator Synthesis', *Journal of Guidance, Control, and Dynamics*, Vol. 14, Issue no. 4, pp. 791–796, August 1991.

[138] Wongsathan, Chaipron and Sririma, Chanapoom, 'Application of GA to Design LQR Controller for an Inverted Pendulum System', in *Proceedings of the 2008 IEEE International Conference on Robotics and Biomimetics*, Bangkok, Thailand, February 21–26, 2009, pp. 951–954.

[139] Li, Young, Liu, Jianchang and Wang, Yu, 'Design Approach of Weighting Matrices for LQR Based on Milti-objective Evolution Algorithm', in *Proceedings of the 2008 IEEE International Conference on Information and Automation*, Zhangjiajie, China, June 20–23, 2008, pp. 1188–1192.

[140] Clarke, Arthur C., 'Extra-Terrestrial Relays', in *XII th International Astronautical Congress*, Washington DC, USA: October 2–7, 1961, originally appeared in *Wireless World*, Iliffe Electrical Publications, Ltd., pp. 305–308, London, October, 1945.

[141] Clohessy, W. H. and Wiltshire, R. S., 'Terminal Guidance System for Satellite Rendezvous', *Journal of the Aerospace Sciences*, pp. 653–658, 64, September 1960.

[142] Tschauner, J. and Hempel, P., 'Rendezvous with a Target in an Elliptical Orbit', *Acta Astronautica*, Vol. 11, Issue no. 2, pp. 104–109, 1965.

[143] Walker, J. G., 'Continuous Whole Earth Coverage by Circular Orbit Satellite Patterns', Royal Aircraft Establishment, London, 77044, March 1977.

[144] Mortari, Daniele, Wilkins, Matthew P. and Bruccoleri, Christian, 'The Flower Constellations', *The Journal of the Astronautical Sciences, Special Issue: The John L. Junkins Astrodynamics Symposium*, Vol. 52, Issue no. 1 and 2, pp. 107–127, January–June 2004.

[145] Schaub, H., 'Stabilization of Satellite Motion Relative to a Coulomb Spacecraft Formation', *Journal of Guidance, Control, and Dynamics*, Vol. 28, Issue no. 6, pp. 1231–1239, November–December 2005.

[146] Carpenter, Russel J., Leitner, Jessie A., Burns, Richard D. and Folta, David C., 'Benchmark Problems for Spacecraft Formation Flying Missions', in *AIAA Guidance, Navigations, and Control Conference and Exhibit*, Austin, Texas, August 11–14, 2003, AIAA-2003-5364.
[147] Lawden, D. F., *Optimal Trajectories for Space Navigation*, London, UK:, Butterworths, 1963.
[148] Capó-Lugo, Pedro A. and Bainum, Peter M., 'Strategy for Satisfying Distance Constraints for the NASA Benchmark Tetrahedron Constellation', in *Proceedings of the AAS/AIAA Space Flight Mechanics Conference*, Vol. 120, Part I, Copper Mountain, Colorado, January 23–27, 2005, AAS 05–153, pp. 775–793.
[149] Capó-Lugo, Pedro A. and Bainum, Peter M., 'Implementation of the Strategy for Satisfy Distance Constraints for the NASA Benchmark Tetrahedron Constellations', in *Proceedings of the AAS/AIAA Astrodynamics Specilist Conference and Exhibit*, Vol. 123, Part II, Lake Tahoe, California, August 7–11, 2005, AAS 05–344, pp. 1463–1482.
[150] Capó-Lugo, Pedro A. and Bainum, Peter M., 'Solar Pressure Effects for a Constellation in Highly Elliptical Orbit', *Journal of Guidance, Control, and Dynamics*, Vol. 32, Issue no. 2, pp. 675–679, March–April 2009.
[151] Capó-Lugo, Pedro A. and Bainum, Peter M., 'Effects of Satellite Dimensions in the Solar Pressure Formulation for Formation Flying', *Proceedings of the Institution of Mechanical Engineers, Part G, Journal of Aerospace Engineering*, Vol. 223, Issue no. G4, pp. 453–463, June 2009.
[152] Guzman, Jose J., 'Primer Vector Optimization: Force Models Consideration', in *AAS/AIAA Space Flight Mechanics Conference*, Vol. 120, Part I, Copper Mountain, Colorado, January 23–27, 2005, AAS 05–153.
[153] Capó-Lugo, Pedro A. and Bainum, Peter M., 'Active Control Schemes to Satisfy the Separation Distance Constraints', *Journal of Guidance, Control, and Dynamics*, Vol. 30, Issue no. 4, pp. 1152–1156, July–August 2007.
[154] Mauro, Massari, Armellin, Roberto, and Finzi, Amalia E., 'Optimal Trajectory Generation and Control for Reconfiguration Maneuvers of Formation Flying Using Low-Thrust Propulsion', in *AAS/AIAA Space Flight Mechanics Conference*, Maui, Hawaii, February 8–12, 2004, AAS 04–258.
[155] Tsuda, Yoichi, 'Global Optimization of Maneuver Schedule for Multiple Spacecraft Flying in Formation', in *International Astronautical Congress*, Vancouver, Canada, 2004, IAC-04-A.2.01.
[156] Bainum, Peter M., Tan, Zhaozhi, and Duan, Xiaodong, 'Review of Station Keeping Strategies for Elliptically Orbiting Constellation in Along-Track Configuration', *International Journal of Solids and Structures*, Vol. 42, Issue no. 21–22, pp. 5683–5691, October 2005, PACAM VIII Special Issue.
[157] Carter, T. and Humi, M., 'New Form of the Rendezvous Equations Near a Keplerian Orbit', *Journal of Guidance, Control, and Dynamics*, Vol. 13, Issue no. 1, pp. 183–186, January–February 1990.
[158] Marec, J. P., *Optimal Space Trajectories*, New York, USA: Elsevier Scientific Publishing Company, 1979.
[159] Inalhan, Gokhan, Tillerson, Michael, and How, Jonathan P., 'Relative Dynamics and Control of Spacecraft Formation in Eccentric Orbits',

Journal of Guidance, Control, and Dynamics, Vol. 25, Issue no. 1, pp. 48–59, January–February 2002.

[160] Eishima, Takashi, 'Determination of Relative Positions for Fomation Flying, -Sensors and Geometrical Considerations-', in *23rd International Symposium on Space Technology and Science*, Matsue, Japan, May 26–June 2, 2002,ISTS 2002-d-49.

[161] Campbell, James B., *Introduction to Remote Sensing*, 2nd ed., New York, USA: The Guilford Press, 1996.

[162] Sengupta, Prasenjit and Vadali, Srinivas R., 'Relative Motion and the Geometry of Formation in Keplerian Elliptical Orbits', *Journal of Guidance, Control, and Dynamics*, Vol. 30, Issue no. 4, pp. 953–964, July–August 2007.

[163] Yao, Yu, Xie, Ruiqiang, and He, Fenghua, 'Flyaround Orbit Design for Autonomous Rendezvous Based on Relative Orbit Elements', *Journal of Guidance, Control, and Dynamics*, Vol. 33, Issue no. 5, pp. 1687–1692, September–October 2010.

[164] Huntington, Geoffrey T. and Rao, Anil V., 'Optimal Reconfiguration of a Tetrahedral Formation via a Gauss Pseudospectral Method', in *AAS/AIAA Astrodynamics Specialist Conference and Exhibit*, Vol. 123, Part II, Lake Tahoe, California, August 7–11, 2005, AAS 05–338, pp. 1337–1358.

[165] Huntington, Geoffrey, Benson, David A. and Rao, Anil V., 'Post-Optimality Evaluation and Analysis of a Formation Flying Problem via a Gauss Pseudospectral Method', in *AAS/AIAA Astrodynamics Specialist Conference and Exhibit*, Vol. 123, Part II, Lake Tahoe, California, August 7–11, 2005, AAS 05–339, pp. 1359–1377.

[166] Williams, Paul and Trivailo, Pavel, 'Numerical Approach Designing and Deploying Formations', in *25th International Symposium on Space Technology and Science*, Kanazawa, Japan, June 4–11, 2006, 2006-d-40.

[167] Dow, John M., Dow, Roberta M., Schmidt, Michael, and Warhaut, Manfred, 'The Implementation of the Cluster II Constellation', *Acta Astronautica*, Vol. 54, Issue no. 9, pp. 657–669, May 2004.

[168] Bainum, Peter M., Strong, Avaine, Tan, Zhaozhi, and Capó-Lugo, Pedro A., 'Techniques for Deploying Elliptically Orbiting Constellation in Along-Track Formation', *Acta Astronautica*, Vol. 57, Issue no. 9, pp. 685–697, November 2005.

[169] Boutonnet, A., Martinot, V., Baranov, A. and Escudier, B., 'Optimal Invariant Spacecraft Formation Deployment with Collision Risk Management', *Journal of Spacecraft and Rockets*, Vol. 42, Issue no. 5, pp. 913–920, September–October 2005.

[170] Capó-Lugo, Pedro A. and Bainum, Peter M., 'Deployment Procedure for the Tetrahedron Constellation', *Journal of Mechanics of Materials and Structures*, Vol. 4, Issue no. 5, pp. 837–854, May 2009.

[171] Williams, Paul, 'Jacobi Pseudospectral Method for Solving Optimal Control Problems', *Journal of Guidance, Control, and Dynamics*, Vol. 27, Issue no. 2, pp. 293–297, March–April 2004.

[172] Tan, Zhaozhi, Bainum, Peter M. and Strong, Avaine, 'A Strategy for Maintaining Distance between Satellites in an Orbiting Constellation', in *AAS/AIAA Space Flight Mechanics Meeting*, Vol. 102, Part I, Advances in

Astronautical Sciences, Breckenridge, Colorado, February 7–10, 1999, AAS 99–125, pp. 343–354.

[173] Capó-Lugo, Pedro A. and Bainum, Peter M., 'Fuzzy Logic Control System for Reconfiguration Procedure Applied to Formation Flying', *International Journal Actual Problems of Aviation and Aerospace Systems: Processes, Models, Experiment*, Vol. 12, Issue no. 3(25), pp. 31–54, 2007, ISSN 1727–6853.

[174] Rubensson, Marcus and Lennartson, Bengt, 'Stability of Limit Cycles in Hybrid Systems using Discrete-Time Lyapunov Techniques', in *Proceedings of the 39th IEEE Conference on Decision and Control*, Sydney, Australia, December, 2000.

[175] Edwards, Christopher, Lai, Nai One, and Spurgeon, Sarah K., 'On Discrete Dynamic Output Feedback Min-Max Controllers', *Automatica*, Vol. 41, Issue no. 10, pp. 1783–1790, October 2005.

[176] Ioannou, P. A. and Sun, J., *Robust Adaptive Control*, New Jersey, USA: Prentice Hall, 1996.

Index

Kolibri 360 GmbH
Ferdinand-Jühlke-Straße 7
99095 Erfurt, Deutschland
produktsicherheit@kolibri360.de